Mammals
of the Holy Land

Mammals of the Holy Land

Mazin B. Qumsiyeh

Texas Tech University Press

This book was set in Caslon and printed on acid-free paper that meets the guidelines for permanence and durability of the Committee on Production Guidelines for Book longevity of the Council on Library Resources. ∞

Jacket design by Lisa Camp

Library of Congress Cataloging-in-Publication Data
Qumsiyeh, Mazin B.
    Mammals of the Holy Land / Mazin B. Qumsiyeh.
        p. cm.
    Includes bibliographical references and index.
    ISBN 0-89672-364-X (cloth : alk. paper)
        1. Mammals—Palestine. I. Title.
QL729.P32Q57 1996
599'.095694—dc20   96-33969
                                                            CIP

96 97 98 99 00 01 02 03 04 05 / 9 8 7 6 5 4 3 2 1

Texas Tech University Press
Box 41037
Lubbock, Texas 79409-1037 USA
800-832-4042

"When the Holy One, blessed be He, created the first man, He took him to all the trees of paradise, and told him: see my work, how handsome and fine they are; everything I have created was created for you. Make sure not to spoil and destroy My world, because what you spoil—no one can repair" (Koheleth Rabba, 7:13).

"Greater indeed than the creation of man is the creation of the heavens and the earth" (The Holy Quran, 40:56).

Dedicated to the inhabitants (present and dispossessed) of the Holy Land and all those who keep the faith in peace and the human spirit

Contents

Preface

Mammals of the Holy Land have been the source of popular inspiration, wonder, food, and folktales at least since the Neanderthal inhabitants of northern Palestine were making cave drawings some 200,000 years ago. More than 100 species of mammals roamed the forests, mountains, and deserts of this ancient land and their impact on humans can be gleaned from the writings of all Near Eastern civilizations. Many mammals are unique to the Holy Land, including forms of the leopard, the caracal, the marbled polecat, the jungle cat, and less known but equally fascinating endemic species such as Buxton's jird (*Meriones sacramenti*) and Bodenheimer's pipistrelle (*Pipistrellus bodenheimeri*). Human population explosion and habitat destruction in recent centuries resulted in the local extermination of several species such as the aurochs (wild ox), red deer, onager, and bubale. Other species became extirpated only in the past few decades due to the introduction of modern guns and shooting animals from vehicles. These tactics not only killed animals and their prey, but the introduction of motorized vehicles enabled more people to gain access to fragile, remote areas causing the destruction of these habitats. Recently extirpated animals include the Syrian wild ass, roebuck (roe deer), fallow deer, oryx, Syrian brown bear, cheetah, and perhaps the lion.

A few species of large mammals barely survived the onslaught and the remaining few individuals still can be seen in remote areas. Urgent conservation measures are needed to salvage such large mammals as the leopard, wolf, wild cat, caracal, ibex, and desert gazelles. Habitat destruction and other human activities also have begun to take a toll on small mammals, especially small carnivores, insectivores, and bats. Within a few decades, we have almost destroyed a delicate web of life that has tied us to the land for thousands of years. However, not all is lost, as local education and international efforts have resulted in promoting conservation efforts in the area. Even more encouraging are recent reintroductions to the area of some previously extirpated or endangered animals such as the oryx, gazelles, wild ass, onager, and roe deer.

Conservation efforts are hampered by the lack of a guide to the mammals of the Holy Land. The hundreds of scattered references in many languages make the task of learning about the local mammals formidable except for the most specialized mammalogist. This work is an attempt to gather in one volume this scattered information and present it in a readable fashion with workable keys.

The interactions of our species, *Homo sapiens,* with the land and its other inhabitants have caused considerable changes in just a small number of years. Many of these changes are irreversibly destructive. Many urgent environmental problems must be addressed immediately. However, environmental concerns have received a low priority in this area that has been dominated by conflict for many millennia. This book is about the information we have about ourselves and the other mammals that coinhabit this historic land. More importantly, it is about the interactions between us and these mammals. A major purpose of this book is providing information about mammals that is useful to inhabitants and visitors to the area. Regardless of their political persuasion, these individuals hold the key to salvaging what remains of a fascinating complex of wild mammals at the gate, or rather the crossroad, between Africa and Asia. Another purpose of the present volume is to provide information that will be helpful to mammalogists as well as other scientists (virologists, ecologists, parasitologists, behavioral scientists, physiologists, etc.) who use mammals in their research.

Many individuals have helped me along the way in the arduous journey to completing this work. I owe a great debt to each and every one of them. The interest in mammals was kindled in my childhood years when I accompanied my uncle, Sana I. Atallah, to observe, collect, and study small mammals near my hometown of Beit Sahur in Palestine. Later, I was encouraged and helped in my studies of the mammals of the region by Duane A. Schlitter, Ralph Wetzel, and David L. Harrison. Many individuals contributed directly or indirectly to this work. Some used their personal funds to copy material or provide transportation. Others supplied photographs or information. Yet others opened their homes to me on field trips. These individuals are too many to list here. However, I am especially grateful to Sami Abdel-Hafez; Maher Abu-Jafar; Zuhair S. Amr; Ahmad M. Disi; David L. Harrison; Heinrich Mendelssohn; Sahar B., Samar, and Majed Qumsiyeh; Darwish M. Shafi, B. Shalmon, Duane A. Schlitter, Yehuda Werner, and Yoram Yom-Tov. Support for several field trips to the area was provided by the Carnegie Museum of Natural History, the Museum of Texas Tech University, and family funds. Finally, I would like to thank my parents, my wife (Hsueh-fang), and son (Dany) for their encouragement and for tolerating my distance during the time in which this work was accomplished.

1

Introduction

This work is intended for individuals with diverse interests such as conservation, zoogeography, and taxonomy as well as the amateur naturalist. Because of this, I attempted to make it as easy as possible for the reader to arrive at the information of interest. Introductory chapters provide basic information about mammals and their ecology in the Holy Land. The use of the term Holy Land is intentionally general because boundaries for such countries as Israel, Jordan, and Palestine have not been historically dependable. As understood here, the Holy Land lies at the southeastern corner of the Mediterranean Sea between latitudes 29° and 34°N and longitudes 34° and 38°E and is approximately 70,000 square kilometers in size (Fig. 1). In the west it is bounded by the Mediterranean Sea and the Sinai Desert. The line separating it from the Sinai Desert is an artificial political line running between the northern tip of the Gulf of 'Aqaba to the Mediterranean coast south of the Gaza Strip. From this western border, the area studied extends east to the Syrian Desert and north to a line that runs along the political boundaries between Jordan and Syria and between Israel and Lebanon (excluding the Israeli self-declared "security zone," a strip of Lebanese territory controlled by Israel since the mid-1980s).

Chapter 2 is a review of previous studies of the mammals of the Holy Land. Chapter 3 is a discussion of the tools and techniques of studying mammals (the scientific field of mammalogy) and is mainly for the benefit of those new to this area of study. Included is a brief description of the anatomy of mammals as it pertains to identifying mammals. Also in Chapter 3 is a description of the scientific terminology and techniques of classification. Specialized morphological structures that are important in particular groups of mammals are discussed in the parts dealing with the introduction to these groups (for example, the tympanic bulla in gerbils or the horns in bovids). Chapter 4 is a discussion of mammalian evolution in the Holy Land as inferred from fossil, archeological, and historical records as well as a discussion of human history in the region. A general discussion of mammalian adaptations with special reference to Holy Land mammals is provided in Chapter 5. Chapter 6 includes a brief list of some mammalian parasites and the impact of mammals on human health. Chapter 7 discusses the geography, climate, and vegetation of the area under study. A discussion of the effect of these and other historical factors on the distribution of mammals follows in Chapter 8. Chapter 9 reviews conservation issues in the Holy Land, including a historical perspective and a discussion of conservation efforts by individuals and governmental and nongovernmental agencies. The largest section of this work is the account of mammals of the region (Chapter 10). These are arranged systematically with keys to aid in identification. Chapter 11 includes a brief account of introduced and domesticated mammals.

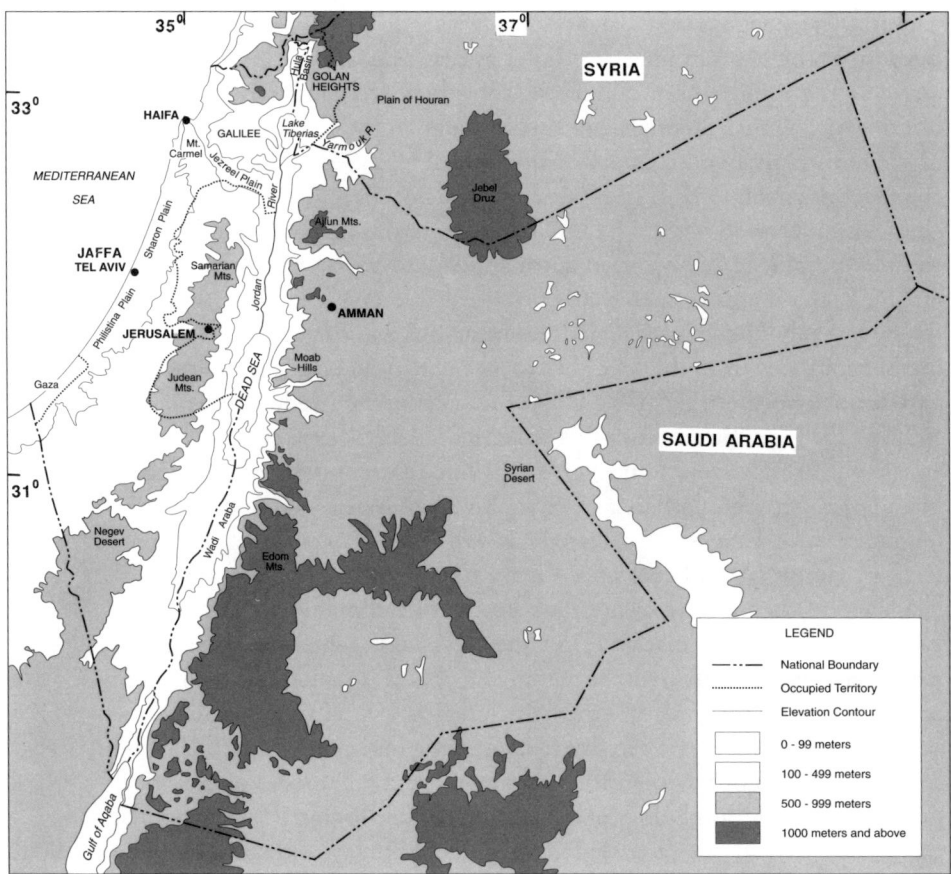

Fig. 1. Map of the area under study showing major geographic features, cities, and present political boundaries in the Holy Land.

The book concludes with a glossary, a list of references, and an appendix listing localities and their coordinates.

In Chapter 10 (synopsis of mammals), the following categories of information are given under each species:

Species names—Scientific names are given followed by common English names. On the same line as the scientific name is a letter indicating the conservation status as categorized in the 1990 IUCN Red List of Threatened Animals (IUCN, 1990). The categories used are: extinct (**X**)—species not found in the wild within the last 50 years; endangered (**E**)—in danger of extinction unless factors causing the decline are reversed; vulnerable (**V**)—likely to become endangered if causal factors continue; rare (**R**)—species at risk, for example, small population or restricted range; indeterminate (**I**)—known to be endangered, vunerable, or rare but not enough data to assign to specific category; insufficiently known (**K**)—not definitely known to belong to any category; out of danger (**O**)—abundant and in no danger at this time.

Nomenclature and synonyms—These are listed to allow readers to relate the present scientific name to names in the older literature. Also included is the "type locality" (the geographic locality from which the name was first described) for each name given. No attempt was made to provide a list of all synonyms for the species but rather only those that are relevant to the area under study.

Diagnosis—These are general comments on the morphology of the species and the best characters that can be used to identify the species and distinguish it from other closely related taxa. This diagnosis complements the keys provided for families, genera, and species. Chapter 3 includes basic terminology and methods to identify different groups of the mammals of the Holy Land. For non-mammalogists, I suggest a reading of that chapter prior to attempting to use the keys and the diagnosis sections. If local variation (subspecies) exist, this is also discussed in the diagnosis section.

Range of species—The general range of the species is briefly given.

Local status—Information is given on the distribution of the species relative to major habitats of the Holy Land. Confirmed locality records are given with references except where these are too numerous to list (as for common species). In the latter case, marginal localities and localities representing diverse habitats are given. This information can be used in conjunction with the distribution maps and the appendix of localities and their geodetic coordinates. In this section, I also comment on the conservation status of endangered or threatened species.

Biology—Comments are made on habits, habitat preferences, feeding, reproduction, and other available biological information.

Genetics—Genetic information is becoming an integral part of describing mammals. Such information helps resolve systematic problems and elucidate patterns of evolution. I thus provide relevant information on the chromosome constitution and any other genetic information that is available for each species. Much of the information from outside the Holy Land may not be applicable to species in the Holy Land because of genetic variation within a species. However, I included such data from nearby areas for comparative purposes, especially when data were not available from the Holy Land.

Human interactions—Interactions and importance of the particular species to previous and present human inhabitants are discussed. These include use as food by humans; impact on agriculture and disease; animal folklore, superstitions, and myths; specific methods of study, capture, and protection of the species; and origin of the species name(s). Quotations from holy books of the three major religions of the area are provided for reference.

I decided not to burden the text with a list of specimens examined. Under the "Local status" category, I report localities for the species from the area under study by citing the person who originally reported these specimens or where specimens are available. If specimens are available without a report in the literature then I cite these with the collection acronym. I have visited or examined specimens from the following collections. I am grateful to the individuals listed in parentheses who allowed me to examine specimens and for their generous hospitality. BM—British Museum of Natural History, London. This museum has collections from Jordan and Palestine that are of historic significance such as those made by Tristram, Carruthers, and Buxton and types of animals described by Oldfield Thomas (John E. Hill). BZM—Berlin Zoological Museum, Berlin, Germany. This museum has the historic collections of such individuals as Hemprich and Ehrenberg, Schmitz, Bodenheimer, and I. Aharoni. CMNH—Carnegie Museum of Natural History, Pittsburgh, Pennsylvania. Collections here have specimens from Jordan and the West Bank collected by M. B. Qumsiyeh in 1981 and 1991 (Duane A. Schlitter). HUJ—Hebrew University of Jerusalem, Jerusalem. This museum has collections from the Holy Land (Eitan Tchernov). HZM—Harrison

Zoological Museum, Sevenoaks, Kent. Collections from Jordan and Israel are housed here (David L. Harrison). JNHM—Jordan Natural History Museum, Yarmouk University, Irbid. Specimens at this museum are primarily from northern Jordan (Darwish M. Shafi). JUNHM—Jordan University Natural History Museum, Amman. Mammal specimens at this museum were collected by M. B. Qumsiyeh, A. M. Disi, Z. Amr, and others (Ahmad M. Disi). MCZ—The Museum of Comparative Zoology, Harvard University, Cambridge, Massachusetts. This collection was made by John C. Phillips in 1914 in a trip from Sinai to Mount Hermon. It is reported on by Allen (1915) (John Kirsch). SANHC—Sana Atallah Natural History Collection, Beit Sahur. Specimens collected by Sana I. Atallah and M. B. Qumsiyeh are housed here (Issa A. Atallah). TAUM—Tel Aviv University Museum, Tel Aviv. This museum has collections from Israel and elsewhere in the Holy Land and Sinai (Heinrich Mendelssohn and Yoram Yom-Tov). TTU— The Museum of Texas Tech University, Lubbock, Texas. This museum has specimens mainly from Jordan and the West Bank collected by M. B. Qumsiyeh in 1985 (Robert J. Baker). USNM—United States National Museum, Washington, D.C. This museum has an extensive collection from Egypt and a few specimens from the Sinai (Michael D. Carleton).

Transliteration of Arabic and Hebrew names into English (or for that matter any European language) is difficult. This problem is encountered for names of localities and common names of species. Many letters do not have corresponding sounds in English and many English letters are pronounced in more than one way in Semitic languages. An interesting example is that T. E. Lawrence ("Lawrence of Arabia") used different spellings for Arabic names in writing his autobiography (Lawrence, 1926). This prompted the publisher to insert a "disclaimer note," which reads in part:

> It seems necessary to explain that the spelling of Arabic names throughout this book varies according to the whim of the author. The publisher's proof-reader objected strongly . . . [but] the author's attitude can best be judged from the following extracts . . . [:] Q. I attach a list of queries regarding inconsistencies in the spellings of proper names . . . will you annotate it in the margins. A. [Lawrence] not very helpfully perhaps. Arabic names won't go into English, exactly, for their consonants are not the same as ours and their vowels, like ours, vary from district to district. There are some 'scientific systems' of transliteration, helpful to people who know enough Arabic not to need helping, but a wash-out for the world. I spell my names anyhow, to show what rot the systems are.

The remaining inquiries by the publisher were answered sarcastically by Lawrence. About the spelling of *Jeddha* or *Jedhah*, the she-camel, Lawrence answered "She was a splendid beast." At any rate, I bring this up only to show the reader that, although I will attempt to have a uniform spelling throughout this book, the reader will find other spellings in references that refer to the same name. When a confusion may arise, I have placed some synonymous spellings in parentheses. In addition, the locality index in the Appendix includes other spellings used for the same locality. I would best advise the reader who looks for identity in words to use the similarity of pronunciation rather than identity in spelling.

Another problem encountered is the change in locality names since the establishment of the State of Israel. A good reference that discusses the changes in locality names is the book by Khalidi (1992). A few of the new names were based on old Hebrew names or on Hebrew translations of Arabic names but many were newly created names. Examples of old and new names are Abde (Avdat), Wadi Karkara (Nahal Bezet), Ein Husub (Hatseva), Wadi Khabra (Nahal Hever), Jezreel (Emeq Yizrael), Wadi El Abiad (Nahal Lavan), and Bir Rekhme (Yeroham). These changes created a complex situation for trying to confirm older locality records or for comparison of new and older records. Because of this, I provide a locality index as an appendix and attempt to standardize names (including spelling in the text). The complexity of this task should not be underestimated and a few inconsistencies may have escaped detection.

2

Historical Background

The mammals of the Holy Land have been a continuous source of fascination and interest to the human inhabitants of the area. The earliest known depictions appear at the time Stone-Age Neanderthal man occupied the area (Repenning and Fejfar, 1982). As Semitic civilizations emerged that depended less on hunting (for example, Canaanites and Assyrians), the inhabitants continued to rely on a keen knowledge of the land and animals. Only cursory review of writings of these eras is needed to document the interest in, and the fascination with, wild animals. Recent Nebatean discoveries from Petra include texts and mosaics with many local animals depicted. The modern Arabic language is derived from the Nebatean language. Another example is the frequent reference to animals in the Bible and the Torah. Quotations from these ancient texts are provided under the accounts of species. In such a wild and untamed wilderness, knowledge of the area and its animals was essential to the inhabitants. Later, Roman and Greek civilizations extended to the eastern Mediterranean region and their ships brought back to Rome and Greece not only pelts of wild mammals, but also knowledge and tales of the wildlife. The Romans hunted many local mammals for sport and food, and depictions of these animals are still seen in the mosaics the Romans left behind. Examples include lion, cheetah, deer, wild ox, gazelle, and other game animals in ruins at places like Jericho, Caesarea, and Madaba.

The Umayyad Empire spread from Arabia and flourished at a time when Europe was in the Dark Ages. Muslim scholars translated many Greek and Roman books and expanded and transferred knowledge to Europe. The Islamic contribution to civilization is best known in the fields of mathematics, chemistry, astronomy, and medicine. However, some scientists devoted their life to studying animals and were perhaps the first specialized zoologists. In the 9th century AD, Al Jahiz (Abu Uthman ibn Bahr al-Jahiz) studied

mammals in the Arabian and Syrian deserts and published a huge monograph titled *Kitab al-Hayawan* (*The Book of Animals*). In the 12th century
AD, Usama ibn Munkidh published a similar treatise titled *Kitab al-Itibar*
(*The Book of Knowledge*), and in the early 1300s AD, Zakharia Al-Qazwini
published *Nuzhat al-Qalb* (*Journey of the Heart*) and *Aja'eb al-Makhloukat*
(*Wonders of the Living*). These earlier works were mostly anecdotal descriptions of animals and their habits. However, some information was remark-
ably detailed and accurate. For example, Al-Damiri (who was born in AD
1349) in *Hayat Al-Hayawan* (*Life of Animals*) spoke of bats as lactating
mammals at a time when many thought they were birds. He spoke of two
"types" of bats, one large and fruit-eating, the other small and insect-eating.
These are divisions that were later "discovered" and are now followed in
formal taxonomy by dividing bats into Megachiroptera and Microchiroptera. The Umayyads, like the Romans and Greeks, also hunted animals and
made depictions of these animals as part of the architecture of palaces and
castles. These palaces were located not only for strategic/military purposes
but also to be used as hunting lodges. An example is Qasr Amra near Azraq,
which was built in the 8th century AD.

   Interest in mammals of the Holy Land was rekindled with the revival of
European science after the Middle Ages. As early as the 16th century, some
European travelers observed and recorded mammals during pilgrimages to
the Holy Land (Belon, 1553; Rauwolff, 1582). Such contributions to the
study of mammals continued to be made on short pilgrimages until today.
Especially profitable were those studies done after the revolution in systematics begun by Linnaeus in the second half of the 18th century. One of
the earliest systematic works done following the tradition of Linnaeus was
Hasselquist's *Iter Palaestinum* (published 1757). Other works were those by
Geoffroy St.-Hilaire, who described and catalogued many mammals from
Egypt and the Near East (including Sinai and Palestine). The 19th century
was the golden age for knowledge of mammals throughout the world and
the Holy Land was no exception (examples include Wagner, 1839, 1840;
Heuglin, 1861; Hart, 1885, 1891). The most important from a local stand-
point was the work of the Reverend Henry Baker Tristram, Canon of
Durham, who spent many years studying and reporting on mammals
(Tristram, 1866a, 1866b, 1867, 1876, 1877, 1884). His culminating work,
published in 1884, was titled *The Survey of Western Palestine: the Fauna and
Flora of Palestine*. It included 30 pages on mammals, and was the first detailed
study of the mammals of the Holy Land.

   At the turn of the century Father (Pater) Ernst Schmitz (Fig. 2) collected
and observed animals in the Holy Land during his tenure (beginning in

Fig. 2. Father Ernst Schmitz from a portrait in the Schmitz collection in Jerusalem.

1908) as head of the Holy Land missions of the Deutscher Verein des Heiligen
Landes, a German Catholic organization (Schmitz, 1912; Anonymous,
1946*b;* Leshem, 1979). The amount of material he amassed during this time
was significant. Although Schmitz was not keen on studying mammals, his
collection, mostly sent to Germany, formed the basis of many scientific
publications by renowned systematists. Some of the Schmitz collection

remains in Jerusalem at the Schmitz Girls College where I taught biology and chemistry in 1978–1979. Allen (1915) reported on the collections made by I. Phillips and others on a trip from Sinai to the base of Mount Hermon. These specimens are now at the Museum of Comparative Zoology, Harvard University.

The interest in local fauna was also maintained by immigrants at the turn of the century. Israel Aharoni and his daughter Bathscheba Aharoni were the first of the immigrant naturalists to report on the mammalian fauna (I. Aharoni, 1917, 1930; B. Aharoni, 1932, 1944). I. Aharoni was born in 1882 in Lithuania and immigrated to the Holy Land in 1902, where he initially settled in the new town of Rehovot. During World War I, he was appointed as army zoologist by the Turkish authorities and later took a government position with the British Mandate. For more than a quarter century he studied the local fauna, mostly reporting on the mammals. He gave Hebrew names to many of the animals he studied (usually based on Arabic or on European common names). Most of these Hebrew names are still in use today. I. Aharoni's studies were followed by less detailed studies of Bodenheimer (1935, 1937, 1958), who did not give many specific locality data nor did he indicate what specimens were at his disposal. This made it difficult to use his data in a study of mammalian taxonomy.

Over the last two centuries, many other travelers made reference to the fauna of the Holy Land. However, most of these references were overdramatized and anecdotal remarks with little scientific value. An example of such inaccurate reports is Howells's (1956) *A naturalist in Palestine*. By contrast, the mammals of Arabia including the Holy Land have been, and continue to be studied by the eminent British naturalist David L. Harrison. His studies culminated in the publication of three volumes (Harrison, 1964*a*, 1968*a*, 1972). This work was more extensive than intensive because it covered the whole of the Arabian Peninsula. A new condensed edition in one volume has been published (Harrison and Bates, 1991; reviewed by Qumsiyeh, 1992).

Among native zoologists, Sana Issa Atallah was a leader in the 1960s (Wetzel and Schlitter, 1970). His career was cut short by a tragic accident at the young age of 27 (Kumerloeve, 1973). His Ph.D. dissertation was the most extensive treatment of the small mammals of the eastern Mediterranean region (Lebanon, Syria, Jordan, Israel, and Palestine). Atallah's dissertation did not benefit from potential changes and editing or proof reading by Atallah, because although he wrote it as a dissertation in 1969, it was not published until eight years after his death (1977, 1978). Many errors in the proof would clearly have been discovered had Atallah lived to publish this work himself. For example, a page mix-up occurs that affects the published pages 317–320 (Atallah, 1977). Page 320 should be followed by page 319 and these two sections should be inserted before "*Arvicola terrestris*" on page 317. Of the living scientists who published and continue to publish on various aspects of the mammalian fauna of the Holy Land, the following deserve special mention (in alphabatical order): Zuhair Amr, Ahmad M. Disi, David L. Harrison, Heinrich Mendelssohn, Eviator Nevo, Eitan Tchernov, and Yoram Yom-Tov.

It is with this background and with a lack of a guide to the mammals of this region that I proceeded to work in this area. The Holy Land is undergoing rapid habitat destruction, human overpopulation, and extensive destruction of wildlife. This prompts an urgency for understanding the factors that resulted in the local extinction of so many species and for the protection of the remaining threatened and endangered taxa. I hope that this text will help disseminate our knowledge of the mammals and the need for their protection to governments and private sectors in the countries of the region. It is also hoped that this humble effort will stimulate additional detailed studies of the different species. It is encouraging that Jordan, Israel, and Palestine have nature protection societies and laws to protect wildlife. However, much more needs to be done, especially because the current priorities in these governments are dominated by political and economic objectives and because many of these objectives are in direct conflict with nature conservation.

3

Mammalogy:
The Study of Mammals

Mammalogy is the discipline within biology dedicated to the study of all aspects of mammals (from the Latin *mamma,* meaning breast). This work deals with land mammals and as such the definitions provided here may not apply to marine mammals (for example, whales, dolphins, seals). Mammals are characterized among vertebrates by the usual possession of body hair rather than scales (as in reptiles) or feathers (as in birds); mammary glands for the production of milk in females; a single bone (dentary) in each lower jaw as opposed to a complex of bones as in the jaws of other vertebrates; replacement of at least part of the deciduous "milk" teeth; the presence of three middle ear bones (ossicles); and the ability to maintain the body at a relatively constant temperature. As in any other discipline, mammalogy entails descriptive terms that are not familiar to most people. However, a lay person can quickly acquire skills for identification and study of mammals. In this chapter, I will introduce information that an individual with little or no background in the field will need to be able to use the keys and species accounts to identify mammals and proceed to study them. The reader also can refer to the Glossary for further definitions of terms. This will circumvent the need to refer continuously to this chapter or to dictionaries of scientific terms.

Systematics

The field of taxonomy concerns itself with the theory and practice of classification or arranging living organisms into groups. The science of studying the relationships and diversity of organisms is the science of systematics. Each animal is classified into a group of similar individuals with subsequently more inclusive groups. The names given to these groups are uniformly accepted in all languages as the "scientific names" as opposed to colloquial or common names. Scientific names are derived from Latin, Greek, or other languages that are then modified to a Latin or Greek structure. Conventional arrangement for the human species illustrates this system:

Kingdom Animalia: the animal kingdom.

Phylum Chordata: from *khor?*, L., spine; animals with a spine.

Class Mammalia: from *mamma*, L., breast; mammals.

Order Primates: from *primus*, L., first or foremost; humans, apes, and relatives.

Family Hominidae: from *homo*, L., genitive *hominis*, a man; *idae*, L., suffix added to form a family name; man and his relatives.

Genus *Homo*: from *homo*, L., genitive *hominis*, a man; genus of man, single living species: *Homo sapiens*.

Species *Homo sapiens*: *sapiens*, L., meaning wise, sensible.

A species is defined as a group of interbreeding natural populations that is effectively reproductively isolated from other such groups (Mayr, 1969). A subspecies is a uniform subset of a species that is genetically distinct and has its own evolutionary tendencies (Lidicker, 1962). As such, human populations in the world are not divided into subspecies, whereas, for example, there are several subspecies of the leopard, *Felis pardus*.

The *scientific names* given for species are underlined or italicized because they are Latin, Greek, or Latinized names and should be separated from the text. These names are recognized internationally, whereas *common* or *vernacular names* can vary among different geographic areas or different languages. Scientific names may sound quite complicated to the nonbiologist, but in most cases the origin of the name is a logical one derived from Latin, Greek, or other languages and usually has a meaning pertaining to the animal of interest. These names can be derived from: the animal's description (*Taphozous nudiventris* meaning "naked abdomen"), the region of the first collection (*Rousettus aegyptiacus* from Egypt), the common name (*Felis leo* from Latin for lion), or a person's name (*Otonycteris hemprichi* after von Hemprich). For each species discussed in this work, name derivations are provided under the section of "Human interactions." The derivation of the genus name is discussed either under the genus or under the first (or only) species in the genus.

Morphology

In describing the external morphology of a land mammal, such designations as ventral (toward the lower side), dorsal (toward the upper side), lateral (to the side), anterior (toward the front), and posterior or caudal (toward the tail) are used.

A mammal's pelage serves such functions as insulation, protection, defense, and signaling (both to members of the same as well as other species). Other specialized functions of skin evolved in conjunction with major morphological and physiological changes such as those involved in burrowing, flying, or jumping on sand. The bat's wing is a modified foreleg with skin stretched between elongated finger bones. The mole-rat's fur is unique in that, unlike that of other mammals, it is not directional, enabling this fossorial (burrowing) animal to reorient quickly in any direction when it is in tight spots. The jerboa has elongated hairs on the hind legs for easy movement on loose sand. Other desert adaptations are discussed in Chapter 5.

The measurements of the body of a mammal are important in many cases for identification purposes. The most common measurements and those used in this book are: head and body length, tail length (base to tip), total length (head and body plus tail), ear length, hind foot length (heel to longest claw or nail), and forearm length (in bats only, elbow to wrist or external measurement of the radius bone). In large mammals, the height at the shoulders may be used as a general measure of size. Information on the weight in mammals is also useful. However, for Holy Land mammals, basic data on variations in weights (by season, sex, reproductive status, etc.) are yet unavailable. When such data become available, weight may be used as a taxonomic character in some cases.

The mammalian skull is distinctive and reflects the specializations of feeding, hearing, vocalization, and smell that are unique to each species. Because of this, the anatomy of the skull provides an excellent source of information for classifying mammals. A knowledge of the various parts of the skull is essential for identifying many species of mammals.

Figure 3 presents the main features of an idealized mammalian skull. The skull is seen to consist of two main parts, the *cranium* (the major part of the skull) and the *mandible* or lower jaw. The main bones of the cranium are paired and meet in a midline. From the dorsal view, the *nasals* are followed by the *frontals*. The frontals also form part of the eye socket and their posterior margin may extend to form the *postorbital process*. More posteriorly, the *parietals* and the *occipital* form the main roof of the braincase. These bone

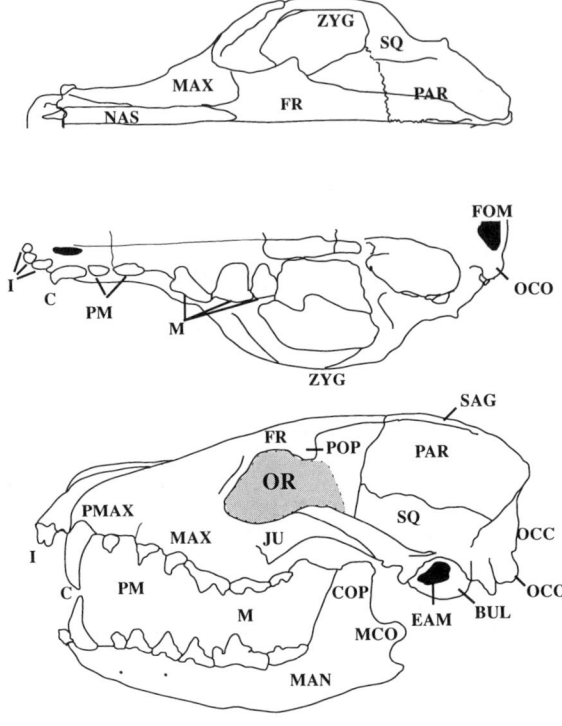

Fig. 3. A skull of a typical mammal, Rüppell's fox (*Vulpes rueppelli*). Dorsal, ventral, and lateral view of skull and lateral view of mandible are shown. Abbreviations are: BUL, bulla; C, canines; COP, coronoid process; FOM, foramen magnum; FR, frontals; I, incisors; JU, jugular bone; M, molars; MCO, mandibular condyle; MAN, mandible; MAX, maxilla; NAS, nasals; OCC, occipital bone; OCO, occipital condyle; OR, orbit; PAR, parietal; PM, premolars; PMAX, premaxilla; POP, postorbital process; SAG, sagittal crest; SQ, squamosal; ZYG, zygomatic bone (arch).

may form a thick joint in the midline that is elevated to support the mastication muscles. This is called the *sagittal crest*.

Ventral to the parietals are the *squamosals*. The parietals and squamosal join the most posterior bone of the brain case, the *occipital*. The occipital bones surround the large *foramen magnum*, which is the opening through which the spinal cord emerges. The vertebral column has a flexible connection to the skull by attachment to the two *occipital condyles* at the sides of the foramen magnum.

The upper jaw bones are also paired. They are the *premaxilla* supporting the anteriormost teeth, called the *incisors*, and the *maxilla* supporting all other teeth. The premaxilla, maxilla, and *palatine* bones also form the roof of the mouth cavity (the so-called hard palate) and serve to isolate the nasal passage from the mouth cavity. The *zygomatic arch* is formed by processes from the squamosal and the maxilla linked by a thin bone, the *jugal*. This arch supports the orbits and the large opening for the mastication muscles, the *temporal fossa*. The paired *auditory bullae* are located lateral and ventral to the occipital. The auditory bullae are bony capsules that support the inner and middle ear bones. In some mammals, the auditory bullae are enlarged and provide important identification characters. The structure of the auditory bulla in one such group of mammals, the gerbils, is discussed in that section.

The mandible is a relatively simple structure in mammals and is composed of the paired solid *dentaries*. The *mandibular condyles* articulate with the cranium. The dentary expands into a *coronoid process* to provide a surface for muscle attachment (dorsal to the condyle). Ventral to the condyle, the *angular process* can be seen.

The teeth represent complex evolutionary adaptations in mammals. Four kinds of teeth are recognized. The incisors are the anteriormost teeth and are attached to the premaxilla. The maxillary teeth are the *canines*, the *premolars*, and the *molars*. Frequently some of these teeth have been reduced during evolution or completely eliminated creating *diastemas* (spaces). Also, premolars and molars may be difficult to distinguish, and one can then term them cheekteeth. Cheekteeth in primitive mammals possessed distinct cusps and these are still found in many primates, bats, and insectivores. In some herbivorous mammals, the cusps have fused to form elongate and irregular ridges termed *lophs* (for example, in ungulates). Also, cusps may fuse to form only lateral lines called *laminae* (as in some rodents). In other herbivores, single cusps may elongate to form crescent shaped ridges, a condition termed *selonodont* teeth (as found in gazelles). The cheekteeth of carnivores are modified by reduction of some cusps and elongation of others. The cheekteeth of the upper and lower jaws of carnivores work in combination like a scissors and are collectively termed the *carnassials*.

Dental formulae may be used to describe the number of the different kinds of teeth in mammals. For the wolf, the dental formula is i 3/3, c 1/1, p 4/4, m 2/3. This means that in the upper jaw, each side has three incisors, one canine, four premolars, and two molars and on the lower jaw, each side has three incisors, one canine, four premolars, and three molars.

Genetics

Classical mammalogy depended heavily on gross morphological features for identification and study of taxa. Numerous technical advances now make it possible to use other characters derived from study of the chromosomes, proteins, or molecular structure for taxonomic and evolutionary studies. I will discuss some of the applications of these modern studies (both previous and anticipated) in studies of mammals of the Holy Land.

The number and morphology of chromosomes provide useful characteristics to distinguish some species and populations of mammals. Numerous
earlier studies by Wahrman and colleagues (Wahrman and Zahavi, 1955;
Zahavi and Wahrman, 1957; Wahrman et al., 1988) showed that gerbil
species could be identified by studies of chromosomes. For example, a
chromosome number of 52 is characteristic for *Gerbillus nanus,* 60 for *G.
dasyurus,* and 42 or 43 for *G. gerbillus.* Chromosome number differences also
exist among populations of the same species. For example, Wahrman and
colleagues (review in 1985) showed chromosomal differences between four
populations of the mole rat, *Spalax leucodon.* Relationships of species are
studied by using chromosome banding homology (similarity due to descent
from a common ancestor). A review of this topic can be found in Baker et
al. (1987). For mammals of the Holy Land, there are few such studies; those
that exist are mainly on gerbils (Qumsiyeh and Chesser, 1988) and bats
(Qumsiyeh and Baker, 1985; Qumsiyeh et al., 1988; Qumsiyeh, 1989).
Similar studies should be performed on other groups of mammals. Those in
the Holy Land that would most benefit from chromosomal banding studies
include shrews, vespertilionid bats, rodents, and gazelles.

Protein electrophoresis is a technique that allows separation and identification of proteins (both enzymes and structural proteins) run on a gel
matrix and stained with specific substrates (chemicals that will be acted upon
and change color). The technique was initially applied to studying differences between human populations. Studies were then extended to other
groups of mammals. Again, such modern studies of mammals in the Near
East lag behind those done in other parts of the world. Examples of the use
of electrophoretic data on mammals of the Holy Land include: gerbils (Nevo,
1982; Qumsiyeh and Chesser, 1988; Qumsiyeh, 1989), mole rats (Nevo,
1978, 1988), dormice (Filippucci et al., 1988*a*), field mice (Filippucci et al.,
1989), and bats (Qumsiyeh et al., 1988). However, it is necessary in such
studies to have large sample sizes that include nearby regions outside of the
Holy Land. Using protein electrophoresis in combination with chromosome
and/or DNA studies should provide answers to some of the taxonomic
problems still encountered in some groups.

The revolution in molecular biology has yet to impact on zoological
studies in the Holy Land. Doubtless, such techniques as the polymerase
chain reaction (for amplifying pieces of DNA) and sequencing of pieces of
DNA (from the mitochondria and/or the nucleus) will become important
genetic study tools.

Methods for Observing and Studying Mammals

The tools used in studying mammals are numerous and a detailed description is not appropriate here. I provide a brief description here of methods for observing and studying live mammals. Most large mammals can be observed directly in the field. Diurnal mammals (those that are active during daylight hours) are much easier to observe than those that are nocturnal (active at night). A pair of binoculars is all that is needed to observe many diurnal mammals such as gazelles, hyraxes (conies), and even some of the carnivores (foxes, for example). Observing nocturnal carnivores is more difficult and requires establishing observation posts at drinking sites and feeding stations.

Small mammals such as rodents, bats, and insectivores are more difficult to observe under field conditions. However, hedgehogs easily can be located in many areas by taking an evening stroll with a flashlight. Bats can be observed in flight and a rough determination of the genus or family may be possible by observing the pattern of their flight and by listening to their flight vocalizations. Using a sonar detector may allow species identification in some cases but this requires a knowledge of the specific pattern of the sonar of each species. Such information is available for some species of European bats that also occur in the Holy Land. More information on echolocation signals of local bats is needed. Bats may also be observed by studying them in their roosting places. However, this should be limited to essential scientific observation because of the extreme disruption (which can cause death of many bats) that such visits cause. This is especially true for maternity colonies and for bats in hibernation. Some bats can be collected for scientific study using specially designed nylon nets (mist nets). Shrews and small rodents can be studied by trapping them in special live traps.

Valuable information can be deduced from sources other than direct observation of live animals. Footprints provide an excellent record of the activity of mammals and can be used for species identification. Valuable information on the local mammalian fauna may be obtained by studying owl pellets. Owls feed on many small mammals (bats, shrews, rodents); undigested bones and hairs are regurgitated, usually near the owl's nest. Other specimens may be found killed by cars on the road or dead by other means in the wild. Observation or collections all add to our information about the local fauna. Museums provide a repository for specimens of mammals and are very important in providing the basis for future study. Any specimen collected for the purpose of scientific study should have the following data recorded if possible: the locality where it was obtained, the sex, the name of the collector, the reproductive condition, and measurements in the flesh. The specimen then can be preserved.

Identification of Orders of Mammals

So far I have provided a brief description of the morphology of mammals and other aspects for their study. In the account of each group of mammals, I provide keys for the identification of the mammals of the Holy Land. A key is a tool for identification that is designed in a bifurcating system. When reading a key, a person decides which of two statements best describes the animal of interest. Once chosen, the option may lead to further subsets of choices until an identification is reached.

Keys in this book are provided under each order of mammals. Thus, the first level of our identification process is assigning the mammal of interest to an order. The following is a brief description that allows this to be done.

Order Primates
Primates

The orbits are usually directed forward. Teeth usually with low crowns or conelike cusps. The postorbital bar or plate is always present. In advanced groups, the brain is enlarged and the rostrum shortened. There are flexible fingers on the hands and feet with an opposable thumb at least on the hands. One extant species lives in the Holy Land, humans or *Homo sapiens*.

Order Insectivora
Hedgehogs and shrews

These animals are difficult to characterize as a group. Hedgehogs have a spiny pelage and range in size from 150 mm to 300 mm in length. Shrews are smaller animals with soft pelage. Both hedgehogs and shrews are distinguished from other mammals of the region by their pointed (rather mobile) snouts and the morphology of the teeth (Fig. 4), and the simple plantigrade structure of the feet.

Order Chiroptera
Bats

These are flying mammals and have elongated finger bones with a skin membrane extending between these finger bones and also to the hind legs. In some species, this membrane also extends between the hind legs and may include part or all of the tail.

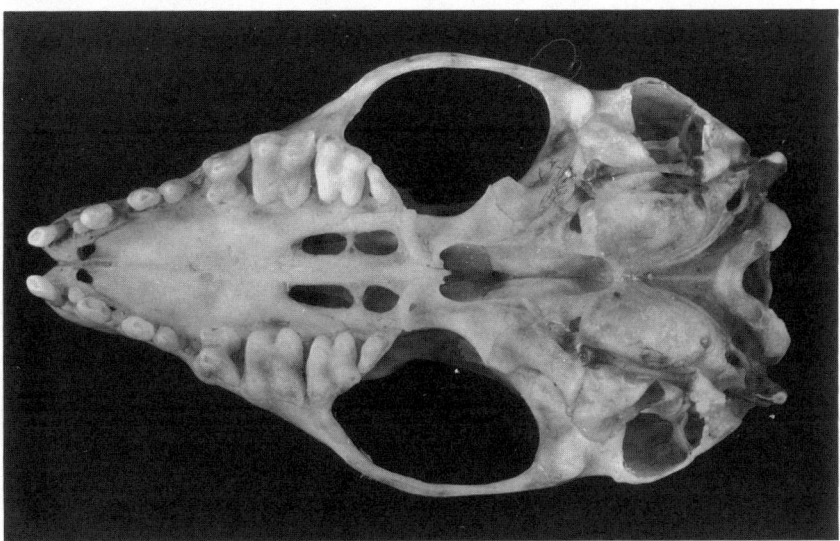

Fig. 4. Ventral view of the skull of a desert hedgehog (*Paraechinus aethiopicus*), from Jordan. The upper teeth from anterior to posterior are two incisors, one canine, three premolars, and three molars.

Order Lagomorpha
Hares and rabbits

These animals need no description here. Everybody is familiar with the soft pelage, long ears, and characteristic longer hind feet. Lagomorphs have a saltatorial (jumping) mode of locomotion. The skull is characterized by the presence of two small incisors hidden behind two large anterior incisors. A gap or diastema separates the incisors from the cheekteeth.

Order Rodentia
Mice, rats, jerboas, gerbils, jirds

This is a complex and diverse group of mammals. The best distinguishing features are in the skull and include one pair of upper and one pair of lower incisors that are chisel-shaped. These incisors are continuously growing. They are separated from the cheekteeth by a gap (canines are absent).

Order Carnivora
Cats, dogs, jackals, wolves, foxes, hyenas, mongooses

Like rodents, this group has a wide diversity in external form. The skull is characteristic, with well-developed canines that are used to tear flesh or pierce snails and insects. The mandible has a transverse condyle (Fig. 3). In doglike and catlike carnivores, the cheekteeth (upper fourth premolar and lower first molar) are modified into carnassials, with sharp scissorlike action to cut flesh.

Order Hyracoidea
Hyraxes

There is only one species that enters this region. These animals are rabbit sized. They have a continuously growing pair of upper incisors that differ from those of rodents and lagomorphs in being triangular in cross section and tusklike. There are two pairs of lower incisors.

Order Perissodactyla
Horses and asses

Each foot has a single functional weight-bearing toe. This digit carries the hoof and is termed an unguligrade type. The skull is elongate anteriorly. The incisors are prominent; the upper canines are small, followed by a gap, and then the premolars and molars. The cheekteeth have a flat surface (hypsodont) with a complex pattern of cusps and ridges and are used to crush plants.

Order Artiodactyla
Pigs, gazelles, deer, antelopes, sheep

The third and fourth digits are the large, functional, weight-bearing digits. Two other rudimentary digits may be present on either side of the third and fourth digits, or may be absent.

4

Mammalian Evolution and Human History

The Mesozoic Era (230 million years ago [MYA] to 70 MYA) is the age at which mammals first appeared in the fossil record. Mammals first evolved from a group of reptiles termed the cynodont reptiles (order Therapsida). Mammalian evolution from then was a spectacular series of adaptations, mainly occurring in the Cenozoic Era (70 MYA to the present). In paleontological terms, eras are divided into periods and periods are further divided into epochs. Some of the earliest mammalian fossils were collected from Europe and are dated to the late Triassic Period some 200 MYA (Mesozoic Era).

Changes in dentition and jaw articulation were some of the most important adaptations that occurred in the transition from reptiles to mammals. The more familiar mammalian characteristics of lactation, live birth, and so on came much later. Indeed some of the present mammalian groups still have so-called primitive characteristics: for example, the Australian duck-billed platypus and spiny anteater lay eggs. Large reptiles (the familiar dinosaurs) dominated the Mesozoic Era, whereas mammals were small and inconspicuous. These small mammals lived in the underbrush and were mostly insectivorous. The end of the Mesozoic and the demise of the dinosaurs was the beginning of the spectacular success of mammals in the Cenozoic Era. The Cenozoic Era is divided into two periods: the Tertiary (65 MYA to about 2 MYA), which is characterized by rapid radiation of mammals; and the Quaternary, which is characterized by widespread loss of species of large mammals and the speciation events that led to most modern mammalian species.

The Tertiary Period is split into five epochs: Paleocene, Eocene, Oligocene, Miocene, and Pliocene. The Paleocene, the first epoch in the Tertiary, was the time that the major split between the marsupial and the placental mammals was established. Most orders of mammals originated in the Eocene Epoch, which began some 50 MYA. Many mammalian orders were already differentiated in the Eocene records: hedgehogs, bats, primates, various rodent groups (Gliroidea, Muroidea), carnivores, hoofed mammals, and primates. Most families of mammals are first seen as Oligocene fossils.

The fauna of the Holy Land was fairly rich in large mammals in the Pliocene Epoch. The Pleistocene Epoch witnessed major climatic shifts corresponding to the advancement and retreat of ice caps in the Northern Hemisphere (termed glacial and interglacial periods). The Holy Land was not covered by ice caps. However, the glacial periods in the north caused cool and moist climates in the Near East (termed pluvials). Northern animals extended their ranges south with the southward extension of the cool and moist climates during these pluvial periods. When the glacials retreated, massive extinctions of northern animals took place. The reverse occurred for animals adapted to drier climates in the south: they expanded their ranges north during interpluvials and were decimated during the pluvials. Thus, it is not surprising that the Pleistocene Epoch was characterized by numerous extinctions and faunal replacements. However, the climatic changes were not the only source of loss of many species of mammals in the Holy Land. Human impact was clearly responsible for the destruction of many species that could have survived in isolated forests. In the moist times of the Pleistocene, the grasslands, marshes, and forests of the Holy Land were rich in such animals as the lion (*Felis leo*), rhinoceros (*Rhinoceros hemitoechus*), hippopotamus (*Hippopotamus amphibius*), horse (*Equus caballus*), several deer species, and several bovid species. In prehistoric times, two species of hyena lived in the Holy Land, the striped *Hyaena hyaena* and the spotted *Crocuta crocuta* (Kurtén, 1965). Some mammals, such as wild camels, became extinct more than 35,000 years ago. Others, such as the hippopotamus, survived until the Neolithic (10,000 years ago). A good example of the Neolithic fauna occurred in finds at Hayonim cave in the western Galilee (Bar-Yosef and Tchernov, 1966). This fauna included most of the species living today in the area but also the rodents *Myomimus roachi* and *Mesocricetus aramaeus*, and hoofed animals such as *Capra*, *Bos*, and *Cervus elaphus*. Still other species were lost only after further human development in the Iron Age and beyond. Animal remains at Jawa in the basalt desert of Jordan indicate a rich history of animal life, including cattle and gazelles (Hatough-Bouran and Disi, 1991; Kohle, 1981). Further excavation at this site may yield even more promising finds. The impact of human activity on the land is further discussed in the chapter on conservation.

Large-scale agriculture and animal husbandry (especially of sheep and goats) were relied upon in the 7th century BC, apparently arising after large populations of gazelles were decimated by mass hunting (Legge, 1972; Legge and Rowley, 1987). It appears that in the 9th to the 7th century BC, gazelle populations of northern Jordan migrated to Syria seasonally (in the late spring). After the young were born, gazelles returned to northeastern Jordan at summer's end. Evidence excavated at Tel Abu Hareireh in Syria indicates that gazelles were captured upon their arrival in massive structures called "desert kites" (because of the resemblance to a child's kite when viewed from the air). Remnants of these desert kites litter most of the Syrian Desert, including many areas of Jordan.

An area of special interest for study would be the investigation of faunal changes in the area since the Pleistocene. For example, three fossil forms of *Apodemus* were reported in deposits at the Tabun caves near Mount Carmel (Bate, 1942). These three forms perhaps correspond to the three species of *Apodemus* currently living in that area. Similarly, an extinct *Cricetulus, C. demetros*, was described from early Pleistocene levels at Wadi El Maghara and this may be the forerunner of the present day *Cricetulus migratorius* (Bate, 1943*b*).

Human History

Clearly "higher" primates (hominoids) occupied the Near East since the Oligocene Epoch, as evidenced by the rich fauna found at the Fayyum deposits in Egypt (Simons et al., 1978; Simons, 1990). Most of the early evolution of humans is now believed to have taken place in Africa. Human evolution in the Holy Land followed a path similar to that reported in Europe and northern Africa. The hand-ax tradition (called Acheulean culture) of early Stone-Age humans was documented from many parts of northern Palestine. Sometime after 100,000 years ago, a new technology of making stone tools emerged and spread to the Holy Land. The new culture (termed Mousterian) was more sophisticated and the chipped stones made were much more varied. The people who used these tools lived primarily in caves in the Holy Land and were classical hunters-gatherers.

In morphological terminology, the early Stone-Age man of the caves was given the name of Neanderthal (after the Neander Valley in Europe). Neanderthals were replaced gradually by more modern-looking humans sometimes between 30,000 and 50,000 years ago. Most scientists believe that these humans were direct descendants of Neanderthals. This is because Neanderthals stood erect, made tools, used fire, and buried their dead comrades. At that time, humans evolved further technological advances and underwent increasing social and cultural complexity. Evidence of the cultural sophistication of early cave-dwellers was documented at Paleolithic sites (between 17,000 and 10,000 years ago) in the form of dramatic paintings of hunting scenes and of rituals (examples are at Carmel Caves). Animal domestication and agriculture became inevitable advancements during the warming trend that followed. With sedentariness and agriculture came the development of centralized states (kingdoms) and the dawn of written history.

The history of the Holy Land is perhaps the most complicated and intensively studied of any region on earth. Civilizations flourished and perished in a succession that is most fascinating. One fact that would be agreed upon by all scholars is that the area was continuously populated at least since the Pleistocene (Neanderthals at Mount Carmel). Evidence of primitive sedentary cultures has been uncovered in many areas of the Holy Land. One of the earliest civilizations (7000–10,000 years ago) was the Assyrian culture centered around the Euphrates and Tigris rivers in Meso-potamia. This culture soon spread throughout the so-called fertile crescent (fertile agricultural lands that include Mesopotamia, Syria, Lebanon, and the Holy Land).

As complex societies evolved, centralized cities and complex communi-cation systems became necessary. These were first seen with the Sumerian culture of southern Mesopotamia. Some 5000–6000 years ago Sumerians were using wedges to make written tablet communications (a writing known as cuneiform). The first evidence of agricultural activity in the old world is also found in this area. Agricultural activity was first based on the domesti-cation of wheat, barley, peas, and lentils (Zohary, 1973). Nissen's (1988) book *The Early History of the Ancient Near East, 9000–2000 BC* is an excellent introduction to the study of early societies. The history of the societies of the Near East can be summarized in four phases: early settlements (9000–3200 BC), early civilization (3200–2800 BC), city-states (2800–2350 BC), and territorial/rival states (2350 BC onward).

It appears that throughout the past several thousand years, conquerors from nearby areas dominated the scene in the Holy Land for variable periods of time and the culture of the inhabitants was molded accordingly. The Philistines were a coastal people who established ports and a thriving economy, especially in the southeastern part of the Mediterranean. The area has since been referred to as the Land of the Philistines or Palestine. The name is misleading because the Philistine culture was short-lived. The more appropriate "Land of Canaan" reflects the dominant Semitic culture in the history of the area, that of the Canaanites. The northern Canaanite groups

(Aramaic and Assyrian) were especially dominant and built most of the ancient cities. The southern Canaanite groups (Hebrew and Arabic) became more influential after the establishment of the dominant religions of the region (Judaism and Islam). Most cities still populated today are "north" Canaanitic cities. Examples include Urusalim (Jerusalem) and Beit Lahm (Bethlehem) (Abel, 1938). The Phoenicians were a particularly influential group of northern Canaanites who excelled in trade, exploration, and sailing. Modern day Palestinians are a mixture of northern Canaanites with inter-marriage and influence by adjacent tribes and conquering groups (Samarites, Philistines, Babylonians, Hyksos, Egyptians, Amorites, Hebrews, Romans, Persians, and Arabs). This racial mixture started as tribal kingdoms began to collapse in the face of expanding empires. Besides these groups, significant tribes/kingdoms of the area were the Ashorians, Hebrews, and Akadians. During the Hebrew, Roman, and Arab conquests of the Holy Land, many tribal kingdoms were destroyed, assimilated, migrated, and/or continued to live as a minority in the Holy Land. Thus at the time of Christ and for a short period thereafter, the Roman governor was in charge of collecting taxes and keeping order in an area with diverse cultures and religions. The diversity of ethnic and religious backgrounds in the Holy Land continued with some loss of small minority cultures due to more powerful cultures. The Jewish revolt in AD 71 and the destruction of the Temple in Jerusalem marked a beginning of the Jewish diaspora during the Roman rule. The Roman rule did not last much longer than the 6th century AD when the Holy Land came under the control of the Muslim Caliphates. During that time Islam, Judaism, and Christianity were the only permissible religions.

For a brief period in the 12th century, Crusaders occupied the Holy Land under the guise of restoring Christian domination there. After the defeat of the Crusaders, a period of Islamic rule resumed. The Turks dominated the scene until the First World War. During World War I, France and Britain promised independence to "Arabia" in exchange for Arab alliance in defeat-ing the Turks. The Turks were removed from the Holy Land by Arab armies instigated by the famous British officer T. E. Lawrence ("Lawrence of Arabia"). However, Britain and France reneged on their promises and carved up the areas of the Near East as protectorates. France controlled Lebanon and Syria, and Britain gained control of Jordan and Palestine, as mandated by the newly formed League of Nations. The British allowed independence to Jordan under the rule of the Hashemite family (forming the Hashemite Kingdom of Jordan).

The situation in Palestine was more complicated. The Zionist idea espoused in the late 1800s resulted in massive organized Jewish settlement in the area beginning at the turn of the century. The British had given the green light to establish a Jewish homeland in the famous Balfour Declaration of 1917. The massive immigration enraged local Palestinians and a major revolt ensued in 1936. The United Nations (UN) recommended a two-state solution in 1947 with Jerusalem as an international city. Following uncertain times and bloody incidents, the British decided to withdraw and the State of Israel was established in 1948. Mass exodus of Palestinian refugees took place from the areas controlled by the State of Israel (an area larger than that in the UN resolution of 1947).

Some areas of Palestine were kept under the control of the Arab armies, which engaged in war in 1948. Jordan annexed the so-called "West Bank," and Egypt kept control of the "Gaza Strip." In 1967, Israel launched a war and gained control of these areas, as well as the Golan Heights from Syria and Sinai from Egypt. More Palestinians became refugees in Jordan, Syria, Lebanon, and other countries. Sinai was returned to Egypt following a peace treaty in the 1970s. Israel invaded Lebanon in 1982, then withdrew, but kept control of a strip of land in southern Lebanon as a "security zone." The Palestine Liberation Organization (PLO) was established in 1964 for the purpose of establishing a secular state in Palestine free of foreign influence. In 1974 the PLO formally recognized the 1947 two-state recommendation made by the UN. In 1993, Israel and the PLO exchanged formal recognition and agreed to work out their differences by negotiations. The reconciliation now underway between the Palestinians (three million outside Palestine and two million under Israeli rule) and the Israelis (four million with more immigration every year) will be a challenging task and must involve sharing natural resources and the small land area.

The various conflicts during this bloody history clearly had massive adverse ecological and human consequences. The standing countries of the Near East use enormous amounts of their gross national product for military spending. Wars takes tolls of lives on all sides not only among soldiers but also among civilians caught in the middle (as well as due to acts of terrorism). Yet, environmental destruction and human population growth continue at a rapid pace. Recent examples of the horrendous toll of war can be seen in the 1991 Persian Gulf war and the shelling of areas in Lebanon in 1996 and the devastating ecological and human consequences. The environment and inhabitants other than humans should not be ignored in the future balance of power in the region. This is because when peace and justice do prevail in the Holy Land (and they will), the local inhabitants—regardless of their political persuasion—must have a salvageable area to rebuild.

5

Mammalian Adaptations

The deserts, the Jordan Valley, and the mesic highlands (with a Mediterranean climate) form the three major habitats for mammalian species in the Holy Land (Fig. 5). Conditions in each habitat require that mammals have certain adaptations in behavior and morphology (form). For example, most areas have dry, hot summers and rainy winters. The Jordan Valley is hot most of the year with cultivation occurring only near sources of freshwater. The desert climate is dry for most of the year with little rainfall in the winter (average less than 100 mm). The different habitats from desert to forests and from mountains to seashores impose distinct requirements on mammals. Evolutionary change allowed mammals to acquire unique adaptations to their habitats. These adaptations are sometimes remarkable morphological modifications.

Behavioral and Physiological Adaptations

Many desert animals are nocturnal, thus avoiding the harsh heat of the day. Others are crepuscular, foraging only during the cool hours of the morning and evening. Animals in general also dig or seek sheltered areas during times of adverse conditions. For example, many desert gerbils build elaborate burrow systems with storage, nest, and escape areas. In the desert, the difference between the temperature one meter below ground and the surface temperature can be dramatic—at midday, the outside temperature can be more than 40°C, but the temperature one meter below can be 20–25°C. Some jerboas also plug their burrows during the day, thus trapping a higher humidity inside and conserving water.

Mammals may avoid times of food scarcity or adverse conditions by special adaptations allowing them to reduce their metabolism (for example, decrease heart and respiration rates) for long periods of time. This process

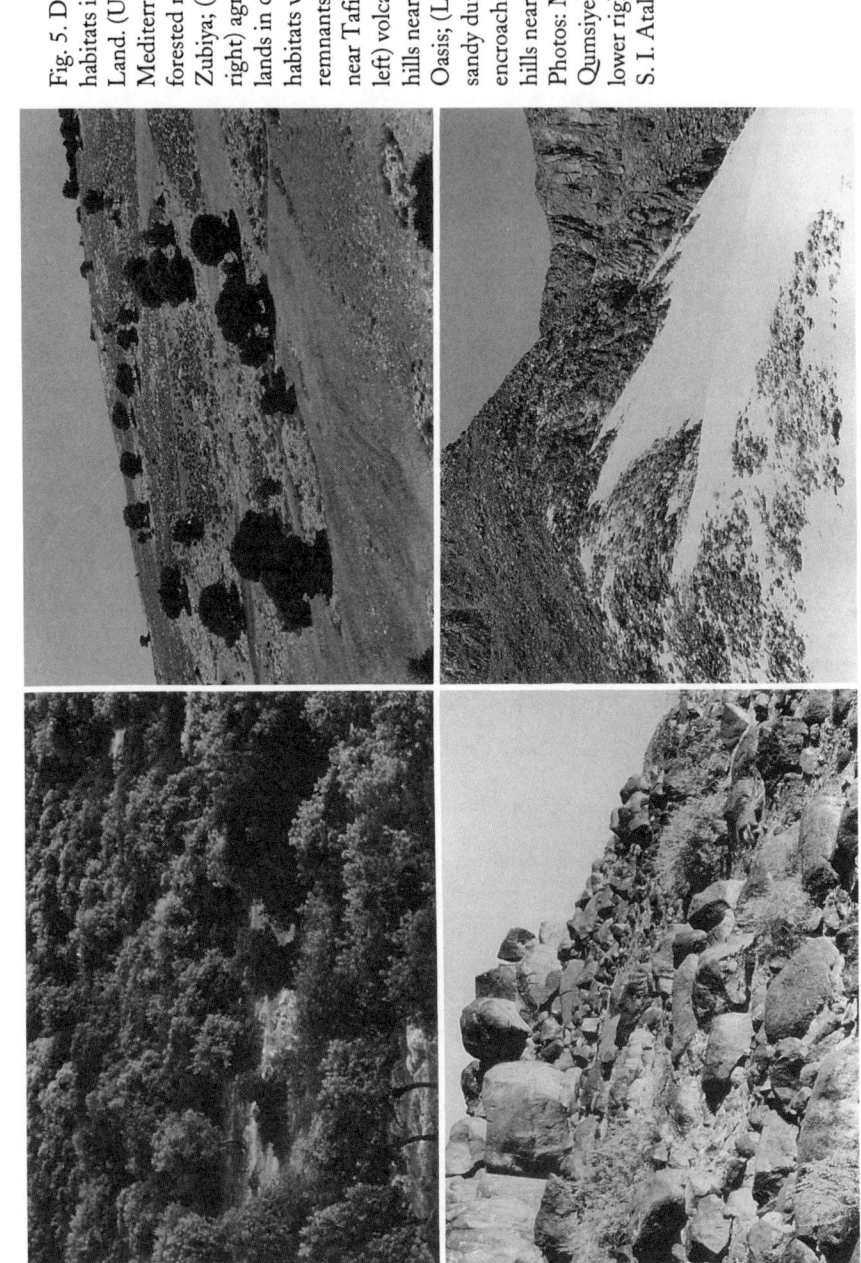

Fig. 5. Diversity of habitats in the Holy Land. (Upper left) Mediterranean forested region near Zubiya; (Upper right) agricultural lands in disturbed habitats with remnants of forests near Tafila; (Lower left) volcanic rock hills near Azraq Oasis; (Lower right) sandy dunes encroaching on arid hills near Rum. Photos: M. B. Qumsiyeh except lower right by S. I. Atallah.

is termed hibernation (for winter months) or estivation (for summer months). Some mammals living in the northern regions of the Holy Land hibernate during the winter; examples include the Mount Hermon vole (*Microtus nivalis hermonis*), and many species of horseshoe bats (genus *Rhinolophus*). In these animals, the body temperature falls significantly during hibernation (almost to the level of the roost temperature). They do not wake for several months and their pulse can barely be measured. If bats are picked up during hibernation and warmed, they will take several hours to awaken. However, this process will deplete a significant portion of their energy reserve and they usually will not survive the winter following such disturbance. Other mammals show partial hibernation, whereby they wake up occasionally to drink or even eat. Examples of that group in the Holy Land include the forest dormouse (*Dryomys nitedula*) and the Syrian squirrel (*Sciurus anomalus*). Desert mammals also may reduce their metabolism due to a lack of water for short periods of time (in rare cases for a whole season). As an example, many bat species estivate during the prolonged periods of drought in the summers.

Studies on the jerboa illustrate some of the physiological adaptations of desert mammals (Kirmiz, 1962, 1965). The jerboa has a resting metabolic rate and evaporation rate about half that of the common rat (*Rattus*). Its body temperature is also lower by 0.75°C (rat 37.55°C, jerboa 36.8°C). The jerboa and many other desert rodents can tolerate temperatures of 45°C by going into a state of lethargy (torpor). In addition, these animals can produce the water they need to survive from the food they eat. In the lab, individuals have lived for up to 3 years on a dry diet.

Many other physiological adaptations allow mammals to survive in desert conditions. For example, the limited water supply imposes significant demands for water conservation. Many desert mammals have evolved physiological mechanisms by which the kidneys can concentrate urine so as to discard the least amount of water. The concentration of urea in the urine of desert mammals, such as jerboas, jirds, and bats, is many times greater than that found in such mammals as humans.

Shrews of the genus *Sorex* that live in the mountains of the Holy Land posess a number of adaptations that enable them to cope with a cold seasonal climate, such as a high metabolic rate, fat storage during the warm season, no torpor, a large home range, solitary behavior, and a large litter size. Shrews of the genus *Crocidura* that live in the Jordan Valley likewise posess a different set of adaptations for life in a warm climate. For example, these shrews have a lower metabolic rate, little fat storage, torpor, small home range, social behavior, and small litter size with nest sharing (Genaud, 1988).

Size and Shape

Desert animals can lose more heat than their relatives from moderate climates because they have a greater surface area-to-volume ratio. The surface to size ratio is greater for smaller animals than for larger species. It is not surprising then, that animals from deserts and other warmer areas are generally smaller than animals from more moderate or cool climates. This relationship is termed Bergmann's rule. For example, foxes in the Holy Land are larger in size in the north (Mediterranean habitat) than in the south (arid habitat). Additionally, the surface area-to-volume ratio is larger in animals with elongated (relatively larger) appendages (tail, legs, ears) than in those with smaller appendages. Thus, one would expect longer extremities in animals from warm regions as opposed to those from colder regions. This relationship is termed Allen's rule. An example is offered by the wild hares, which have relatively larger ears and longer legs in the south than in the north of the Holy Land.

Locomotary Modifications

The soles of the feet of most sand-dwelling desert mammals are covered with hair. This is especially noticeable in such species as the lesser Egyptian jerboa (*Jaculus jaculus*), the lesser Egyptian gerbil (*Gerbillus gerbillus*), and the fennec fox (*Vulpes zerda*). The long balancing tail and long hind legs of the jerboas (both *Allactaga* and *Jaculus*) provide for a ricochetal (jumping) method of movement on desert sand.

In contrast to sand-dwelling groups, mammals inhabiting rocky areas have bare soles, allowing them to move easily without slipping in such habitats. Examples include the hyrax (*Procavia capensis*), the bushy-tailed jird (*Sekeetamys calurus*), Wagner's gerbil (*Gerbillus dasyurus*), and spiny mice (genus *Acomys*).

Pelage

As a rule, most animals possess coloration that closely matches their surroundings. Such camouflage occurs in desert animals with regard to the substrate on which they move (sand, soil, or rock). These substrates are surprisingly variable (for example, sand is not always light in color). Camouflaging pelage is especially advantageous in deserts where there is little cover and where considerable nighttime lighting is produced by the moon and the stars (there is usually little if any cloud cover).

The most striking example of the matching of the the color of the animal to the substrate in this region can be seen in the golden spiny mouse (*Acomys russatus*). Tristram (1877:41) made the following, and one of the earliest, comments on the desert coloration of a Palestinian rodent: "*Acomys dimidiatus*, a beautiful little isabel-coloured mouse, with the coat of a Hedgehog, I watched for an hour or two among bare gravel near the Dead Sea. Its color so perfectly harmonized with the pebbles that I could not have detected it had I not caught its eyes as it was nibbling a root, puffed up like a little ball." It is most likely that the animal Tristram described was *Acomys russatus*, the diurnal species and not *A. dimidiatus* (a synonym of *A. cahirinus*). A new subspecies was described almost a hundred years later from the same mountains that Tristram visited: *A. r. harrisoni* (Atallah, 1970*a*). More dramatic is the dark, almost black coloration of *A. r. lewisi*, which closely resembles its habitat of dark lava rocks in the Azraq Oasis (Atallah, 1967*c*; Harrison, 1975). Specimens of this species obtained elsewhere in the Holy Land are golden colored and also closely match their rocky habitats.

Dark coloration in desert animals, especially diurnal species, increases heat absorption. The almost black *Acomys russatus lewisi* is diurnal, although not active during midday periods. It was reported, based on studies of the paler-colored subspecies, *A. r. russatus*, that this species is more tolerant of heat (ambient temperature to 42°C) than the closely related nocturnal species, *Acomys cahirinus* (Shkolnik and Borut, 1969). Based on these data, we would expect that *A. r. lewisi* either has even higher temperature tolerance than other taxa of *Acomys* or that it compensates by decreased internal transfer of heat (by fat layer, convection, decreased transmission through hair [Cloudsley-Thompson, 1977]), or both. It would be interesting to investigate such problems further.

In addition to coloration, the pelage provides physical protection. Mammals at higher altitudes with cooler climates may have a thick pelage, which protects and insulates them against cold temperatures. In desert areas, mammals that are exposed to the sun develop a different adaptation. The camel's fur is glossy, short, and dense, which allows it to reflect heat and maintain a cooler temperature than that of its surroundings. Without this and several other adaptations, the camel would not survive the desert midday temperatures that can reach 70°C on the surface.

Hearing

Many desert mammals, including rodents, foxes, cats, mongooses, and weasels, have enlarged tympanic and mastoid portions of the middle ear.

These modifications are presumed to increase the animal's ability to hear in the conditions of wide open spaces and low wave conductance characteristic of hot, dry air (Legouix et al., 1954; Wisner et al., 1954; Lay, 1972). Petter (1968) showed that the sand jird, *Meriones crassus,* which has extremely enlarged tympanic bullae (ear bones), was able to home successfully from several kilometers even when deprived of sight. This may indicate a role for these specialized hearing organs in orientation.

Diet

The abundance of rodents in desert environments has been noted by many travelers and researchers. The logical explanation has been the habit of desert plants of storing nutrients in roots and bulbs during the dry season. Such plants provide an excellent source of stored food for the many rodents of these desert regions (Tristram, 1877).

Many desert animals drink rarely and conserve water by various means. Others seem to derive all their water from their diet. The fennec fox (*Vulpes zerda*) derives moisture from the plants, insects, and small vertebrates on which it feeds (Schmidt-Nielson, 1964).

Fat Storage

In cold climates, mammals such as polar bears and seals store fat in an even layer under the skin and this acts as an effective insulator. Desert animals also have to store fat to cope with times of food shortages. However, the distribution of fat is significantly different in that it is usually localized to certain areas of the body. The remainder of the body is then free to dissipate heat. The most dramatic example of this is in the mouse-tailed bats of the genus *Rhinopoma.* These bats roost in shallow caves in desert regions where the temperatures, even in the deepest areas of the cave, can reach 32–40°C. They also need large fat deposits because they do not migrate and food shortages can be significant in certain periods of the year. The fat deposits are localized in a small area of the lower abdominal region.

6

Mammalian Parasites and Human Health

Some mammals in the Holy Land, as elsewhere in the world, are vectors of human disease. Some of these diseases have been with humans at least through recorded history. Plague, which produces severe septicemia, is caused by the bacillus bacterium *Yersina pestis* (formerly *Pasteurella pestis*). The bacteria are transmitted from the blood of rodents to man through flea bites. The disease is rarely transmitted from one infected human to another. This is because human blood carries less bacteria than that of rodents (humans succumb more quickly). Fleas reported to transmit this disease include the human flea *Pulex irritans* and such rat fleas as *Xenopsylla cheopis* and *Leptopsylla segnis* (review by Twigg, 1978).

Lice are known to transmit epidemic typhus, a disease caused by *Rickettsia prowazakii*. The species of lice involved include body lice (*Pediculus humanus*) and crab lice (*Phthirius pubis*). The latter is named because it primarily infects hair of the pubic region.

There is some inconclusive evidence that bats may act as reservoirs for human disease. The common Kuhl's pipistrelle (*Pipistrellus kuhlii*) was found to carry a spirochete infection similar morphologically to that of human relapsing fever (Zulueta et al., 1971). Rabies has been found in a small percentage of bats in Europe and North America. However, the rate of infection among bats is lower than among such domestic animals as dogs. Also there is little chance of human contact through bats. Many other bat parasites are known in the Holy Land (Theodor and Moscona, 1954) but are harmless to humans (most can only live on bats).

Cutaneous leishmaniasis is a disease caused by the protozoan parasite *Leishmania tropica*. The disease causes ulcers and lesions of various degrees, which can produce permanent disfigurement. The parasite life cycle includes sand flies of the genus *Phlebotomus* as well as reservoir rodents such as the fat sand rat (*Psammomys obesus*), jirds (*Meriones*), and rats (*Rattus rattus* and *Rattus norvegicus*) (Morsy et al., 1981; Oumish et al., 1982; Saliba et al., 1985).

Murine typhus was also known in the Holy Land (Kligler and Comaroff, 1936; Royen et al., 1944). Fortunately, with medical advances and with increased education, many of the transmitted diseases in the Holy Land have been controlled. Diseases transmitted by mammals now contribute to a negligible proportion of hospitalizations in the Holy Land with deaths from such cases being quite rare. On the other hand, animal parasites are important for economic reasons as they may cause significant losses in both domestic and wild mammals. Thus, a knowledge of these parasites is important.

7

Geography and Ecology
of the Holy Land

The history of the Near East (the "Middle East") is such that political
boundaries cannot be reliably used for distributions of animal or plant
species. Two main reasons for this are the frequent political changes and the
fact that political boundaries and names do not correspond to geographic
areas. Boundaries and their accompanying names have changed throughout
history. Examples include the Land of Canaan, and the kingdoms of
Samaria, the Philistines, the Hebrews, Israel, Judea, and Palaestina. In the
past century, the land west of the Jordan Valley was known as Cisjordan or
Western Palestine, and the land east of the Jordan Valley as Transjordan or
Eastern Palestine. These terms were no longer used by the time of the British
Mandate on this area (1919–1947) and the areas simply were referred to as
Palestine and Jordan, respectively. More recent disputes over the boundaries
of Israel, Jordan, and Palestine prevent the general use of these terms for the
area under study. Thus, and for a lack of a better term, I have chosen "the
Holy Land" as a name for this area, which represents a unique geographic
region at the crossroads between Africa and Asia.

The Holy Land has a surprising variety of habitats for such a small area.
In fact, it may be true that no other land on earth offers such a variety of
scenery and climate in such a small place. Topography and geography allow
division of the region into major longitudinal areas and latitudinal division
within those areas into biotopes. From west to east, the areas that will be
discussed are as follows: the coastal plains, the western foothills and highlands, the Jordan Rift Valley, the eastern highlands, and the Syrian Desert.

The Coastal Plains

For most of its length in the Holy Land, the Mediterranean coast is smooth and featureless. Only between the cities of Haifa and Acre is there an inlet (gulf). In most coastal areas, the land contacts the sea in sandy smooth beaches, but in places the rocky foothills slope westward to contact the sea in small overlooks. The hills of Galilee come closest to the sea at Ras en Naqura. From there, the coastal plains slowly expand southward. From north to south, the coastal plains increase in width from about 4–5 km near Acre to about 40 km near Gaza. Three major subdivisions of the coastal plains are recognized. The Plain of Acre is about 32 km long and extends from Ras en Naqura to Mount Carmel near Haifa. This area shows varied hilly overlooks and small sandy beaches. The Sharon Plains extend from there south to the Yarkon River (a distance of about 130 km). Finally the Philistia Plains are about 110 km long.

The Western Foothills and Highlands

The western highlands slope gently into the coastal areas by a series of foothills, noticeable especially in the south. These foothills (known as the shephelah, meaning lowlands) consist mainly of limestone and average about 15 km wide. They were heavily forested in historic times by *Pistacia* and other trees. For example, the Bible contains statements such as "and he made cedars as plentiful as the sycamore trees that are in the shephelah" (1 Kings, 10:27). The western mountain ranges can be divided from north to south into the Galilee, Samaria, Judea, and Idumea.

The Galilee Mountains rise from the Nahr El Litani in south Lebanon to reach their highest peak at Jebel Jarmaq (Mount Meiron) at 1200 m. To the south, these mountains slope gently to the Jezreel Plain (the biblical Esdraelon). This plain is a branch of the Great Rift Valley and extends from the Jordan River area to the seashore near Haifa. Because of this topography, the Jezreel Plain is a fertile strip of land reaching 20 km wide in some areas. The mountains bordering the Jezreel Plain to the southeast near Haifa are the Carmel Mountains (highest peak at 550 m). Further south, the mountain range around Nablus and as far south as Ramallah is known in the ancient Canaanite language as Samaria. These mountains are continuous with the mountains surrounding Jerusalem and Bethlehem. However, the latter hills are more arid and are known as the Judean Hills. Some authors further recognize the mountain range around Hebron to the south as separate (Idumea). These mountain ranges gradually merge with the arid mountains of the Negev and then the Sinai. Because of lower rainfall, the southern mountains are less productive in agriculture than the Galilee and Samaria mountains.

The western faces of the mountain ranges in the north collect significant precipitation and are fertile. The major rivers generated drain to the Mediterranean coast. The Qishon River enters the Mediterranean near Haifa and the Yarkon River enters near Jaffa-Tel Aviv. In the south, the mountain ranges taper into arid regions in the west and even more arid regions in the south (the Negev Desert plateau). The northern limit of the Negev Desert is near Beersheba and Sodom. Deserts include varied biotopes and the distributions of mammals reflect these biotopes. Rocky mountains are common in the Negev, in many areas on the rim of Wadi Araba and in the lava hills of the Eastern Desert. Rocky desert mountains have sparse vegetation and it is sometimes surprising to find an abundant and diversified mammalian fauna. In some instances, rodents on these slopes depend on prevailing winds to bring them seeds from the wadi beds. Examples of rodents found in such rocky habitats are bushy-tailed jirds (*Sekeetamys calurus*), Wagner's gerbils (*Gerbillus dasyurus*), spiny mice (*Acomys russatus*, *Acomys cahirinus*), and Sinai dormice (*Eliomys melanurus*). In more vegetated mountain areas or those overlooking permanent water sources, such larger animals as hyraxes, ibexes, and leopards are found.

The Jordan Rift Valley

The Great Rift Valley was formed gradually in geologic time by the contact between the Arabian and the African tectonic plates. In the Holy Land, this geologic feature encompasses the Hula Basin, the Jordan River Basin, the Dead Sea, and Wadi Araba with minor branches such as the Jezreel Plain to the west. It further extends south through the Gulf of 'Aqaba and the Red Sea, then into Ethiopia and Kenya. The northern horn of the Rift Valley in the Holy Land opened recently in geologic times (Gridler and Southern, 1987). The first stage occurred some 25–15 MYA and the second some 4.5–0 MYA; the very northern tip was the most recent to open. Thus, there is much similarity between the biotopes and mammal distributions on the mountains east and west of the Jordan Valley.

Significant snow and other precipitation fall on Mount Hermon during the winter months. Mount Hermon, the Galilee, the Basan, and the Golan Heights drain into the Hula basin, a depression representing the northern branch of the Rift Valley in the Holy Land. It is this significant amount of water that resulted in the formation of Hula Lake and many marshes and swamps. Here a unique assemblage of plants and animals is found that to this day has been poorly studied. Extensive settlement and study of this unique area were hindered by the prevalence of malaria and the physical barrier of the swamps. A major engineering project was initiated by the State of Israel to drain this basin and open it for development. The project, completed in 1957, involved building two canals running from north to south; the swamps quickly dried up exposing fertile agricultural lands. It is unfortunate that the environmental impact of such a project was not carefully evaluated, although a small area was saved in its natural habitat (the Hula Nature Reserve). However, experience in other parts of the world indicates that many species cannot be saved in such small areas (especially those species that are large or highly mobile). The extent of the environmental damage in that area may never be known.

The water from the Hula Basin follows the depression in the Rift Valley to empty into the largest freshwater lake in the region—Lake Tiberias (Galilee Lake). The Jordan River stretches from Lake Tiberias to the Dead Sea in the south. The rich agricultural valley surrounding the Jordan River is commonly referred to as Al Ghor, which literally means the "depression." Many valleys both on the east and the west empty into the Jordan River. Several of these valleys carry permanent streams of great economic and ecological significance. The Yarmouk Valley and River in the east form a natural boundary between Jordan and the Golan area. In the west, Israel established a canal from the mouth of the Yarmouk River to irrigate lands controlled by settlements west of the Jordan River. Similarly, the Jordanian government built a canal about 75 km long that supports agriculture in the Jordan Valley east of the Jordan River. These and various other draining projects resulted in a drop in the water level of the Dead Sea and in the formation of large salt plains. In the south the depth of the Dead Sea Basin fluctuates from 0 to 20 m. Desiccation also can be seen clearly by an increase of the area of the Lisan, a stretch of land surrounded by water on all but its eastern side. In fact, the lines depicting the shores of the Dead Sea in Figure 2 are only those for maximum status in the last 20 years. The adverse impact on the ecology of the Yarmouk and Jordan valleys remains to be assessed. Another river that has been affected by human activities is the Zarqa River (the ancient Jabbok River), which was tapped at King Talal Dam for hydroelectric power. Again, little ecological assessment was performed to study the impact on the environment. Another major water source is the joining of the Heidan and Mujib tributaries to pour into the Dead Sea in the area of Ras el Ghor.

After collecting many tributaries and springs, the Jordan River finally empties into the Dead Sea, which lies at some 400 m (1315 ft) below sea level. Several springs (for example, Ain Fashkhah, Ain Gedi, Zarqa Ma'in) also pour freshwater directly into the Dead Sea. However, the high rate of evaporative water loss of the Dead Sea over thousands of years has resulted in an extremely salty inland sea (more appropriately a lake), which supports no life with the exception of some extremely halophilic (salt-adapted) algae. The rate of incoming water has dropped significantly in recent years because of extensive diversion of freshwater for agricultural purposes. The largest diversion project is one discussed earlier that was constructed by the Israeli government near the Yarmouk River.

The southernmost section of the Rift Valley in the Holy Land is the Wadi Araba. This is primarily an arid hot region about 160 km long and reaching widths of 30–32 km in some areas. The soil consists of alluvial sands and gravel deposited by ancient rivers and recent floods from valleys both from the west and the east.

The Eastern Highlands

The Hermon Mountains (with highest elevation at about 2800 m) slope gradually to the south forming the extreme northern tip of the area under study. From the foothills of Mount Hermon in the southeast rise the Basan and Golan heights. They abruptly terminate at the Hula Basin in the west and at the Yarmouk River in the south. However, they slowly merge into the Syrian Desert and the Houran Hills in the east. Further south, the mountain ranges are traversed by deep valleys that empty into the Jordan Valley, the Dead Sea, or Wadi Araba. In Jordan, the northern mountain ranges of Ajlun (the ancient Gilead) run between the Yarmouk and Zarqa river valleys. They abruptly terminate in the west at the Jordan Valley, whereas in the east they slowly slope into the Syrian Desert. The Yarmouk River is the largest river in the north and empties into the Jordan River south of Lake Tiberias. It forms a natural boundary between the Ajlun Mountains and the Golan Heights. The Yarmouk spreads east into the Plain of Houran. Further east, this plain rises steeply to the Druz Mountains, which were formed by ancient volcanoes. The highest peak of this range is at Jebel ed Druz (2000 m).

Between the Yarmouk River in the north and the Zarqa River in the south lie the Ajlun Mountains (the biblical Gilead). The highest elevation is at Jabal Um Al Daraj (1247 m). Further south, the Hills of Amman (ancient Ammon and Moab hills) are bordered by the Zarqa River to the north and Wadi El Hasa in the south. These and other rocky mountain ranges in arid regions slope steeply into the Araba Valley in the west and more gradually to the Syrian Desert in the east. The mountainous regions are traversed by many valleys, which have been carved by water and wind over thousands of years. Valleys that carry river beds, especially those that are dry in some part of the year, are referred to as wadis. Wadis are common throughout the Negev Desert, the overlooks of the Araba, and the Syrian Desert. In most areas, the rock formations are igneous or metamorphic, with limited water porosity. By the nature of such formations, wadi beds receive significant amounts of water flow when there is rain, even with as little as a few millimeters of rain. Many such wadi beds are flooded quickly, and because of this flooding, plants survive only on the banks and shelves. The major valleys and wadis east of the Rift Valley are (from north to south): the Zarqa Valley (water throughout the year); Wadi Shuaib extending to Wadi Nimrin as it enters the Ghor (Ghor Nimrin) near Shuna; Wadi en Nusariyat joining Wadi Es Sir and Wadi Na'ur; Wadi Zarqa Ma'in (continuous water at least in the upper part, including hot springs); Wadi El Mujib; Wadi El Karak; and Wadi El Hasa.

The Edom Mountain region is bordered in the north by Wadi el Hasa and on the north and east by the Arabian Desert. The Jabal El Ataita in the north reaches an elevation of 1640 m, the Shera Mountains reach a height of 1850 m, and the southern Edom Mountains reach an elevation of 1590 m at the mountain of Shafat Ibn Jad. Many wadis provide drainage to Wadi Araba, with Wadi Finan being one of the major tributaries.

The Syrian Desert

The Syrian Desert forms the northern segment of the great Arabian Desert. The deserts of the eastern parts of the Holy Land receive limited precipitation (20–80 mm/year) and in many areas are covered by sandy dunes. Sandy areas support such animals as hairy-footed gerbils, jerboas, desert jirds, and desert foxes. Hard-soil deserts can include wadi beds that support more vegetation and are rich in mammals that require more moisture. Another type of hard-soil desert is the hammada. This is the Arabic name for deserts that are covered with gravel and large rocks but have soil underneath that supports some vegetation. The hammada habitat is used by many rodent species including the Baluchistan gerbil (*Gerbillus nanus*), the Libyan jird (*Meriones libycus*), the three-toed jerboa (*Jaculus jaculus*), and the less common five-toed jerboa (*Allactaga euphratica*). This habitat is also rich in birds, lizards, snakes, insects, and other life. This combination of prey species supports carnivores such as the honey badger (*Mellivora capensis*), and the striped hyena (*Hyaena hyaena*). In the not too distant past, the hammada and other hard-soil deserts (especially near wadis) supported gazelles, cheetahs, and wolves.

Climate

Rainfall varies significantly in the Holy Land from more than 1000 mm (about 40 inches) in the Lebanon Mountains and on Mount Hermon to less than 20 mm in certain desert areas of the southeast. In addition to rain, many areas receive dew precipitation. In the coastal plains, dew falls about 250 nights per year (much less occurs in the hills). The Holy Land receives rainfall between November and April with peaks in January and February. August is the warmest month with high temperatures of about 30–32°C in the hills near Jerusalem and near Amman. In Jericho and deserts of the east, the high temperature can be 37–39°C. January is the coldest month with lows of 0–5°C in the central mountains.

Snowfall is rare south of Galilee and Mount Hermon. Snowfall near Jerusalem and Amman averages 1–3 days a year and usually melts within a few hours. Because of this rather mild climate, few European species of mammals occur farther south than the northern highlands. In the forested regions of the Galilee, and on Mount Hermon, small mammals are forced to hibernate in the winter months. Examples include voles, dormice, and field mice.

The mild to cold climate in the mountain region sustains (or used to sustain) a Mediterranean forest that is dominated in many places by oaks. The primary trees are *Quercus calliprinos*, *Quercus ithaburensis*, *Pistacia* sp., and *Ceratonia siliqua*.

8

Zoogeography of Mammals

Patterns of Distribution

Zoogeography is the field of knowledge concerned with the study of the distribution of animals on earth. Distributions of animals occur in distinct geographical clusters. These areas are recognized by biologists as distinct biogeographical regions. The biogeographical regions are as follows: the Australian (Australia and New Guinea), Oceanic (Pacific Islands), Nearctic (North America), Neotropical (South America), Palearctic (Europe, North Africa, southwestern Asia, and most of Asia north of the Indian and Malaysian subcontinent), Ethiopian or Afrotropical (Africa south of the Sahara and southwestern Arabia), and Oriental (India, Southeast Asia). The latter three regions are relevant to the Holy Land. Most mammals in the Holy Land have Palearctic affinities; a few have affinities to the Ethiopian and Oriental regions.

Early zoogeographic treatments of the mammals of the Holy Land (Nehring, 1902*a;* Aharoni, 1932; Kosswig, 1955; Bodenheimer, 1958; Niethammer, 1987) provide an excellent introduction to the subject but are limited by the lack of proper distribution maps for the species. I previously tabulated species of mammals in 14 areas in the southern Palearctic region including the Holy Land and Sinai (Qumsiyeh, 1985). Results of numerical analyses on these data were intriguing in that the Holy Land was closest to Iran and Afghanistan when analyses were limited to bat distributions. On the other hand, the Holy Land grouped closely with Sinai and Egypt when land mammals were analyzed without the bats. Other than this disagreement, the relationships of areas based on bat distributions were similar to those based on land mammal distributions. The explanation for the discrepancy may be due to one or more of the following factors. Many bats are widespread in the Palearctic region and thus occur in Iran, Afghanistan, and the Holy Land. On the other hand, land mammals appear to be more restricted in their distributions. Many north African land mammals have Ethiopian affinities and many of these reach their northern limit in Egypt and only occasionally enter the Holy Land.

A similar study of distribution and area relationships was carried out on the rodents of Arabia and Asia Minor (Turkey) (Neronov et al., 1987). Results of this study and my earlier study suggest that the mammals of the Holy Land can be divided into three main zoogeographical groups:

1. Mediterranean. This is a distinct subregion within the Palearctic region. It includes mountain areas of southern Europe, northwestern Africa (i.e., the Maghreb), the Jabal Akhdar region in Libya, and the mountain ranges of the eastern Mediterranean. Examples of Mediterranean mammals in the Holy Land include: *Crocidura suaveolens, Rhinolophus ferrumequinum, R. euryale, R. mehelyi, Myotis blythii, Spalax leucodon, Eliomys melanurus, Microtus guentheri,* and *Apodemus mystacinus.*

2. Saharo-Sindian (also referred to as the Saharo-Arabian and Irano-Turanian phytogeographic region by Zohary [1973]). This is another subregion within the Palearctic and includes the Sahara Desert, the Arabian Desert, and the deserts of southwestern Asia to Sind in southern Pakistan. Examples of these mammals are numerous: *Paraechinus aethiopicus, Hemiechinus auritus, Rhinopoma hardwickei, R. microphyllum, Asellia tridens, Eptesicus bottae, Otonycteris hemprichi, Gerbillus pyramidum, G. gerbillus, G. dasyurus, G. henleyi, Sekeetamys calurus, Meriones crassus, M. libycus, M. sacramenti, M. tristrami, Psammomys obesus, Acomys russatus, Jaculus jaculus, J. orientalis, Allactaga euphratica, Vulpes rueppelli, V. zerda,* and *Gazella dorcas.*

3. Ethiopian or Afrotropical. Many species of African mammals reach their northern limit in the eastern Mediterranean region. Examples are: *Rousettus aegyptiacus, Nycteris thebaica, Rhinolophus clivosus, Pipistrellus rueppelli, Herpestes ichneumon, Procavia capensis,* and *Alcelaphus buselaphus* (the latter extirpated in the Holy Land).

In extreme northwestern areas of the Holy Land (north of the Jezreel Plain), one finds an almost pure Mediterranean fauna. In Wadi Araba and the deserts of the south, the fauna is predominantly Saharo-Sindian (also known as the Eremian subregion) with a few Ethiopian or Mediterranean species. Between those two areas are areas occupied by a mixture of mammalian faunas.

I previously commented (1985) that it is noteworthy that many of the Saharo-Sindian species have their ranges centered around the northern Red

Sea. This observation may explain the numerous endemic forms in this region including *Pipistrellus ariel*, *P. bodenheimeri*, and *Eptesicus bottae innesi*.

Factors that Affect the Distribution of Mammals

The distributions of mammal species in the Holy Land, as elsewhere, are determined by habitat preferences, interspecies competition, history, and distributional barriers.

Habitat.—Evolution and adaptation to a particular habitat are important in determining distributions. Clearly, a species that is not adapted to live in desert areas will not be found in those areas. Thus, and as a rather simple example, the Syrian squirrel is found in the Holy Land only in areas with remaining rather dense forests that can support this arboreal species. It follows also that the range of a species will change with changing environments unless the species can adapt (evolve) to meet the new requirements of the new habitat. This is one way in which subspecies and species evolve. For example, the expanding Sahara desert forced many species (for example, voles, deer, hamsters) to retreat northward (their ranges decreased). However, some northern species remained and through time adapted to the increasingly arid conditions. Depending on the degree of adaptation and the length of time since they split from their ancestral species, these forms are now either new subspecies, species, or even genera. The Palestine mole-rat, (*Spalax leucodon*), has formed four chromosomal forms in the Holy Land and the southernmost form is the one with the highest diploid number and with the most xeric adaptation (Nevo and Bar-El, 1976; Nevo et al., 1990). *Pipistrellus kuhlii* and *Pipistrellus aegyptius*, on the other hand, are closely related species with the latter found only in the arid regions of the Sahara (Qumsiyeh, 1985). Three species of jirds (*Meriones tristrami*, *M. shawi*, and *M. libycus*) also probably evolved in this way (Qumsiyeh and Chesser, 1988).

Interspecies competition.—When the ranges of two closely related species are adjacent to each other (usually in two unique habitats), it is difficult to judge whether the lack of overlap in their distribution is due to habitat preference, interspecific competition, or both. Indeed the two variables should not be discussed independently. This is because each can lead to the other. Species A could split into two populations (B and C) in response to an environmental gradient. If the environment keeps changing, population—and now species—B could advance and replace species C in certain areas. This would be considered interspecies competition in those areas. In very few instances we can clearly demonstrate that there is a limiting factor (food, space) that is shared between the two species. Studies of limiting factors would be most productive in cases where species come close together in intermediate habitats. For example, the distributions of the European and the Ethiopian hedgehogs in the Holy Land are allopatric (that is, occurring in different areas). However, the two species come very close together in certain areas near the Mediterranean coast (Fig. 7) and it would be interesting to find an area where both species occur or to study food utilization in experimental enclosures. Similarly, the distributions of the mountain gazelle and the dorcas gazelle are also allopatric in the Holy Land.

A strikingly similar distribution of arid- as opposed to mountain-adapted forms occurs in many closely related species of mammals in the Holy Land. Three groups are recognized: group A, which is adapted to a Mediterranean climate and occurs in areas with an average annual rainfall of more than 100 mm; group B, which is a desert-adapted group restricted to areas with an average rainfall of less than 100 mm; and group C, which is intermediate and overlaps A and B but avoids the extremes of both habitats. The following is a partial list of representatives of these groups in the Holy Land.

Group	A: Mediterranean	B: Desert	C: Intermediate
Hedgehogs	*Erinaceus concolor*	*Paraechinus aethiopicus*	*Hemiechinus auritus*
Horseshoe bats	*Rhinolophus ferrumequinum*	*Rhinolophus clivosus*	*Rhinolophus blasii,* *R. mehelyi*
Jirds	*Meriones tristrami*	*Meriones crassus*	*Psammomys obesus*

Historical factors and distributional barriers.—In many cases mammals are not found in certain areas that are suitable for their survival simply because they have not been able to reach those areas. The presence of barriers seems to have the most significant effect on distributions of small and poorly mobile animals. The short-tailed bandicoot rat (*Nesokia indica*) could thrive in marshes and wet-soil habitats in the north but is found only around the Dead Sea and oases to the south. Similarly, the water vole (*Arvicola terrestris*) could potentially thrive in other areas of the Holy Land with marshes or swamps but is now found only in the Hula Valley. The water vole is found in Egypt in habitats very similar to those of oases in the Wadi Araba.

The significance of climatic shifts in the Pleistocene on the distribution of mammals cannot be overemphasized. The Pleistocene in the Near East was characterized by periods of cool moist climates (termed pluvials) alternating with dry hot climates (interpluvials). Pluvials and interpluvials correlated with the glacials and interglacials in the Northern Hemisphere. In pluvial times, northern species expanded their ranges southward, reaching in some instances areas as far south as Egypt and the Sudan. When the glaciers receded, these species were forced back north, with occasional remnant populations left isolated on mountain ranges. The most striking example of a remnant fauna since the last pluvial is the fauna of Jabal Al Akhdar (the Green Mountain) in Libya. Many typical European mammals occur there surrounded by a typical North African desert fauna. Such European species include field mice (*Apodemus*), voles (*Microtus*), dormice (*Eliomys*), noctules (*Nyctalus*), common pipistrelles (*Pipistrellus pipistrellus*), long-winged bats (*Miniopterus*), and genets (*Genetta*) (Ranck, 1968; Qumsiyeh and Schlitter, 1982). Similar remnant populations of field mice and mole-rats (*Spalax*) are found in the mountains of south Jordan. Another example may be the distribution of the mountain gazelle (*Gazella gazella*), which extends south in Jordan along the mountain ranges; a remnant population occurs in the south of Wadi Araba apparently surrounded by the more xerophilic species, *Gazella dorcas*.

9

Conservation

"These are the animals which you may eat: the ox, the sheep, the goat, the deer, the gazelle, the roebuck, the wild goat, the ibex, the antelope and the mountain sheep" (Deuteronomy, 14:5).

"And the earth—we have spread it out wide, and placed on it mountains firm, and caused life of every kind to grow on it in a balanced manner, and provided means of livelihood for you as well as for all living beings whose sustenance does not depend on you" (The Holy Quran, 15:18–19).

In prehistoric times (late Pleistocene), the Holy Land harbored an abundant rich fauna that included rhinoceros, hippopotamus, spotted hyenas, and even elephants. Most of these were lost before recorded historical times. The first documented massive exploitations of large mammals occurred with the earliest reported settlements of hunter-gatherers. These first villages were the sites of the first destruction of the large populations of gazelles that used to roam the area (Legge, 1972; Legge and Rowley, 1987). Travelers in the last century, such as Hart (1885) and Tristram (1866b, 1876, 1884), reported an abundance of large mammals. Hart (1885:251) reported talking to Towarah Bedouins near 'Aqaba who "knew of leopards on Serbal and Umm Shaumer; wolves in Wady Lebweh and neighborhood; hyenas, ibexes, gazelles, hares, jerboas, and mice." The fertility of these plains near 'Aqaba, where even prides of lions occurred, was referred to in the classical text of Diodorus Siculus and quoted by Bodenheimer (1960). Reminders of these animals can be found in names of places in the Holy Land. Examples include Beersheba (Lion Spring), Ain El Asad (Lion Spring), Ain Ghazal (Spring of the Gazelle), Jebel Ghazaleh (Mountain of the Gazelle), Ghor Nimrin (Leopard Depression), Wadi Namr (Leopard Wadi), and Wadi Yahmur (Roe Deer Wadi). Many of these creatures became extirpated, and the few remaining ones are endangered. In this chapter, I will review the causes of environmental damage, the history of conservation efforts, and the successes and failures of protecting the endangered mammals of the Holy Land.

The major environmental concerns for mammals of the Holy Land are habitat destruction and massive exploitation (hunting, poisoning, trapping). As in most of the world, "civilization" in the Middle East meant environmental destruction. As the first agricultural settlements were established in the Natufian Period, clearing of forests began. The scale of deforestation increased at the turn of the 20th century. For example, the Turkish army destroyed much of the forests in Jordan to construct the railroad and to provide fuel for the locomotives of the Hejaz railway linking Turkey and Syria with the holy cities of Mecca and Medina in Saudi Arabia (Hatough-Bouran and Disi, 1991).

The first law intended to protect wildlife was enacted in 1924 by the British Mandatory government of Palestine and was titled "Game Preservation Ordinance." This law was a step in the right direction, but there was little enforcement. In fact, the three decades that followed were characterized by a breakdown of law and order in the region, where small groups of armed individuals (British and government troops, local Arabs, and Jewish settlers) roamed the wilderness. This situation was conducive to massive destruction of large mammals as hunting became a respite from occasional skirmishes. When the State of Israel was established in 1948, hunting was prohibited for one year (although not primarily as a conservation measure). After a request by a group of zoologists, hunting in the Negev Desert was forbidden in 1951. In 1953, concerned zoologists established a private society named the Society for the Protection of Nature in Israel (SPNI). The SPNI is a public organization receiving most of its funding through donations and is concerned primarily with education and public awareness issues. Lobbying by this group and other concerned citizens resulted in the "Wild Animal Preservation Law" enacted in 1954. This law afforded some wildlife protection and hunting regulation, but as in previous laws, little enforcement existed. Further pressure from SPNI resulted in the establishment of a government body, the Nature Reserve Authority (NRA) in 1964, which began to prosecute offenders. Hunting and poisoning of many animals such as wolves continued and perhaps increased in the 1960s due to lax enforcement and the consideration of predators as nonessential even by government agencies (Mendelssohn, 1982). Presently, all mammals are protected with the following few exceptions: fruit bats, hares, Tristram's jirds, nutria, hyraxes, commensal (human-associated) rats and mice, and wild boars.

The NRA is attached to the Ministry of Agriculture and is the body that runs the reserves and enforces nature protection laws. The SPNI is the main public awareness organization in Israel and is successful in its educational programs. The SPNI established the Mammal Information Center in 1984 to collect data and distribute information about mammals and mammal conservation. Other organizations are primarily concerned with creating a better environment for man's benefit. For example, Keren Kayemeth Le'Israel is a society that raises funds and coordinates efforts to establish parks and perform other activities that create a better environment for human settlement. Their efforts includes large-scale forestation, usually with *Pinus halepensis,* even though this may not be an ecologically wise choice (Tchernov, 1980, unpublished report).

Several reserves were established in Israel (starting around 1963) to protect wildlife and four of these are now established national parks: Ain Avdat (Abde), Gan Hashlosha (Sachne), Hurshat Tal, and Ma'ayan Herod. There are many other important areas for the conservation of mammals, of which a few are of special interest:

1. Mount Meron: This area (9600 ha) in the Upper Galilee is one of the best run reserves. It affords protection to such forest-dwelling mammals as mountain gazelles, wild boars, jackals, hyraxes, polecats, martens, and many others.

2. Mount Hermon: A small reserve of about 186 ha with a rich mammalian fauna including wolves, martens, wild boars, hyraxes, squirrels, hamsters, and voles.

3. Yehudiya Forest Park: Located in the Golan Heights, Yehudiya Park encompasses 6620 ha and harbors rich populations of gazelles, wild cats, jackals, wild boars, and other temperate-climate mammals.

4. Beit Zayda: A wetland habitat near Lake Tiberias with such mammals as otters, Egyptian mongooses, and badgers.

5. Mount Carmel: This area includes steep cliffs with caves that protect many bat species as well as such mammals as hyraxes, badgers, porcupines, and foxes.

6. Ein Gedi: The largest (2780 ha) reserve in the Judean Desert located on the western shores of the Dead Sea. The fauna includes leopards, Blanford's and Rüppell's foxes, ibexes, gazelles, bats, and many other desert-adapted species.

7. Hanehalim Hagdolim: A Negev reserve with large wadi systems. Protected mammals include Blanford's and Rüppell's foxes, wolves,

hyenas, caracals, wild cats, desert hedgehogs, and jerboas. Syrian wild asses were reintroduced here.

8. Eilat Mountains: This is the southernmost reserve with populations of the Dorcas gazelle and an endemic subspecies of the mountain gazelle as well as wolves, foxes, caracals, ibexes, porcupines, hyraxes, and desert rodents and bats. The proposed area of protection is about 40,000 ha.

9. Yotvata (Ain Ghidyan) in the Araba Valley: This arid land reserve harbors populations of gazelles, foxes, and caracals as well as introduced oryx, Syrian and African wild asses, and other large mammals.

In many regions, feeding centers for carnivores were established and perhaps helped their recovery. This practice was recently discontinued because of cost and the fear of the spread of rabies. Reintroductions were carried out in some of the nature reserves. Some of the previously extirpated or endangered mammals (onagers, wild asses, oryxes) have been bred at centers such as the Tel Aviv University Zoo and Yotvata (Ain Ghidyan) Reserve. The main obstacles facing conservation efforts in Israel are continued massive immigration and the human population explosion versus the small size of the country and the utilization of large areas of the land by the army. The preservation of habitats and nature took a distant back seat to establishing new settlements and to the numerous political and economic disruptions that followed. Although many settlements were on previously inhabited areas, many others were established on previously unsettled land that was then cleared for agriculture and settlement. Large scale irrigation and the use of large areas by the expanding settlements left few places undamaged.

Jordan in the 1950s and 1960s did not fare any better in conservation than did Israel of that period. The war of 1948 and the establishment of the State of Israel resulted in burdening Jordan with a large population of refugees and massive economic troubles. The creation of many refugee camps contributed to the decline of the environment by clearing many previously uninhabited areas in a situation similar to that seen for some of the Jewish settlements. Again the economic and political situation was such that conservation issues were not addressed. King Hussein of Jordan invited British expeditions to conduct an extensive study of the desert and mountains of Jordan and to make recommendations on the protection of wildlife. This developed into the International Jordan Expeditions of 1963 and 1965. Results of the survey and the recommendations were presented and accepted by the government of Jordan. The problems seen were summarized by Guy Mountford, leader of the expedition (Mountford, 1964). A more expanded report was published as a book (Mountford, 1965). The two major problems cited are: (1) extensive deforestation for fuel and the subsequent loss of wildlife and desertification, and (2) extensive hunting and decline of many large mammals with several species becoming locally extinct in this century.

In 1973, a law to protect wildlife was introduced by the Ministry of Agriculture and was adopted by the government (Law 20, Section 3). This law regulated hunting and trapping of mammals and birds including prohibiting the use of military guns, vehicles, and lights in hunts. This law was unfortunately little enforced. Even if properly enforced, the fines proposed were very low. For example, for illegal hunting of a desert gazelle, the fine was 15 Jordanian Dinars (at the time 1 JD = U.S. $3.30) and for a mountain gazelle or ibex 10 JD. Hunting was permitted and continued under the direction of the Royal Society for the Conservation of Nature (RSCN, established in 1966), previously the Royal Hunting Club. This is a quasi-governmental body charged with all aspects of conservation in Jordan. Funding for this organization is derived from international agencies (for example, the World Wildlife Fund) and national donations (both governmental and private). The RSCN was charged by the Ministry of Agriculture to enforce the nature conservation laws and in this effort receives the full cooperation of the police, justice system, and other government agencies. The society also establishes and operates nature reserves in several areas in Jordan, of which the following are the most advanced and are of significance to the mammalian fauna:

1. Shawmari: This is the first and best-run reserve in Jordan. It is a small (22 km^2) area of desert wadis and harbors such mammals as foxes, hyenas, caracals, wild cats, desert hedgehogs, and jerboas. Oryx, Syrian wild asses, and gazelles have been reintroduced here. The reintroduction of the oryx is a success story in Jordan. The herd now numbers over 70. This reintroduction was made possible by generous gifts (of money or animals) from the World Wildlife Fund, the Zoological Society of San Diego, the Zoological Society of London, and the Fauna Preservation Society.

2. Azraq Wetland: This swampy area of some 12 km^2 is mainly intended to protect waterfowl but also harbors mammals such as hyenas, jackals, foxes, hares, and a few gazelles.

3. Zubiya: This woodland reserve of some 13 km^2 harbors gazelles, wild boars, jackals, hyraxes, polecats, martens, forest bats, and many others.

4. Wadi El Mujib: This area on the eastern shores of the Dead Sea (about 212 km^2) is mountainous desert. Mammals possibly existing there

include leopards, Blanford's and Rüppell's foxes, ibexes, gazelles, bats, and many other desert-adapted species.

5. Dhana: This is a mountainous forested region of about 150 km^2 with martens, wild boars, hyraxes, squirrels, hamsters, and voles.

6. Wadi Rum: This area of about 560 km^2 supports such mammals as Blanford's and Rüppell's foxes, wolves, hyenas, caracals, wild cats, desert hedgehogs, and jerboas.

Much of the land for these reserves was partially protected from development by the Ministry of Agriculture. When a reserve is to be established, private property within the reserve may be purchased and fencing may result in better conservation. The effects of protection can be quite dramatic; for example, plant cover and diversity significantly increase inside a reserve (Hatough et al., 1986). Reserves also act as centers for reintroduction of wildlife. The oryx has now been successfully reintroduced in the Shawmari Wildlife Reserve (Talbot, 1960; Anonymous, 1975, 1978; Clarke, 1979; Lamb, 1984; Nelson, 1985).

Both Israel and Jordan have institutions designated to protect nature and both are doing well at decreasing the pace of environmental destruction. There also have been good wildlife reintroduction programs in the area. Proposed protected areas, if enacted in both countries, would result in the protection of more than 5% of the Holy Land (Green and Drucker, 1991). However, wildlife destruction continues, though at a decelerated pace. Areas where improvements are needed are numerous. First and most obvious is the need for adequate education in conservation issues. Second, considerable economic factors result in the allocation of minimal funds for natural resources, especially in Jordan. Although this is a problem shared by many third-world countries, it is exacerbated in the Middle East because of the almost continuous war effort. In Jordan the resources devoted to wildlife conservation at present are insufficient (Green and Drucker, 1991). Third, overpopulation results in habitat destruction and an increase in illegal hunting. Both western Palestine (Israel) and eastern Palestine (Jordan) have a very high rate of population increase. The population of these areas at the end of World War I was estimated at one million, in 1950 at two million (Ives, 1950). In 1983, the estimated (from various government publications) population was eight to nine million. The last five wars had little impact on population size. Most of the casualties in the Arab–Israeli wars have been on the Egyptian and Syrian sides, and more recently in Lebanon.

Ives (1950:103) warned about the economic/agricultural consequences of a population explosion four decades ago and concluded that: "the oft-repeated dream of a Jewish Homeland in Palestine, supporting all of the perhaps 20,000,000 Jews in the world, is, and must remain, only a wild and unworkable dream. Under no economic system now known or proposed can Palestine support ten times its present population. There is valid reason to doubt that the country can ever support more than twice its present population on any tolerable level."

The present population in that area is over five million and may have to accommodate more if refugees are allowed to return. If not, the population is still growing because of the high birth rate and continuous Jewish immigration. Thus, with the political settlement of the Israeli– Palestinian conflict, it seems the population will continue increasing by one or more of the following: Jewish immigration, return of Palestinian refugees, high birth rate, and improved health care. The adverse effects of this increase could strain limited natural resources and could also foster continued habitat destruction and a general decrease in the quality of human life.

An example of previous ecological disasters can be cited with regard to the decline of insectivorous bats. A campaign for the destruction of fruit bats was started in Israel in 1958. Many caves were sealed and fumigated with ethylene dibromide and later (beginning in 1982) with lindane (hexachlorobenzene) by Israeli farmers and agricultural organizations (Makin, 1979; Makin and Mendelssohn, 1986, 1989). Lindane is a an organochloride similar to PCB, DDT, and other chemicals that linger and accumulate in the food chain. The Egyptian fruit bat (*Rousettus aegyptiacus*) was specifically targeted by these fumigations. Indeed, on a visit to the caves near Bitan Aharon, accompanied by David Makin and a member of the NRA, we found the caves littered with skeletons of fruit bats but also of insectivorous bats such as horseshoe bats and common bats. One fumigated cave had 30 dead and more than 60 live *Rousettus*. Another fumigated cave had more than 2000 live fruit bats. It thus appears that the smaller insectivorous bats were more vulnerable than were the target animals. Insectivorous bats are very useful animals because of their consumption of huge numbers of harmful insects and their deposition of guano (bat feces), which is a very valuable fertilizer (Nelson, 1926). Replacement of insectivorous bats by fruit bats also occurred because of the spread of agricultural activity and thus the availability of food for fruit bats. Fruit bats are very noisy, and if they inhabit a cave where insectivorous bats occur, they displace them. In Jordan, where there was no fumigation, I visited a cave near the Yarmouk River that held 3000 fruit bats. I was told by the locals that this cave had only "small bats" (that is, insectivorous bats) the year before

A similar example can be cited with attempts to control rodent populations by means of poisoning. The populations of the colonial Mediterranean vole (*Microtus guentheri*), fluctuate dramatically and they are combated as dangerous agricultural pests in Israel (Bodenheimer, 1949*b*, 1959; Atallah, 1978; Ilani and Bouskila, 1982). Thallium sulfate poisoning was used in Israel for many years (Bodenheimer, 1949*b;* Ilani and Bouskila, 1982). However, such measures are ecologically harmful when such substances enter into the food chain.

Human population explosion, habitat destruction, and mistakes of human interference in nature are problems that are found worldwide. These ecological dangers seem to be exacerbated in the Near East. The almost constant military conflicts make habitat destruction much more dramatic. Sophisticated war machinery makes obliterating fragile habitats much easier. Threats of nuclear and chemical attacks are taken more seriously here than anywhere else in the world. Fortunately, worldwide interest in conservation is now extending to reach people of the Near East. It is very encouraging that Palestinians started to organize nature protection acivities even before self-rule was established. A group calling itself "Children for the Protection of Nature" is already active in more than 70 schools (as of 1995). The proposed Palestinian Authority has also shown interest in establishing nature reserves and enacting laws to protect natural resources. Much has been done to protect wildlife and much needs to be done yet. More people of all races and religions from the region have come to realize the importance of preserving the land and the wildlife. It is now very clear that, regardless of how long it takes to resolve political conflicts, future generations of all inhabitants will have to live in this area and they cannot live on a barren desert created by their ancestors. We should learn from our past errors and move forward to conserve what remains of the magnificent life of the region. Within this generation lies the key to the future quality of life in the Holy Land.

10

Synopsis of the Mammals of the Holy Land

For the arrangement and order of the genera and species of mammals in the Holy Land, I followed that used by Wilson and Reeder (1993). Departures from that classification are discussed as needed. On the same line as the scientific name is a letter indicating the conservation status as categorized in the 1990 IUCN Red List of Threatened Animals (IUCN, 1990). The categories used are: extinct (**X**)— species not found in the wild within the last 50 years; endangered (**E**)—in danger of extinction unless factors causing the decline are reversed; vulnerable (**V**)—likely to become endangered if causal factors continue; rare (**R**)—species at risk, for example, small population or restricted range; indeterminate (**I**)—known to be endangered, vunerable, or rare but not enough data to assign to specific category; insufficiently known (**K**)—not definitely known to belong to any category; out of danger (**O**)—abundant and in no danger at this time.

ORDER INSECTIVORA   Insectivores

The Insectivora (from insect and *voro*, L., I devour, meaning insect eaters) of the Holy Land fall into two families that are easily distinguished, hedgehogs and shrews. Hedgehogs or Erinaceidae (from *erinaceus* or *ericius*, L., meaning a hedgehog) are larger animals with stout bodies recognized by the presence of spines covering most of the back, and with a short tail. Shrews or Soricidae (from *sorex*, L., genitive *soricis*, meaning the shrew) are smaller (skull length less than 35 mm) insectivores with no spines on the back and with a slender body and small inconspicuous eyes.

FAMILY ERINACEIDAE   Hedgehogs

These are robust animals with a protective body covering of spines. The tail is short and the body can be curled in a protective posture. A layer of muscles found under the skin aids in erecting the spines and in the protective posture. The hedgehog family includes eight genera and 16 recent species. They are found in areas of Africa, Europe, and Asia. The family was recently reviewed (Corbet, 1988).

Key to the Genera of Erinaceidae of in the Holy Land (mainly from Atallah, 1977)

1. Anterior spines parted over the head creating a bare dorsal patch; pterygoid fossa inflated and connected to tympanic cavity dorsally; four sacral vertebra. . . . . *Paraechinus*

   Spines over head without a conspicous bare patch; pterygoid fossa shallow and not connected to tympanic bulla; three sacral vertebrae. 2

2. Smaller, adult skull length less than 48 mm but with relatively longer ears (ear-to-skull length ratio > 0.6). First upper incisors point forward (proodont). . . . . *Hemiechinus*

   Large, adult skull length more than 50 mm but with relatively short ears (ear-to-skull length ratios < 0.4). First upper incisors point down (orthodont). . . . . *Erinaceus*

Genus *ERINACEUS*

The dental formula is i 3/2, c 1/1, p 3/2, m 3/3 = 36.

Erinaceus concolor   East European hedgehog, common hedgehog   O

Erinaceus concolor Martin, 1918. Proc. Zool. Soc. Lond. 1918:103. Type from near Trebizond, Turkey.

Erinaceus roumanicus sacer Thomas, 1918. Ann. Mag. Nat. Hist. ser. 2, 9: 212. Type from near Jerusalem, Palestine. Perhaps a valid subspecies: *E. c. sacer*.

Diagnosis.—This is the largest of the three species of hedgehogs in the region. Head and body length ranges from 200 to 260 mm in adults with a skull length over 55 mm. This hedgehog weighs about 550–700 g, with males generally larger than females. The ears are small (usually less than 30 mm long) and are thick and covered with dense fur (Fig. 6). The overall

Fig. 6. External appearance of hedgehogs of the Holy Land (photo credit in parentheses). (Upper) European hedgehog (*Erinaceus concolor*) (S. I. Atallah); (Middle) desert hedgehog (*Paraechinus aethiopicus*) (D. Shafi); (Bottom) long-eared hedgehog (*Hemiechinus auritus*) (D. Shafi).

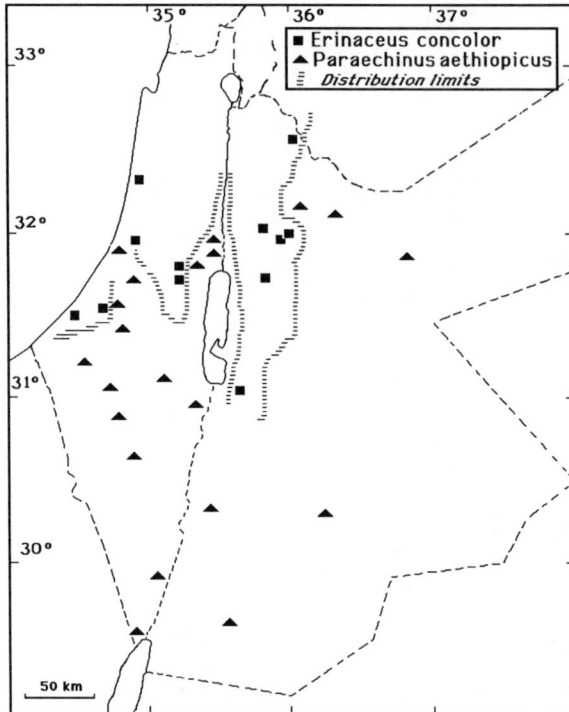

Fig. 7. Distribution of the European hedgehog (*Erinaceus concolor*) and the desert hedgehog (*Paraechinus aethiopicus*). Only marginal localities are given for the European hedgehog. Note that the distributional limits of the European hedgehog are not expected to overlap those of the desert hedgehog.

coloration is usually uniform brownish to gray. Compared with *Paraechinus aethiopicus,* the other large hedgehog of the area, this species lacks the bare patch between the spines of the head, has no conspicuous dark and white markings on the face and the underside, has more coarse hairs on the underside and the head region, and has smaller tympanic bulla. Females have 10 nipples.

The eastern European *E. concolor* recently has been recognized as a species distinct from *E. europaeus* because the two forms overlap in distribution in eastern Europe. Most authors consider the Holy Land form as *E. concolor* (Corbet, 1988; Harrison and Bates, 1991). The uniform dark color of the underside of our hedgehog is similar to that of *E. europaeus. Erinaceus concolor* has a bicolored ventral pelage with white on the breast region. It is thus possible that the characters used to distinguish the two species in Europe break down outside of the regions of sympatry. However, recent chromosome studies from Jordan (Qumsiyeh, 1991) document a karyotype that is indistinguishable from typical *E. concolor.* Thomas (1918) remarked that "*Erinaceus roumanicus*" and "*E. r. sacer*" differ from *europaeus* and *concolor* by the shorter maxillary bones not reaching the level of the "muscular fossa."

Range of species.—Europe across southern Russia to northern and north-western China. Near East as far south as Iran, Iraq, and the Holy Land.

Local status.—The east European hedgehog is found in the more mesic areas of the Holy Land (Fig. 7). Records are available from the mountain regions (receiving more than 100 mm rain/year) and the coastal plains south to Gaza. In the mountains of Jordan, hedgehogs occur as far south as Tafilah.

Where it occurs it is common and records from the Holy Land are abundant. Important earlier and marginal locality records follow. Thomas (1918) reported a specimen from "near Jerusalem" that he designated as a separate subspecies "*E. roumanicus sacer.*" Both Bodenheimer (1958) and Aharoni (1930) reported common hedgehogs as abundant in Palestine extending as far south as Ruhama and Gaza. Harrison (1964*a*) recorded this species from Beit Lid (near Natanya), Tivon, and near Lod. Atallah (1977) reported specimens from Bethlehem and Amman and speculated that the southernmost limit may be around Tafilah. Specimens at the JUNHM are from Mahis, Jubeiha, Jerash, Al-Hummar, Ramtha, Tafilah, and Irbid (Amr and Disi, 1988) and those at HZM are from Wadi Zarqa Ma'in (Bates and Harrison, 1989). I also collected specimens from Beit Sahur (SANHM) and 9 km NE Madaba (CMNH).

Biology.—This is a temperate weather species. Atallah (1977) indicated that the southern limit in the Holy Land may be a line connecting Gaza with Tafilah. Populations in some areas of Palestine and Jordan live close to semidry regions but are never found in the Negev or Syrian deserts. This animal is nocturnal and shy and can roll into a defensive posture with only the spines showing. This "ball of spines" proves an effective defense against predators such as foxes or badgers. Common hedgehogs eat almost anything given to them (omnivorous diet). I kept some specimens for months and fed them eggs (raw or boiled), meats, insects, snails, vegetables, and cheese. They are voracious eaters making considerable noise as they tear into their food. In the wild, they feed on small reptiles, insects (Coleoptera, larvae of Lepidoptera and Diptera, and so on), myriapods, gastropods and other mollusks, and plants.

I have found these animals to be common from January through August, especially early in the night and occasionally found them wandering into village streets. Hedgehogs are solitary animals and fight among themselves (especially males in the breeding season). Schoenfeld and Yom-Tov (1985) reported three young born in May in nests constructed under foliage or other objects. In Europe, usually four to five (range two to seven) young are born in June to September (Corbet, 1988) following a gestation period of 31–35 days.

Genetics.—The banded karyotypes (2N = 48, AA = 90) of the east European hedgehog from Jordan are indistinguishable from those seen in eastern Europe and Greece (Qumsiyeh, 1991).

Human interactions.—Hedgehogs are common animals in the folklore of the locals. All hedgehogs are referred to in Arabic as *kunfud* (*kunfuth*) and rarely as *kababet chouk* (spiny creature) or *khlund*. They are eaten by some locals, especially among Bedouins. In many areas, elders attribute medicinal qualities to hedgehog meat (for example, a cure for arthritis and rheumatism).

Genus *HEMIECHINUS*

Hedgehogs of this genus are characterized by large ears and no bare patch on the neck. The tips of the spines are pale. The dental formula is i 3/2, c 1/1, p 3/2, m 3/3 = 36.

Hemiechinus auritus   Long-eared hedgehog   V

Erinaceus auritus Gmelin, 1770. Nova Comm. Acad. Sci. Petrop., 14:519. Type from Astrakhan Prov., USSR.

Erinaceus aegyptius É. Geoffroy St.-Hilaire, 1803. Cat. Mamm. Mus. Nat. Hist. Nat. Paris, 1803:46. Type from Cairo area, Egypt. Possibly a valid subspecies in the Holy Land: *H. a. aegyptius.*

Erinaceus syriacus Wood, 1876. Bible animals, p. 83, from Palestine. Synonym of *aegyptius.*

Erinaceus brachydactylus Tristram, 1884. *In* Fauna and flora of Palestine, p. 95.

Erinaceus calligoni Satunin, 1901. Port. Obsch. Est. Kazan, no. 192, p. 2 and Proc. Zool. Soc. Lond., 1901:284. Type from Aralyk, about 26.5 km S Erivan, Turkey. Synonym.

Diagnosis.—The smallest (weight 250–400 g, skull length 38–48 mm in adults) of the local hedgehogs but with relatively long ears (30–45 mm; longer than adjacent spines). Females have 8–10 nipples. The tips of the dorsal spines are white. From small specimens of *Paraechinus aethiopicus,* this species is easily identified by the absence of contrasting dark and white areas on the face and by the absence of the gap in the spines of the nuchal (back of the neck) area. In the skull, the pterygoids are not inflated nor do they communicate with tympanic cavities.

Range of species.—From the deserts of Mongolia throughout most of the arid and steppe regions of Asia to Iran. Northern parts of the Arabian Peninsula, Turkey, Cyprus, and northeastern Africa (Egypt and Libya).

Local status.—The long-eared hedgehog occupies an area in the Holy Land that is intermediate in rainfall (100–400 mm) (Fig. 8). It avoids

FAMILY ERINACEIDAE   Hedgehogs   65

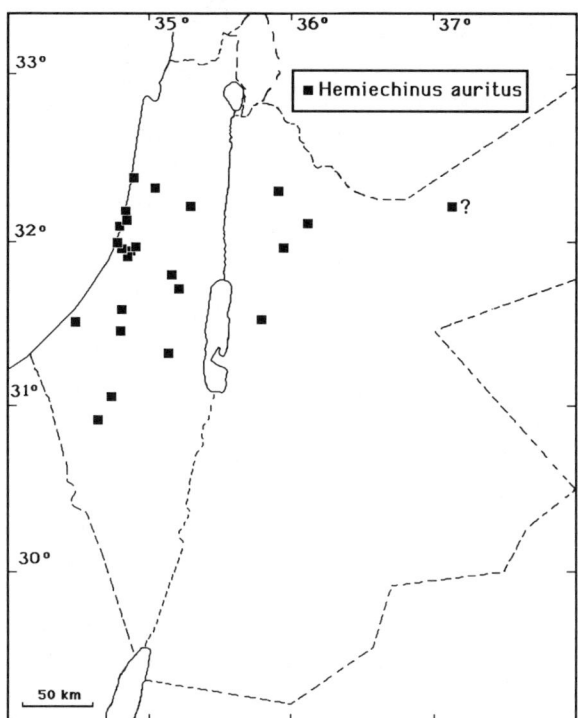

Fig. 8. Distribution of
the long-eared hedge-
hog (*Hemiechinus
auritus*).

extreme desert conditions (where the desert hedgehog, *P. aethiopicus*, is
found) and the northern cold mountain regions (where the east European
hedgehog dominates) (compare with Fig. 7). The range of *Hemiechinus* is
thus partially intermediate and overlaps that of both other species of hedgehogs.

Flower (1932) reported that Major Maurice Portal collected specimens
in 1917 and 1918 from Gaza, Ramleh, and Bir Salem. Bodenheimer (1958)
reported that this species is common along the coastal plains from Tel Aviv
southward to Gaza and Al Arish (northern Sinai) but gave only few localities:
Beersheba, Ramleh, Tel Aviv, Bir Salem (near Lod), and "north to about
Tulkarem." Harrison (1964*a*) reported specimens from Amman, Beersheba,
27 km N Beersheba, 27 km E Beersheba, and near Rishon le Zion. Atallah
(1977) reported specimens at the BM from Jaffa and Lod. I obtained
specimens at Beit Sahur and Bethlehem. The species is also common around
Givatayim, near Tel Aviv (Schoenfeld and Yom-Tov, 1985). Specimens at
HUJ are from 10 km S Beersheba, Beit Hanan, Herzelia, Kfar Vitkin, Moza,
and Revivim and at TAUM are from Tel Aviv, Gevim, Shivta, Zahala, and
Tel Shoqet (western Arad). Atallah (1977) used the name *H. a. calligoni* for

Syrian and Jordanian specimens, such as those reported by Harrison (1964a) from Amman and by Atallah (1977) from Amman, Zarqa, and Jerash. A specimen from West Dhuhayba is at JUNHM (Amr and Disi, 1988) and a report from the "H4–5 region" is given by Saliba and Amr (1985). Bates and Harrison (1989) reported material from owl pellets from the north bank of Wadi Zarqa. Osborn and Helmy (1980) reported specimens from Al Arish (northern Sinai) and Gaza.

Biology.—These hedgehogs are found foraging in the early hours of the night and they occasionally wander into village streets where they are caught easily by locals. Harrison (1964a) suggested that this species may be intermediate in its adaptation to desert habitat between *E. concolor* and *P. aethiopicus*. Long-eared hedgehogs live near available water sources. Food consists of insects, myriapods, gastropods, batrachians, small vertebrates, and plants. Hedgehogs may feed on snakes or other animals and they exhibit an elaborate behavior that allows them to kill such prey. For example, a hedgehog may hold the tail or other parts of an animal, then roll into a sphere to protect itself while chewing away at the hapless prey (Krishna and Prakash, 1956). Hoogstraal (1962) stated that long-eared hedgehogs rest in dry places in buildings, under stones, in sandy cliff sides, and similar places and that they seldom dig their own burrow. On the other hand, Roberts (1977) stated that they are active diggers. Roberts also showed that these hedgehogs hibernate in the Rajasthan Desert from November to early March. These are hardy animals that can go for weeks without food even during the summer, but they have a voracious appetite when food is available.

The species is active throughout the year, but may hibernate for short periods (maximum reported 40 days) (Schoenfeld and Yom-Tov, 1985). In Afghanistan, four to five embryos were found in females collected in July (Hassinger, 1973). The gestation period varies between 35 and 42 days (Herter, 1965). One to five young were reported to be born in May or June in Egypt (Flower, 1932; Schoenfeld and Yom-Tov, 1985). Two to three young are usually the norm. Flower (1932) reported that these animals make a "loud, exacerbating, snarling growl" when they fight. Otherwise they are relatively silent most of the time.

Genetics.—2N = 48, AA = 92 was reported for specimens from Egypt (De Hondt, 1972) and Iraq (Bhatnager and El-Azawi, 1978). The X is metacentric and the Y is acrocentric.

Human interactions.—The name *auritus* derives from the Latin *auris*, ear. Also see comments under *Erinaceus concolor*.

Fig. 9. Desert hedgehog
(*Paraechinus aethiopicus*).
Photo by D. M. Shafi.

Genus *PARAECHINUS*

Wilson and Reeder (1993) tentatively placed this genus as a subgenus of *Hemiechinus* based on Frost et al. (1991). From personal observations, I believe the *Paraechinus* should be retained as distinct genus. The genus includes three species and is found in areas from India to Morocco. These desert-adapted hedgehogs have a naked area on the crown of the head extending to the neck. The dental formula is i 3/2, c 1/1, p 3/2, m 3/3 = 36. Specimens from Jordan of this genus are missing the tiny upper second premolar and thus have 34 teeth (Fig. 4).

Paraechinus aethiopicus   Desert hedgehog   V

Erinaceus aethiopicus Ehrenberg, in Hemprich and Ehrenberg, 1833. Symb. Phys. Mamm., Dec. 2, sheet k, rect (footnote). Type from Dongola Desert, northern Sudan.

Hemiechinus pectoralis Heuglin, 1861. Nova Acta Leopold. Carol., 29:22. Type from Petra, Jordan. Valid subspecies: *P. a. pectoralis*.

Erinaceus dorsalis Anderson and de Winton, 1901. Ann. Mag. Nat. Hist. ser. 7, 7:42. Type from Hadramut, Yemen. Synonym of *P. a. pectoralis*.

Paraechinus ludlowi Thomas, 1919. J. Bombay Nat. Hist. Soc., 26:748. Type from Hit, along the Euphrates, Iraq. Synonym of *P. a. pectoralis*.

Diagnosis.—This hedgehog is intermediate in size between the small *Hemiechinus auritus* and the large *Erinaceus concolor*. Skull length is generally 45–52 mm. It is distinguished from both of these species by the presence of a bare patch of skin between the spines of the neck and by the color. A dark color extends on the legs, the belly, and the face anterior to the eyes (Figs. 6, 9). This contrasts with (usually) a distinct white area on the forehead (between the spines and the eyes). The extent of the dark and white coloration and the intensity of the colors varies geographically and individually. Each of three individuals I collected from Jericho had a distinct color pattern. The ears are large (more than 30 mm in length, longer than spines) and are less clothed with hairs than those of the east European hedgehog. The fur is soft and dense unlike the coarse and rough fur of the east European hedgehog.

Osborn and Helmy (1980), following Setzer (1957), maintained that *P. aethiopicus* is a distinct species from *P. dorsalis*; the latter name would apply to the eastern Mediterranean forms. I agree with Corbet (1988) that this separation appears artificial because the distributions of these taxa do not overlap and because the differences are very slight. In any case the name *dorsalis* is superseded by the earlier name *pectoralis*. Genetic studies would be ideal in this group of hedgehogs to resolve systematic problems.

Atallah (1977) maintained that *pectoralis* and *ludlowi* are two valid subspecies and that both extend their range into Jordan (the former from Petra and El Jafr and the latter east of Amman). I use a single name here because of the variation in coloration discussed by Atallah (1977), which makes color difficult to use as a systematic character and also because of the habitat continuity in the eastern desert of Jordan.

Range of species.—Arid and semiarid regions of northern Africa south to Somalia. Throughout Arabia and in Syria, Jordan, and Palestine to Iraq and Iran.

Local status.—The distribution of the desert hedgehog is peculiar in that it is found in areas not occupied by the east European hedgehog, *Erinaceus concolor* (Fig. 7). It occurs throughout the Syrian Desert, the Negev, and the Wadi Araba.

The first specimen from the area is that from Petra on which the name *pectoralis* was based (Heuglin, 1861). Bodenheimer (1958) reported this species from Beer Sheba. Tristram (1884) and Bodenheimer (1958) reported that this hedgehog is common in southern Palestine. This species was reported from 7 mi S Qasr al Helqum (Harrison, 1959), Wadi Araba, (Harrison, 1964*a*), El Jafr (Atallah, 1967*a*, 1977), and Azraq Shishan (Atallah, 1967*b*, 1977). Specimens are also available at the JUNHM from Wadi Dhulayl and Qasr Al Hallabat (Amr and Disi, 1988). I collected three specimens from Jericho, one from 1 km S Azraq ed Druz, and two from Wadi Rum (at Dieseh). Other localities delineating the distribution of this species are found in specimens at the TAUM from Turabah, Wadi Raman, Revivim, Ein Radian, 10 km S Beersheba, Rehovot, Zeelim, 10 km S Jericho, Jericho, Nahal Zin, 10 km E Jerusalem, Massua, and Ain Auja; and at HUJ from Ahuzam, Beersheba, and Avedat Horvot.

Biology.—This hedgehog is nocturnal and is distributed in desert and dry steppe areas. Little is known of its reproductive biology or habits in the region. Individuals were obtained in the spring and were collected usually in the early hours of the night. A specimen collected at Azraq had remains of frogs and insects in its stomach (Atallah, 1977). Longevity is estimated at up to 6 years.

Genetics.—2N = 48, AA = 92 is found in specimens from Saudi Arabia (Al-Saleh and Khan, 1985). Similar data exist for specimens from Iraq (Bhatnager and El-Azawi, 1978). The X is metacentric and the Y is submetacentric.

Human interactions.—See *Erinaceus concolor.*

FAMILY SORICIDAE   Shrews

Shrews are among the smallest and most secretive of mammals. The scientific knowledge of this widespread group still lags far behind that of other groups of mammals. They have been reported from habitats from desert to subtundra. In some habitats they are so common that they may outnumber rodents. The eyes and ears are small and the snout is elongated. They usually have slender bodies and are extremely active. In the skull, the tympanic bulla and the zygomatic arches are absent. About 20 genera are recognized but the exact number of species is not known (perhaps well over 300).

Key to the Soricidae of the Holy Land

1. Upper unicuspids usually four, the smallest species in the Holy Land (total length less than 85 mm). . . . . *Suncus etruscus*

   Upper unicuspids usually three, specimens of medium size. . . . . *(Crocidura)* 2

2. Skull relatively thick and wide. Leg color whitish. In dorsal view, the space between the upper incisors and the premaxilla is relatively large; 2N = 28, FN = 56. . . . . *C. leucodon*

   Skull relatively long and narrow. Leg color dark. In dorsal view, the space between the upper incisors and the premaxilla is relatively small; 2N = 40, FN = 50. . . . . *C. suaveolens*

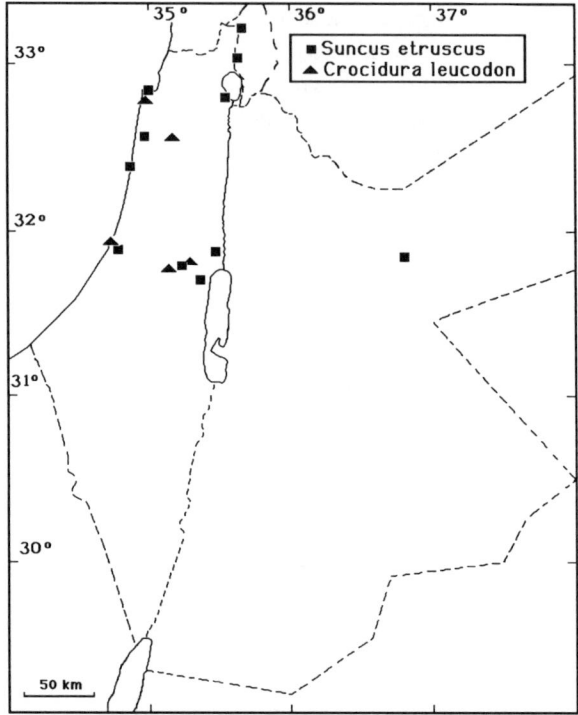

Fig. 10. Distributions
of Savi's dwarf shrew
(*Suncus etruscus*), and
the bicolored
white-toothed shrew
(*Crocidura leucodon*).

Genus *CROCIDURA*

The taxonomy of this large genus (perhaps more than 150 species) is in disarray. Much of the previous work was done using small size differences, which are inaccurate at best. Recent chromosomal and electrophoretic data help resolve some questions regarding species identification (Catzeflis et al., 1985). Unfortunately, identifications are difficult in the field and especially if one relies only on morphological criteria. I list only two species from the Holy Land. *Crocidura lasia* (*lasiura*) was reported from Lebanon but confusion exists about its status (see Atallah, 1977; Hutterer and Harrison, 1988). If it proves to be a valid species, it may exist in the northern regions of Palestine. The dental formula for the genus is i 3/1, c 1/1, p 1/1, m 3/3 = 28.

Crocidura leucodon   Bicolored white-toothed shrew   **R**

Sorex leucodon Hermann, 1780. *In* Zimmerman, Geogr. Gesch. Mensch. Vierf. Thiere, 2:382. Type from Strasburg, Bas Rhin, France.

Crocidura russula judaica Thomas, 1919. Ann. Mag. Nat. Hist. ser. 9, 3:32. Type from near Jerusalem, Palestine.

Diagnosis.—The following characters distinguish *Crocidura leucodon* from the closely related *C. suaveolens/russula*: body more robust, tail length less than half head and body length, tail with thick bristles, tail distinctly bicolored and paler, presence of a sharp demarcation between dorsal and ventral body color, and the cusp on the outside anterior surface of the large upper premolar is higher than the last unicuspid (Harrison, 1963*d*, 1964*a*). According to some authors, in the dorsal view of the skull, the anterior teeth protrude more from the rostrum of *C. leucodon* than they do with *russula/suaveolens*, though the skull of the former is smaller (Felten et al., 1973). This shrew is rather large for the genus with head and body length of 64–80 mm and a skull length of 19.2–21.5 mm in adults.

Range of species.—From France and central Europe to the Volga and south to Palestine, east to Iran.

Local status.—The distribution and status of this species in the Holy Land are poorly known (Fig. 10). Thomas (1919*d*) reported two specimens from near Jerusalem collected by Capt. G. C. Shortridge (types of *C. russula judaica*). Harrison (1963*d*) showed that these actually belong to *C. leucodon* and added localities Mishmar Ha'Emeq, and Carmel Caves, and later (1964*a*) from Nahr Rubin. A specimen at HUJ is from Qiryat Saide.

Biology.—Little is known of the biology of this species in the Holy Land. Harrison (1964*a*) stated that it prefers densely vegetated areas and that in the Holy Land it is less abundant than *C. russula*. In Iran, this species was preyed upon by *Felis silvestris* (Lay, 1967).

Genetics.—2N = 28, AA = 52 was reported for specimens from Europe, which is very different from the sympatric species *C. suaveolens* (2N = 40, AA = 46) (Catzeflis et al., 1985). Electrophoretic differences are also evident and allow clear definition of species in this group.

Human interactions.—*Crocidura* is derived from *krokus*, Gr., meaning, a nap, a pile of cloth, and *oura*, Gr., the tail, in reference to the bristled tail. See also comments under *Suncus etruscus*.

Crocidura suaveolens   Lesser white-toothed shrew   K

Sorex suaveolens Pallas, 1811. Zoogr. Rosso-Asiat., 1:139, pl. 9, fig. 2. Type from Khersones, Crimea, southern USSR.

Crocidura gueldenstaedti Pallas, 1811. Zoogr. Rosso-Asiat., 1:139. Type from USSR, Georgia. Valid subspecies: *C. s. gueldenstaedti*.

Crocidura russula monacha Thomas, 1906. Ann. Mag. Nat. Hist., 17:471. From Scalita, south of Trebizond, Turkey. Valid subspecies: *C. s. monacha*.

Fig. 11. The lesser white-toothed shrew (*Crocidura suaveolens*). Photo: A. Shoob.

Crocidura portali Thomas, 1920. Ann. Mag. Nat. Hist. ser. 9, 5:119. Type from Ramleh, Palestine. Valid subspecies: *C. s. portali.*
   Suncus tristrami, Bodenheimer, 1935. Animal life in Palestine, p. 95. Synonym.

Diagnosis.—Previous authors have allocated specimens from the Near East referred to as *C. r. monacha* or *gueldenstaedti* to subspecific status under *C. russula* and not under *C. suaveolens.* This allocation was mainly based on size differences. For example, according to Felten et al. (1973), Near Eastern suaveolens can be distinguished from russula by the condylobasal length (CBL). In specimens they examined from Israel, females of *"suaveolens"* were those with CBL less than 18.5 mm whereas those of *"russula"* have CBL larger than 19.5 mm. However, these measurements are different in other areas of the ranges of these species.

   This is a small shrew (skull length 17–20 mm, Fig. 11) with a bicolored tail that measures about one-half the length of head and body. There are scattered bristles along the entire length of body. The dorsal color is brownish gray with a light ventral color. There is no sharp demarcation between the upper and under parts. The skull is distinctly convex in the frontal region (above the braincase). The external cusp on the large upper premolar is small (Harrison, 1964*a*).

Range of species.—From Japan and Korea throughout most of the temperate regions of Asia to Russian Turkestan and Armenia. The northern Arabian Peninsula through Turkey and southern Europe and in northern Africa (Egypt, Algeria, Morocco).

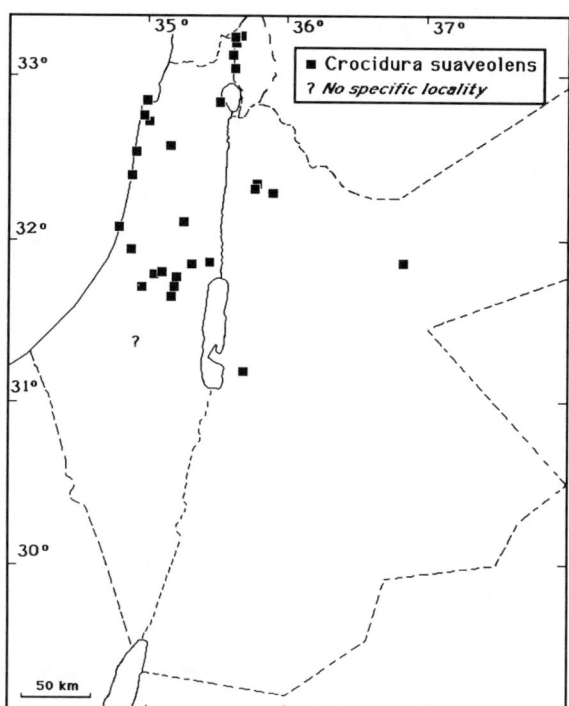

Fig. 12. Distribution of
the lesser white-
toothed shrew
(*Crocidura suaveolens*).

Local status.—The lesser white-toothed shrew is known from areas of the country with enough moisture (vegetation) to support the varied insect diet of this species (Fig. 12). As judged by its presence at Azraq Oasis, it may prove to be more widespread in desert areas near springs or oases. Tristram (1866) reported *Sorex crassicandus* from the Negev, which Bodenheimer (1935) described as *Suncus tristrami*. Later, Bodenheimer (1958) suggested that this was actually *Crocidura portali* described from Ramleh (Thomas, 1920). The latter is a junior synonym of *C. s. monacha*. Bodenheimer (1958) also reported on two specimens, one from the Negev and one from Dan. Harrison (1964a) reported on specimens from Wadi Masrura (near Tel Aviv). Felten et al. (1973) reported specimens from Tel Aviv. The following records were reported as "*C. russula*" but are most likely *C. suaveolens*. Specimens (now at the MCZ) were obtained at Rasheya, Baniyas, Ammik, and Aithenis, all at the base of Mount Hermon (Allen, 1915).

Other specimens were reported from Beit Oren, Haifa, Dan, Beit Meir (near Jerusalem), Hulata, Wadi Tira (15 km S Haifa), Beit Hananya, and Carmel Caves (Harrison, 1963d), and near Kfar Vitkin, Beit Yanani, and Tel el Qadi (Dan) (Harrison, 1964a). Felten et al. (1973) reported specimens from Tel Aviv and Atallah (1967b) reported on the first specimens from an arid locality in Azraq Shishan. Records are also available from Wadi Zarqa, 3 km N Ajloun, 2 km W Sakib, 5 km W Kerak (Harrison and Bates, 1991), Bethlehem, Beit Sahur, and 4 km NW Beit Fajjar (Atallah, 1977; M. B. Qumsiyeh, personal collection). Specimens at HUJ are from Jericho, Tel Aviv, Kefar Zakariya, Aqua Bella, Ain Fara, Ginosar, Horbat Saadim, Hula, Jerusalem, Mishmar Ha'Emeq, Qiryat Saide, Ramleh, and Tel el Qadi (Dan).

Biology.—In Iran the species appears to inhabit the same areas as *C. leucodon* (Lay, 1967). In certain areas of Afghanistan, *C. suaveolens* was common and nests with three juveniles were collected (Hassinger, 1973). It is adapted to more arid conditions than other species of *Crocidura*.

Reproduction in this species (under the name *C. russula monacha*) was studied in captivity (Hellwing, 1971, 1973) and in the field (Zafriri and Hellwing, 1973). With a gestation period of 28 days, in the life expectancy of a shrew (1.5 years in captivity and 1 year in nature), a female may have as many as 10–12 litters, each with one to seven (usually four) young.

Genetics.—Recently, chromosomal, electrophoretic, and hybridization studies documented that populations previously referred to as *C. r. monacha* and *C. r. gueldenstaedti* are actually *C. suaveolens* (Catzeflis et al., 1985; Vogel et al., 1986). The latter authors were able to produce fertile hybrids between specimens from Tel Aviv identified morphologically as "*C. russula*" and western European specimens of *C. suaveolens*. Near East specimens (previously referred to *suaveolens, monacha,* and *gueldenstaedti*) all had 2N = 40, AA = 46, whereas *C. russula* from western Europe has 2N = 42, AA = variable (Catzeflis, 1983; Zima and Král, 1984; Catzeflis et al., 1985). Based on these studies, it is safe to assume that *russula* does not occur in the Holy Land until such time as specimens from the area are collected with 2n = 42.

Human interactions.—The name *suaveolens* comes from *suavis*, L., sweet and *olens*, L., smelling: sweet-smelling, in reference to the musky odor of shrews. See *Suncus etruscus*.

Genus *SUNCUS*

This genus is not well separated taxonomically from *Crocidura* except for the usual presence of four upper unicuspid teeth rather than three. The dental formula is i 3/1, c 1/1, p 2/1, m 3/3 = 30. There is a single species in the Holy Land and it is easily distinguished.

Suncus etruscus   Savi's dwarf shrew, Mediterranean pygmy shrew   I

Sorex etruscus Savi, 1822. Nuovo Giorn. de Letterati, Pisa, 1:60. Type from Pisa, Italy.

Sorex pygmaeus Tristram, 1884. Survey of western Palestine, fauna and flora of Palestine, p. 24. Type from Dir Mar Saba, Palestine. Synonym.

Diagnosis.—A very small shrew (Fig. 13) with an average head and body length of 40–50 mm, tail 20–30 mm, and skull length less than 14 mm. This makes this animal one of the smallest mammals in the world. The dorsal fur is gray to light brown. Ventral regions are whitish.

Range of species.—Southern Europe from Spain through Turkey and the Turkistan and Azarbaijan. Northern Arabian Peninsula with isolated records from Yemen (Harrison, 1964a; Spitzenberger, 1970).

Local status.—The distribution in the Holy Land is poorly known (Fig. 10). Tristram (1884) reported a specimen that he called *Sorex pygmaeus* from Mar Saba, which Bodenheimer (1958) suggested represents this species. Theodor and Costa (1967) reported specimens from 'Emeq and Harrison (1964a) reported others from Tiberias, Dan, Rehobot, Hulata (near Tel Hassan), 3 km W Wadi Zikhron Ya'aqov, and Beit Yanani (near Kfar Vitkin). Atallah (1967b) recorded it from Azraq Shishan. Specimens at HUJ are from Haifa and Jerusalem. I collected a specimen from Jericho.

Biology.—Nothing is known of the biology of this species in the Holy Land. The following are some anecdotal remarks by Roberts (1977). After a gestation period of 28 days, a litter of four to six young are born. At that time they are no bigger than a few millimeters long and are pink and blind. The young develop strength quickly and may soon follow their mother in the typical shrew "caravan" (holding to the tails of each other). They feed on invertebrates such as spiders, small beetles, and termites. Atallah (1977) identified skeletal remains of beetles in stomach contents of two Jordanian

Fig. 13. Savi's dwarf shrew (*Suncus etruscus*). Photo: A. Shoob.

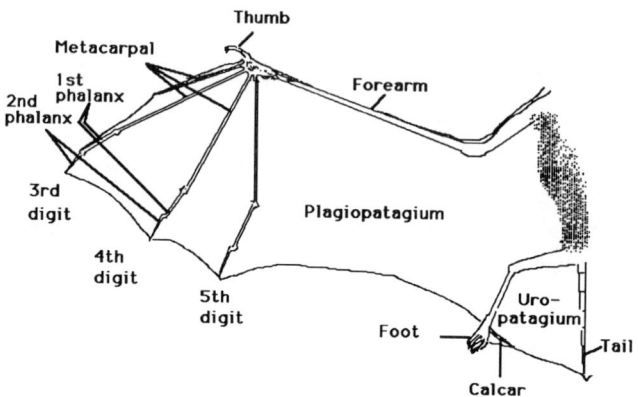

Fig. 14. External anatomy of a bat flight membrane.

specimens. These shrews appear to depend on smell to detect their food. In Iran, this species was preyed upon by *Felis silvestris* (Lay, 1967).

Genetics.—2N = 42, AA = 68 was reported for a specimen from France (Meylan, 1968).

Human interactions.—Most locals are not familiar with shrews and simply refer to them as *firan* (singular *far*), which is the name for mice. In some areas, the locals call the shrew (regardless of species) *salita* or *'urs*. The latter is also a name given to *Mustela*. Hemprich and Ehrenberg (1833) listed "far sunki" as the Egyptian–Arabic name, whereupon came the name of the genus, *Suncus*.

ORDER CHIROPTERA   Bats

Because bats are mammals, their anatomy is basically like that of other mammals. However, these mammals possess unique adaptations allowing them to fly. The forearm and the bones of the fingers (digits) of the hand are elongated and a membrane (patagium) connects the arm, the digits, the side of the body, and the leg, and in some bats also encloses the tail (Fig. 14). The name Chiroptera reflects this (from *kheir*, Gr., the hand, and *pteron*, Gr., wings). The wing membrane is referred to as the plagio-patagium, and that between the legs is the uropatagium or interfemoral membrane. The uropatagium is highly variable among bats; in some families it is rather large, encloses the tail, and augments the flight membranes (for example, common bats or Vespertilionidae). In those cases a bony or cartilaginous support (the calcar) may extend from the ankle to support the membrane. In other bats, the uropatagium is small or absent (for example, tomb bats or Rhinopoma-tidae). The thumb in bats is not significantly modified from other mammals and it also varies in length between species. The flight membranes of bats are tough and resistant to tear despite their delicate appearance. To the touch they feel leathery and dry but are actually composed of skin that is well supplied with blood vessels and nerves.

The remainder of the body of bats is also proportionally adapted to aerial locomotion. The pectoral (chest) muscles are very large. The lung and heart are also enlarged to provide sufficient oxygen and nutrition to the muscles during flight. Unlike most other mammals where the hind limbs are large and powerful to provide the main locomotion power from behind, the hind limbs of bats are reduced and the forelimbs are significantly enlarged and elongated.

All bats can produce sound in the range audible to humans (below 20 kHz). However, one of the most striking adaptations found in many bats is the ability to utilize high frequency sound (ultrasound) for foraging or other purposes. These sounds are produced in the larynx (voice box) and are used by most species for orientation to objects (termed echolocation). The fruit-eating bats (suborder Megachiroptera) lack the ability to produce ultrasonic sounds. However, some fruit bats (for example, the Egyptian fruit bat [*Rousettus aegyptiacus*] in the Holy Land) can orient acoustically by producing lower frequency sounds that are audible to humans. Unlike the Microchiroptera, in *Rousettus*, these sounds are not produced in the larynx but by clicking the tongue.

In some bats, morphological specializations were also acquired for more complex echolocation tasks. Thus, in the families Nycteridae, Rhinolophi-dae, and Hipposideridae, one finds elaborate foliaceous appendages around the nostrils. In other bats, such as certain vespertilionids, the ears are significantly enlarged and specialized.

Bats are found in tropical, temperate, and desert areas throughout the world. They are absent only on some remote islands. There are 19 families and some 170 recent genera. The number of species of the order Chiroptera, about 900, is second only to the rodents. A major subdivision is recognized between the fruit-eating bats and flying foxes (suborder Megachiroptera, containing one family, Pteropodidae) and the remaining mostly insectivo-rous and carnivorous bats (suborder Microchiroptera).

Finally, it is important to describe the method of measuring the forearm of bats. This measurement is important in identification of many species. With the wing mostly folded (not spread), the forearm is the measurement between the elbow and the carpal joints externally.

Key to the Families and Genera of Bats of the Holy Land

1. Second manual digit with claw. Ear simple, with margin
forming a complete ring. Tail vestigial and not enclosed
by the interfemoral membrane. Large bats, forearm
88–100 mm. Fruit-eating bats. . . . . .   (Pteropodidae) *Rousettus*

    Second manual digit without claw. Ear margin does not form a
complete ring. Tail longer and at least partially enclosed by the
interfemoral membrane. Smaller, forearm less than 80 mm.
Insectivorous bats.   2

2. Tail projecting through dorsal surface of uropatagium.
. . . .   (Emballonuridae) *Taphozous*

    Tail not projecting through dorsal surface of uropatagium. . . .   3

3. Tail long and completely or almost completely free from
uropatagium. . . . .   (Rhinopomatidae) *Rhinopoma*

    Tail enclosed by uropatagium or protruding only for a short
distance. 4

4. Muzzle with complex nose leaf. 5

    Muzzle without nose leaf. 6

5. Posterior nose leaf tridentate, anterior nose leaf flat and not
elaborate. (Hipposideridae) *Asellia*

    Posterior nose leaf triangular, anterior nose leaf horizontal and
deeply notched. . . . .   (Rhinolophidae) *Rhinolophus*

6. Upper lip wrinkled. Tragus absent. Tail thick and short and
projecting conspicuously beyond the uropatagium.
. . . .   (Molossidae) *Tadarida*

    Upper lips not wrinkled. Tragus present. Tail longer and enclosed
completely or almost completely by the uropatagium. 7

7. Anterior dorsal region of head traversed by longitudinal furrow
enclosing the nostrils. . . . .   (Nycteridae) *Nycteris*

    Not as above. . (Vespertilionidae, Key to Genera is on page 107)

FAMILY PTEROPODIDAE   Fruit bats

The suborder Megachiroptera (*mega,* Gr., meaning big or large) contains a single family, the Pteropodidae (from *pteron,* Gr., wings, and *pous,* Gr., cognitive *podos* the foot, meaning wing-footed). There are certain aspects of the fruit bat morphology that are very close to the Primates and very different from the Microchiroptera (for example, the eye structure and lack of sonar echolocation). Because they are evolutionarily distinct, scientists have debated whether fruit bats and insect-eating bats should be regarded as separate orders of mammals.

The pteropodids are distinguished from other bats by a number of characters. In most areas, including the Holy Land, fruit bats are much larger animals than insectivorous bats. The tail, however, is rudimentary. The ear is simple and without evidence of a tragus. The postorbital process of the frontal is long and well developed. The dental formula in the single species, *Rousettus aegyptiacus,* in the Near East is i 2/2, c 1/1, p 3/3, m 2/3 = 34.

Genus *ROUSETTUS*

Rousettus aegyptiacus   Egyptian fruit bat   O

Pteropus egyptiacus É. Geoffroy St.-Hilaire, 1810. Ann. Mus. Nat. Hist. Paris, 15:96. Corrected to *aegyptiacus* by É. Geoffroy St.-Hilaire, 1818. Description de L'Egypte Hist. Nat., 2:134. Type from Great Pyramid, Giza, Egypt.

Diagnosis.—This is the largest bat in this area, weighing around 150 g and a forearm greater than 84 mm in adults. These bats are very distinctive in their size and appearance. The ear margin forms a complete ring, which distinguishes the fruit bats from other bats of the region. The wing membrane is connected to the first toe. The tail is very small, projecting less than 1 cm from the dorsal part of the small interfemoral membrane. The fur color is drab gray in young individuals and may become brown in adults. The skull is robust and is distinctive from other bats of this area in the absence of the external bony auditory capsule and the low crowns on the cheekteeth. Males are larger than females.

Range of species.—Most of Africa south of the Sahara, the Nile Valley in Egypt. Cultivated areas of the eastern Mediterranean region, Arabia, Turkey, and east to Pakistan.

Local status.—The Egyptian fruit bat is an African species that has found its way into the eastern Mediterranean region and appears to continue to expand its range in the Holy Land (Fig. 15). It was initially collected by H.

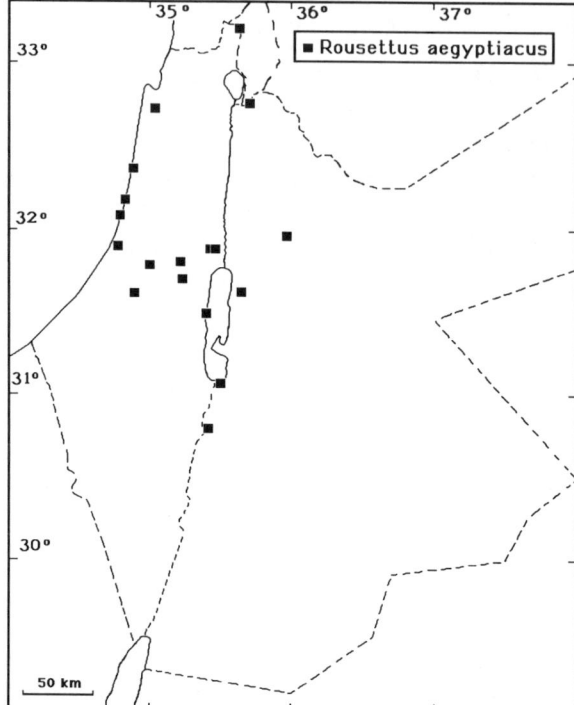

Fig. 15. Selected
locality records of the
Egyptian fruit bat
(*Rousettus aegyptiacus*)
to illustrate its range.
Notice that the only
areas it avoids are the
harsh arid areas of the
Negev in the south and
the eastern deserts.

B. Tristram from Wadi Kern (Dobson, 1878). The following localities were
further reported: Beit Sahur, Beit Guvrin, Bitan Aharon, Dan, Mount
Carmel, Ain Gedi, Herzelia, Hartuv, Mar Jiryis (Wadi Kelt), Jericho,
Jerusalem, Rehobot, Me'arath Hateamim Cave (Hartuv), Tel Aviv (Tris-
tram, 1884; Bodenheimer, 1935, 1958; Dor, 1947; Harrison, 1964a; Makin,
1977; Spitzenberger, 1979; Qumsiyeh, 1985 and personal collection). East
of the Jordan River, this species is present at Wadi Zarqa (Kock, 1969), El
Hemma, near the Yarmouk River (Qumsiyeh et al., 1986; Amr et al., 1987),
and Ibn Hammad near Karak (Qumsiyeh, pers. obs.), and El Mahatta in
Amman (Qumsiyeh, 1980). Additional specimens I examined at the
JUNHM are from Ghor As Safi, Wadi Fidan, and El Hemma (Amr and
Disi, 1988).

Biology.—These bats roost in large colonies in caves, ruins, mosques,
wells, and other structures both in Egypt (Qumsiyeh, 1985) and the Holy
Land. For example, a colony of some 3000 fruit bats was found at a small
well-lit (widely open) cave in El Hemma, near the Yarmouk River. They
seem to prefer relatively humid caves with a certain amount of reflected light
(Lewis and Harrison, 1962). They are diurnal but with peaks of activity in

the early evening and dusk hours. Fruit bats feed on sycamore, mulberries, guava, dates, and figs (*Ficus religiosa* and *F. elastica*) (Anderson and de Winton, 1902; Flower, 1932; Madkour, 1977*a*). They have also been reported to visit flowers of the ornamental silk cotton tree (*Salmalia malabaricum*) (Kaisila, 1966; Roberts, 1977).

Young are reported born from June to August in Lebanon (Atallah, 1977) and from March to May in Egypt (Qumsiyeh, 1985). In northern Jordan, I obtained juveniles in February. There may not be a fixed breeding season for this species (Flower, 1932). The gestation period in this species is 4 months according to Kulzer (1958).

Fruit bats are noisy animals and one can immediately recognize them upon entering a cave by their sharp squeals, squeaks, and flapping wings. They are not particularly good fliers. Unlike insectivorous bats, their echolocation is based on audible clicks, which are produced by a different mechanism than the sonar of the Microchiroptera (Mohres and Kulzer, 1956).

Genetics.—2N = 36, AA = 66 for specimens from eastern Africa (Dulíc and Mutere, 1973).

Human interactions.—The common names in English (Egyptian fruit bat) and Arabic (*khafash el fawakeh*) indicate the preference of this bat for fruits. Fruit bats are agricultural pests in some areas, although the extent of damage they cause to such crops as oranges and dates is unknown (Bate, 1904; Lewis and Harrison, 1962). An extensive campaign was initiated by Israeli farmers and agricultural organizations to destroy these bats by fumigation of caves (Makin, 1979; Makin and Mendelssohn, 1986, 1989). A cave near Bitan Aharon was littered with bodies of insectivorous bats killed by fumigation (*Rhinolophus* and *Myotis*). One cave had 30 dead and more than 60 live *Rousettus*. Another fumigated cave had more than 2000 live fruit bats. Thus, it appears that insectivorous bats were more vulnerable than fruit bats. The replacement of insectivorous bats by fruit bats could also be due to increased agricultural activity and thus the availability of food. Fruit bats are noisy and if they inhabit a cave where insectivorous bats occur, they probably displace them. In Jordan, where there was no fumigation, I visited a cave near the Yarmouk River that held 3000 fruit bats. I was told by the locals that this cave had "small bats" (that is, insectivorous bats) the year before. It is also possible that the numbers of fruit bats fluctuate; in Lebanon, more fruit bats were found in the summer than in the winter (Lewis and Harrison, 1962).

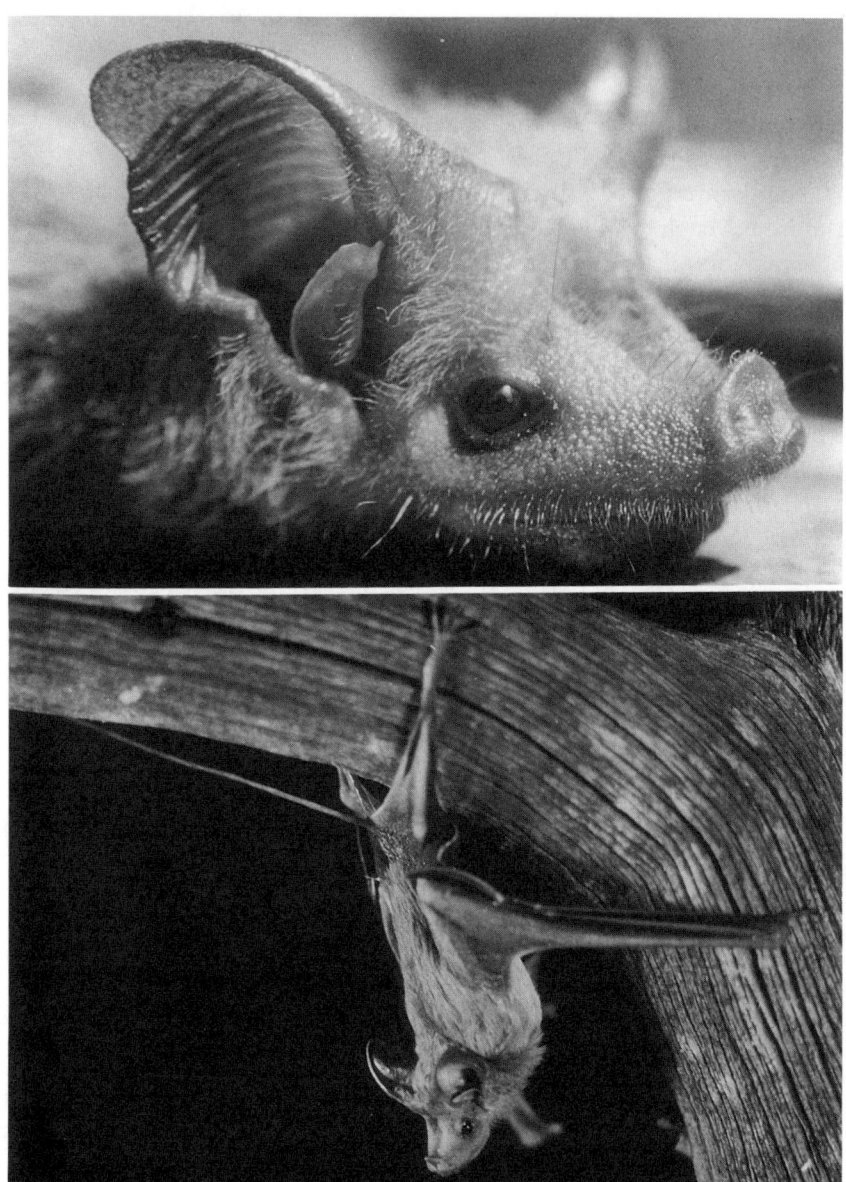

Fig. 16. External morphology of the lesser mouse-tailed bat (*Rhinopoma hardwickei*). Photos: (Upper) M. B. Qumsiyeh and (Lower) C. S. Hood.

FAMILY RHINOPOMATIDAE   Mouse-tailed bats

Genus *RHINOPOMA*

Bats of this family are distinguished by the retention of many primitive features including a long tail with a very small tail membrane (uropatagium) enclosing less than one-third of the tail. The facial area is very distinctive with a swollen muzzle consisting of glandular structures and a pig-like nose (Fig. 16). The numerous glands on the facial region are very sensitive and respond to stimuli (for example, touch, fear) by secreting an oily substance (Madkour, 1961; Kulzer et al., 1985). The skull is distinctive in its rather primitive form and in having conspicuous swellings in the nasal areas. The dental formula is i 1/2, c 1/1, p 1/2, m 3/3 = 28. The scientific names of the family and genus are derived from *rhis*, Gr., genitive *rhinos*, the nose, and *poma*, Gr., genitive *pomotos*, a lid, a cover, in reference to the peculiar skin flap over the nose.

Key to the Species of *Rhinopoma* of the Holy Land

1. Smaller, forearm generally less than 61 mm. Tail usually longer than forearm. Skull delicate but with lacrimal regions inflated. Males do not show strong development of the sagittal crest. Posterior edge of palate rounded. Diploid chromosome number 36. . . . . *R. hardwickei*

    Larger, forearm usually 61–72 mm in adults. Tail usually shorter than forearm. Skull characteristic (Fig. 18); it is more massively built but with less-inflated lacrimal regions; males show strong development of the sagittal crest. Posterior edge of palate triangular. Diploid chromosome number 42. . . . . *R. microphyllum*

Rhinopoma hardwickei
Lesser mouse-tailed bat, small mouse-tailed bat   **V**

Rhinopoma microphyllus É. Geoffroy St.-Hilaire, 1818. Description de L'Egypte Hist. Nat., 2:123, pl. 1, no. 1. Type from Great Pyramid, Giza, Egypt. This is a junior homonym of *R. microphyllus* Brünnich, 1782.

Rhinopoma hardwickii Gray, 1831. Zool. Misc., 1:37. Type from Bengal, India.

Rhinopoma cystops Thomas, 1903. Ann. Mag. Nat. Hist. ser. 7, 11:496. Type from Luxor, Egypt. Valid subspecies (not in the Holy Land): *R. h. cystops*.

Rhinopoma senaariense Fitzinger, 1866. Sitz. Akad. Wiss. Wien, 54(1):547. Used as *R. h. senaariense* by Kock (1969). Nomen nudum (Koopman, 1975).

Rhinopoma cystops arabium Thomas, 1913. Ann. Mag. Nat. Hist. ser. 8, 12:89. Type from Wasil, Yemen. Valid subspecies: *R. h. arabium* in the Holy Land.

Diagnosis.—From the larger species, *Rhinopoma hardwickei* is distinguished by more slender build and an overall smaller size (forearm less than 61 mm, skull length 16–18.5 mm). However, the tail is relatively longer (usually longer than forearm), the lacrimal regions of the skull are more inflated, the posterior edge of palate is more rounded, and the chromosome number is 36 (42 in *R. microphyllum*). In the skull, the most obvious difference is the slender build and the inflation of the lacrimal regions, which are thus separated by a small depression on the nasal region.

Eastern Mediterranean populations were described as *Rhinopoma cystops* (Thomas, 1903) and later assigned to *R. h. cystops* (Harrison, 1964a; Atallah, 1977). Kock (1969) reassigned these to *R. h. senaariense*. This latter name is, however, a nomen nudum (unavailable name) as suggested by Koopman (1975) and thus these specimens should be referred to *R. h. arabium.* Qumsiyeh (1985) examined specimens from different areas of Egypt and Palestine and reviewed the nomenclature. As a result of that review, the northern Egyptian and Palestinian specimens were found to be closer to *R. h. arabium.*

Range of species.—Northern Africa south to Niger, Mali, and Kenya. Arabia and southwestern Asia to India, Burma, and perhaps Thailand.

Local status.—Both species of mouse-tailed bats are found in the Holy Land but older records probably include many misidentifications. The distribution for both species appears to be similar (Fig. 17). The lesser mouse-tailed bat was recorded from Mount Qarantal, Jerash, Tiberias, Dan (Lachish), Arbel, 30 km S Ain Gedi, Ain Gedi, Wadi Amud, Wadi Dalam (Zalam), Wadi Darajah, Wadi Suweinit (HUJ), Bir Hindis, Hanita, Ain Fashkhah, Neot Hakikar, Nahal Amram, and Mar Jiryis (Wadi Kelt) (Thomas, 1913; Harrison, 1959, 1964a; Atallah, 1977; Makin, 1977; Qumsiyeh, 1985; Yom-Tov et al., 1992b). Specimens were also collected recently from near Quraiqira in Wadi Al Fidan (Qumsiyeh et al., 1992) and Al Majdal (JUNHM).

Biology.—Both species of *Rhinopoma* are well adapted to arid conditions. I did not catch any individuals in mist nets near streams or water sources where many bat species come to drink. It is possible that they derive water only from food sources. Food items reported to be taken by these bats include dipteran insects in India (Brosset, 1962, 1966) and beetles in Kenya (Kingdon, 1974). One can see extensive fat reserves in the lower abdominal region

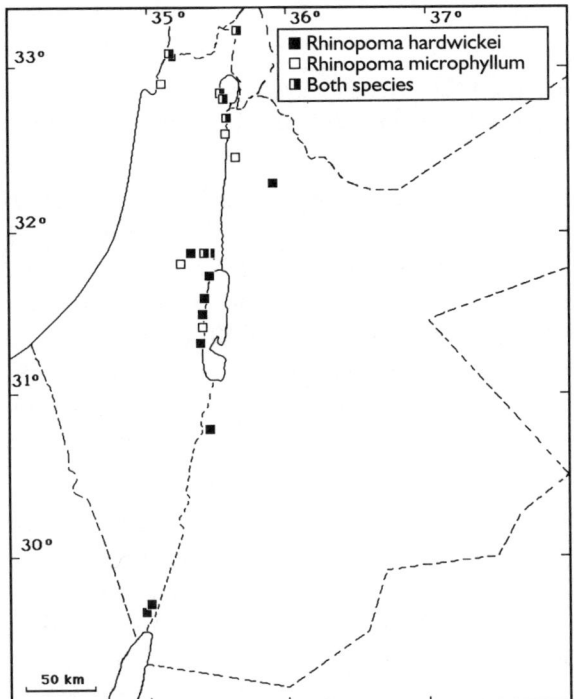

Fig. 17. Locality records of the two species of mouse-tailed bats in the Holy Land.

of these bats and this may explain their year-round activity (Poulet, 1970; Vogel, 1977). A specimen was kept alive for 2 months and it refused food or water until sacrificed (Qumsiyeh and Jones, 1986). These bats prefer dry roosting places in shallow caves, ruins, tunnels, and the like.

Young are born in June and July in India (Prakash, 1960). Pregnant females were found in April in Egypt, indicating a similar late spring parturition (Gaisler et al., 1972). I found immatures and females in colonies near Ain Fashkhah on 19 August 1975, and on Jebel Qarantal on 5 August 1976. Bats in these areas were found in visits from June through November but a visit in January 1981 yielded only one bat observed.

Genetics.—2N = 36, AA = 68 for specimens from India (Ray-Chaudhuri et al., 1968) and Palestine (Qumsiyeh and Baker, 1985).

Human interactions.—The name *hardwickei* is a latinized reference to Hardwicke, one of the first Europeans to collect this bat in Egypt.

Rhinopoma microphyllum
Larger mouse-tailed bat, rat-tailed bat   V

Vespertilio microphyllus Brünnich, 1782. Dyrenes Hist., 1:50, pl. 6, figs. 1–4. Type from "Arabia and Egypt" and later restricted to Giza Pyramids, Egypt (Anderson and de Winton, 1902; Kock, 1969).

Diagnosis.—This species is larger than *Rhinopoma hardwickei.* The tail is, however, relatively shorter and generally does not exceed forearm length. The skull is robust, especially in males, which also have a well-developed sagittal crest (Fig. 18). The nasal inflations and tympanic bulla are, however, less inflated than in *R. hardwickei.* The posterior margin of the bony hard palate is triangular and not rounded as in the smaller species. The lacrimal region of the skull is almost flat. The diploid number in this species is 42, FN = 66, which is distinct from *R. hardwickei* with 2N = 36, FN = 68 (Qumsiyeh and Baker, 1985). Males are larger than females in most measurements.

Range of species.—Not well known due to previous confusion with *R. hardwickei.* Definitely known to occur in sympatry (coexisting) with the smaller *R. hardwickei* in Mauritania (Qumsiyeh and Schlitter, 1981), Morocco (Panous, 1951; also specimens at USNM), Egypt (Qumsiyeh, 1985), Jordan (Qumsiyeh and Baker, 1985; Qumsiyeh et al., 1986), Saudi Arabia (Nader, 1975), Iran, Pakistan, India, and Sumatra (Lay, 1967; Kock, 1969). Thus, the range is perhaps similar to that of *R. hardwickei.*

Local status.—There is much confusion on the identity of this species in the literature prior to the 1960s (Harrison, 1963*b*; Kock, 1969; Atallah, 1977). For example, although Tristram (1884), Bodenheimer (1958), Dor

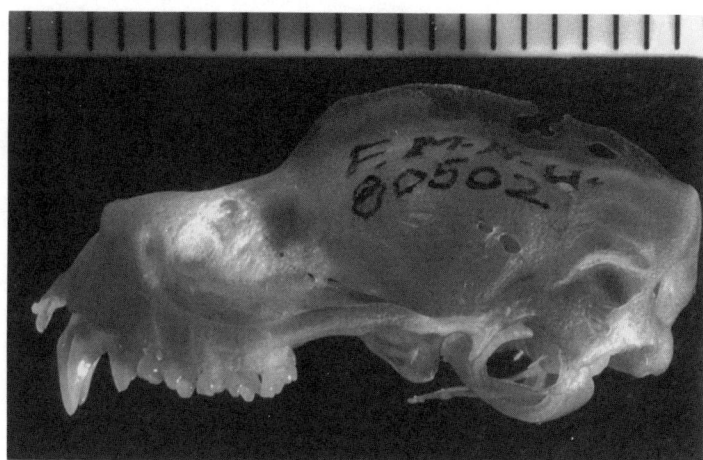

Fig. 18. Lateral view of the skull of a male larger mouse-tailed bat (*Rhinopoma microphyllum*).

(1947; specimens from owl pellets), and Aharoni (1930) report that this species is common in Palestine and Jordan, especially around the Dead Sea, they were probably referring to the smaller species, *R. hardwickei* (Harrison, 1963*b*). Authenticated records of *R. microphyllum* (Fig. 18) are those from Jebel Qarantal (Tristram, 1884; Harrison and Bates, 1991), Ain Gedi (Yom-Tov et al., 1992*b*), Jerusalem, Kineret, Buzieh (Dan) (Harrison, 1963*b*, 1964*a*), and Acre (= Akko) (Kock, 1969). I examined specimens from Hanita, Kineret, Tiberias, Wadi Amud, Jerusalem, Mar Jiryis in Wadi Kelt, Jebel Qarantal, and Wadi Khabra (Qumsiyeh, 1985). Some of these specimens were reported by Makin (1977), who also listed Wadi Bira. I recently collected specimens from Tabqat Fahl in the Jordan Valley (Qumsiyeh and Baker, 1985; Qumsiyeh et al., 1986) and Al Majdal (JUNHM).

Biology.—In its adaptation to desert conditions, this species is similar to *R. hardwickei* (see account of that species). Little is known of the biology of *R. microphyllum*. In India, copulation probably occurs in April with parturition expected in June.

These bats share the same caves as *R. hardwickei* and *Taphozous nudiventris* in areas such as Jebel Qarantal near Jericho and in Jordan. Harrison (1964*a*) and many other workers including myself rarely encountered this species in the Holy Land, unlike the more abundant *R. hardwickei*. Roberts (1977) reviewed what little is known of the biology of the species (mostly from the Indian subcontinent). Rat-tailed bats are colonial and the two sexes occupy separate roosts during the summer and autumn months. They appear to leave the area (in the Punjab) from October to May, probably to hibernate somewhere else. Females give birth in June or July.

Genetics.—2N = 42, AA = 66 for specimens from Jordan (Qumsiyeh and Baker, 1985).

Human interactions.—The name *microphyllum* is derived from *mikros*, Gr., small, and *phullon*, Gr., a leaf, "small leaf nose cover."

FAMILY EMBALLONURIDAE
Sheath-tailed bats, ghost bats, tomb bats

This family is unique among bats due to a combination of characters. First, the tail is enclosed in the uropatagium for a short distance and then emerges from the dorsal surface of the uropatagium. This gives these bats their scientific and common name, the sheath-tailed bats, Emballonuridae (from *emballo*, Gr., I put in, and *oura*, Gr., the tail). The nasal area is simple with no nose leaf. The ear has a tragus that is wide and short. The postorbital process of the skull is long and well developed. For *Taphozous*, the single genus of this family in the Holy Land, the dental formula is i 1/2, c 1/1, p 2/2, m 3/3 = 30.

Genus *TAPHOZOUS*

Key to the Species of *Taphozous* of the Holy Land

1. Fur does not extend posterior to the tail base; this gives the appearance of a naked belly (hence *nudiventris*). Size larger, forearm usually greater than 67 mm in adults. . . . . . *T. nudiventris*

   Fur extends posterior to the tail base. Size smaller, forearm usually well under 68 mm in adults. . . . . *T. perforatus*

Taphozous nudiventris
Naked-bellied tomb bat, naked-rumped bat   **V**

Taphozous nudiventris Cretzschmar, 1830 vel 1831. *In* Rüppell, Atlas Reise Nördl. Afr., Säugeth., p. 70, fig. 27b. Type from Giza, Egypt.

Taphozous magnus Wettstein, 1913. Ann. Naturhist. (Mus.) Hofmus, Wien, 27:466, pl. XX, figs. 1–6. Type from Basra, Iraq.

Diagnosis.—The tail projects through the dorsal surface of the uropatagium. The lower body is not covered with hair and hence the name nudiventris. The radiometacarpal pouch (fold of skin near elbow) is very small. The muzzle is nearly naked. The skull is robust with well-developed sagittal and lambdoidal crests. This species differs from *T. perforatus* in its larger size and the distinct naked skin around the tail base and lower abdomen. The dorsal profile of the skull is also less concave than in *T. perforatus* (Fig. 19). The males are slightly larger than females (Qumsiyeh, 1985) and also have a distinct throat pouch. The available specimens from the Holy Land are slightly larger than the Egyptian specimens (Qumsiyeh, 1985) and may represent intermediates between the nominate subspecies and *T. n. magnus* of Iraq.

Range of species.—Savanna and semiarid regions from Mauritania through Somalia north through Sudan, Egypt, and east through Arabia and southwestern Asia to India and Burma.

Local status (Fig. 20).—Tristram (1866*b*, 1884) reported on large colonies in "ravines in Galilee." Bodenheimer (1958) and Dor (1947) reported it from no exact locality in Palestine. The species was reported from Jebel Qarantal

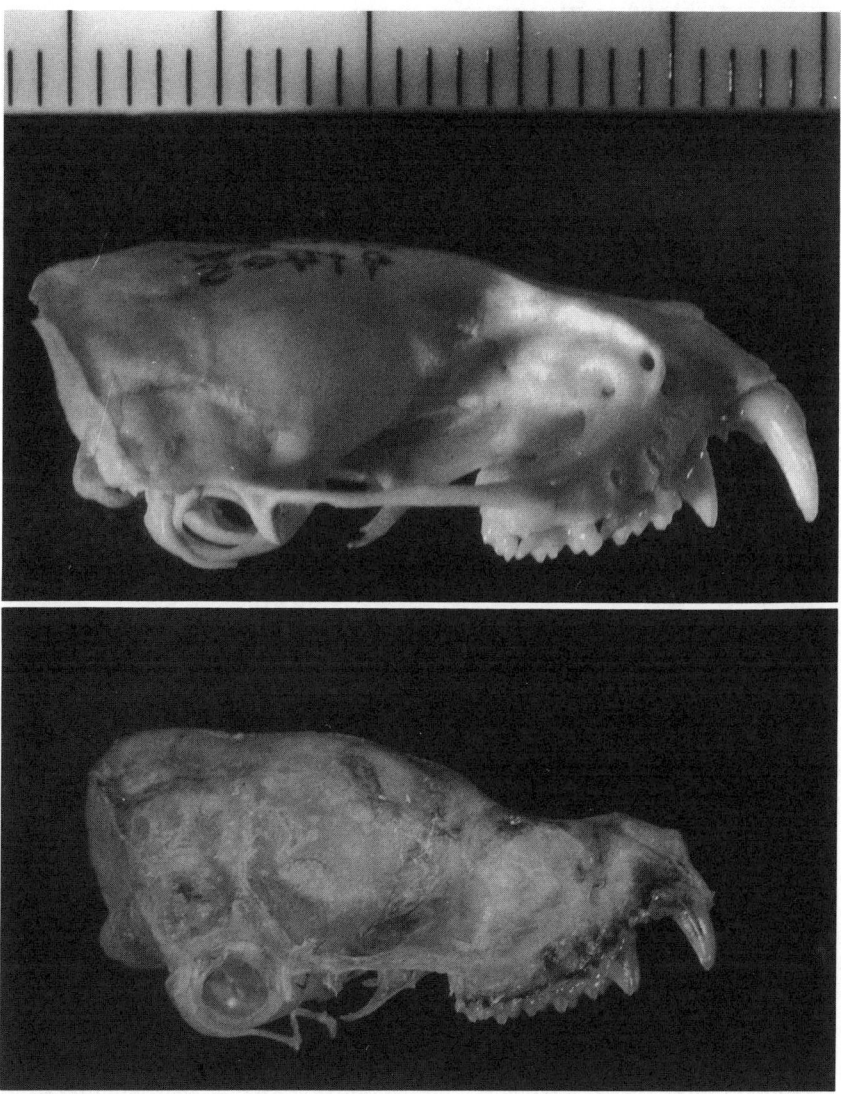

Fig. 19. Lateral views of skulls of (Upper) the naked-bellied tomb bat (*Taphozous nudiventris*); and (Lower) the Egyptian tomb bat (*Taphozous perforatus*). Note that the skull of *T. perforatus* is smaller and has a more concave profile in the forehead.

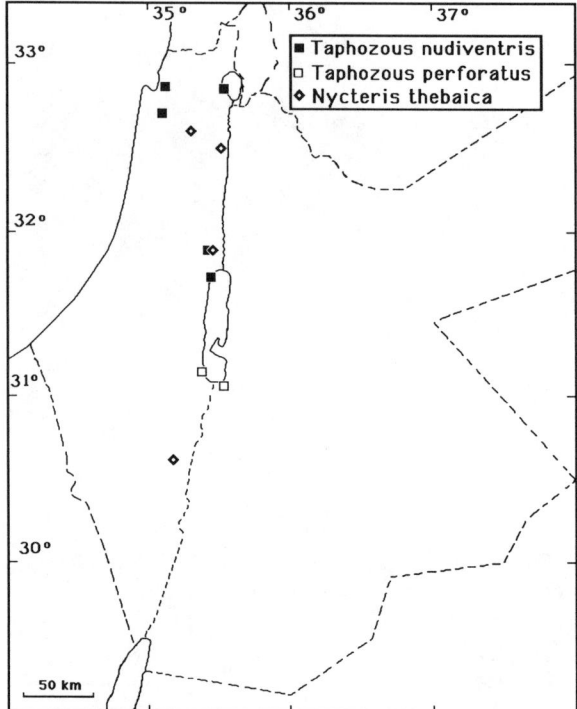

Fig. 20. Locality records of the naked-bellied tomb bat (*Taphozous nudiventris*), the Egyptian tomb bat (*Taphozous perforatus*), and the Egyptian slit-faced bat (*Nycteris thebaica*).

(Jericho) (specimens examined at MCZ) (Allen, 1915), north of Haifa, Lake Tiberias, Sha'ar Ha'amarqim, and Wadi Amud (northern Galilee) (Harrison, 1964*a*), and Galilee (Qumsiyeh, 1985). Specimens at HUJ are available from Ain Fashkhah and Wadi "Zamin."

Biology.—In the Holy Land, this tomb bat roosts in old buildings, small caves that receive a good amount of light (such as those at Jebel Qarantal), and crevices in small sandstone hills. I found males, females, young, and lactating young on 5 August 1976 at Jebel Qarantal. Active individuals were also collected in September. In Iraq, these bats mate in September and October, hibernate from November to March, and the females separate from the males to deliver their young in April (Al-Robaae, 1968).

Although migration has been suggested (Harrison, 1964*a*), there is no direct evidence for this. This bat was found to feed on the cotton leaf moth (*Spodoptera littoralis*) in Egypt (Madkour, 1977*b*). Large deposits of fat are found in the abdominal regions, similar to those found in *Rhinopoma*, and this may be source of reserve nutrition in the hibernating season (Dobson, 1878).

Genetics.—A report from Egypt (Yaseen et al., 1994) showed a 2N of 42 and AA = 64 for both *Taphozous perforatus* and *T. nudiventris*. The X is metacentric and the Y is a small acrocentric.

Human interactions.—The name *Taphozous* comes from *taphos*, Gr., a tomb, and *zoos*, Gr., living, "one that lives in a tomb." The name *nudiventris* is a Latin description of the naked ventral portion (belly) of this species.

Taphozous perforatus   Geoffroy's tomb bat, Egyptian tomb bat   V

Taphozous perforatus É. Geoffroy St.-Hilaire, 1818. Description de L'Egypte Hist. Nat., 2:126. Type from "Ombos and Theben," Egypt, restricted to Kom Ombo by Kock (1969).

Diagnosis.—This species is smaller than *T. nudiventris*. The skull length is generally under 22 mm and the forearm less than 67 mm. The fur is short and dark chocolate brown. Unlike *T. nudiventris*, the bases of the wing membranes near the body as well as the lower body are covered with fur. There is a large gular (throat) sac in males (absent in females). The radio-metacarpal pouch is well developed. The skull is slender and shows a more concave dorsal profile than that of *T. nudiventris* (Fig. 19).

Range of species.—Most of Africa south of the Sahara, Egypt, Arabia, and to Pakistan and western India.

Local status (Fig. 20).—Geoffroy's tomb bat is known only from Ain Boqeq (Ain Um Baghak) (Yom-Tov and Shalmon, 1989), Nahal Amud (Yom-Tov *fide* Harrison and Bates, 1991), and from 2 km E Ghor As Safi (Qumsiyeh et al., 1992). The latter record was of a specimen with a forearm length of 67 mm, which is larger than typical *T. perforatus* from Egypt. It maybe that the Holy Land populations are a distinct subspecies.

Biology.—This bat roosts in caves, ruins, crevices, and tunnels. In India and in Egypt a single young is born in May (Brosset, 1962; Gaisler et al., 1972).

Genetics.—2N = 42, AA = 64 reported from specimens from Egypt (Yaseen et al., 1994).

Human interactions.—The name *perforatus* is Latin meaning perforated; in reference to the fact that the tail membrane is perforated to accommodate the tail.

FAMILY NYCTERIDAE   Slit-faced bats

The face is deeply grooved longitudinally. This groove extends from the nostrils to the area between the ears. The sides of this groove consist of small cutaneous outgrowths. The skull morphology is also similarly affected with a concave interorbital region. The ears are large but the tragus is relatively small. The terminal tail vertebra is T-shaped. The dental formula in *Nycteris* is i 2/3, c 1/1, p 1/2, m 3/3 = 32. Nycterids are found in Africa, Arabia, and Asia.

Genus *NYCTERIS*

Nycteris thebaica   Egyptian slit-faced bat   V

Nycteris thebaicus É. Geoffroy St.-Hilaire, 1818. Description de L'Egypte Hist. Nat., 2:119, pl. i. Type from Egypt and, as indicated by the name (from Thebes), can be fixed to Luxor, southern Egypt (Anderson and de Winton, 1902).

Diagnosis.—This bat is distinguished by the fleshy longitudinal projections on the dorsal region of the elongated head and the shape of the skull, which creates the impression of a slit face or depression in the midfacial region. The ears are long and have a pyriform tragus. The forearm measures 40–52 mm. The skull can be easily distinguished by the presence of a deep concave shield in the dorsal region and the absence of the nasal branches of the premaxilla. The color of the back is brownish in adults, whereas immature specimens have more grayish colors.

Range of species.—Most of the savanna region of Africa throughout northern Africa and as far north and east as Palestine and the the Arabian Peninsula to Oman.

Local status (Fig. 20).—The slit-faced bat was reported from Beisan (Aharoni, 1944), Jericho, Merhavya, Ain Yahav (Qumsiyeh, 1985), Ain Gedi, and Neot Hakikar (Yom-Tov et al., 1992b).

Biology.—In Yemen, this species was found to feed on Orthoptera (Yerbury and Thomas, 1895). It may also feed on scorpions (Felten, 1956). This species was collected year-round in Egypt and nursing young appeared in June and July (Qumsiyeh, 1985). Maternity colonies were noted on 26 April 1969 in Egypt (Gaisler et al., 1972). In southern Africa, animals roost in hollow trees and forage close to the ground and near tree canopies (LaVal and LaVal, 1980; Aldridge et al., 1990).

Genetics.—2N = 42, AA = 78 reported for specimens from Kenya (Peterson and Nagorsen, 1975).

Human interactions.—The names *Nycteris* and Nycteridae are derived from the Greek *nukteris,* genitive *nukteridos* meaning a bat. The name *thebaica* is from the Greek *Thebai* or *Thebes,* which is an ancient city in Egypt now known as Luxor.

FAMILY RHINOLOPHIDAE
Horseshoe bats, Old World leaf-nosed bats

The family Rhinolophidae includes a single genus, *Rhinolophus,* with about 65 species. This group is characterized by the presence of leafy cutaneous outgrowths around the nostrils (Fig. 21). This is the origin of the name *Rhinolophus* (*rhis,* Gr., genitive *rhinos,* the nose, and *lophos,* Gr., a crest). The features of this complex structure are important for distinguishing the species as shown in the Key to the Species. These bats also have other distinguishing characters (for example, in the the structure of the premaxilla bones and in the bones of the pelvic region). The dental formula is usually i 1/1 or 1/2, c 1/1, p 2/3, m 3/3 = 30 or 32. Rhinolophids are found in Africa, Europe, and Asia (to Japan) in tropical, subtropical, and temperate regions.

Genus *RHINOLOPHUS*   Horseshoe bats

Key to the Species of *Rhinolophus* of the Holy Land

1. Small species, forearm shorter than 40 mm, skull length less than 17 mm. . . . . *R. hipposideros*

   Larger, forearm longer than 40 mm, skull length more than 17 mm. 2

2. Large species, skull length more than 22 mm. Connecting process of sella short and rounded (blunt) in side view (Fig. 21B). The upper canine and large premolar are in contact. 3

   Medium-sized rhinolophids. Skull length less than 22 mm. Connecting process of sella elongated and pointed in side view (Fig. 21C). The canine and large premolar are separated by a gap that includes the small premolar (as in Fig. 22). 4

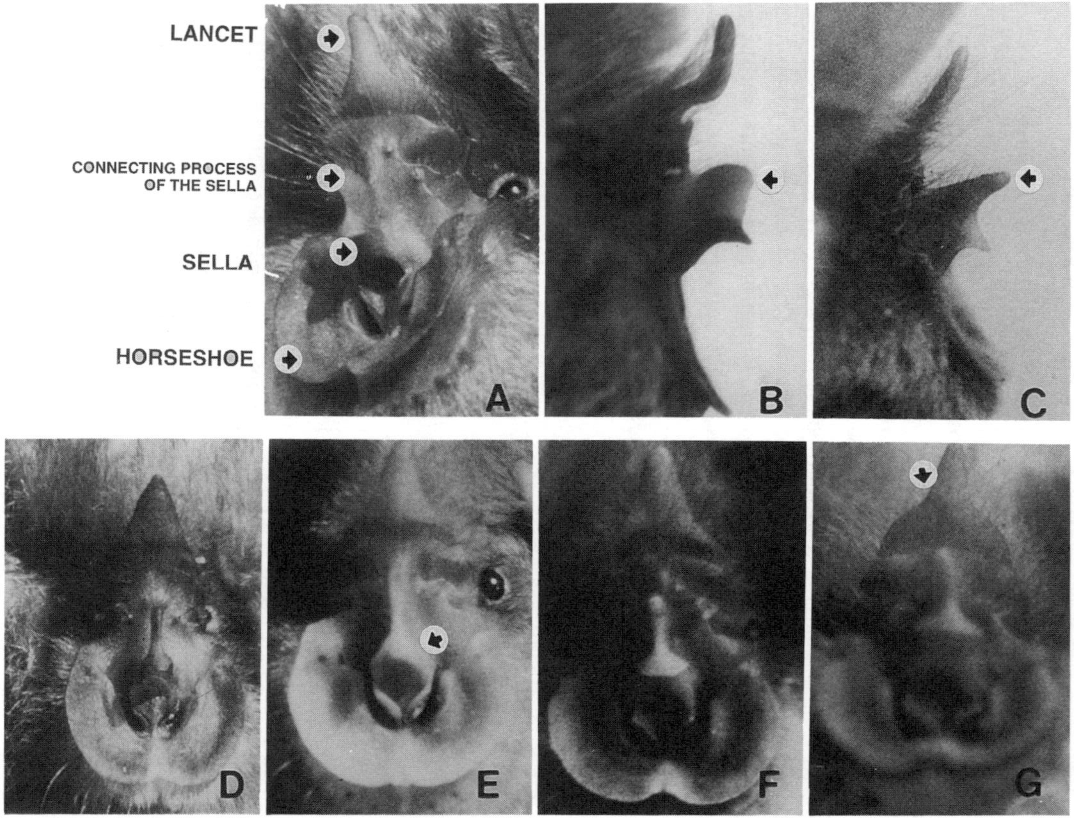
LANCET
CONNECTING PROCESS
OF THE SELLA
SELLA
HORSESHOE
A
B
C
D
E
F
G

Fig. 21. Nose-leaf structures in the horseshoe bats of the Holy Land. Arrowheads point to significant distinguishing features. (A) *Rhinolophus ferrumequinum*, (B) *R. clivosus*, (C) *R. euryale*, (D) *R. hipposideros*, (E) *R. blasii*, (F) *R. euryale*, (G) *R. mehelyi*.

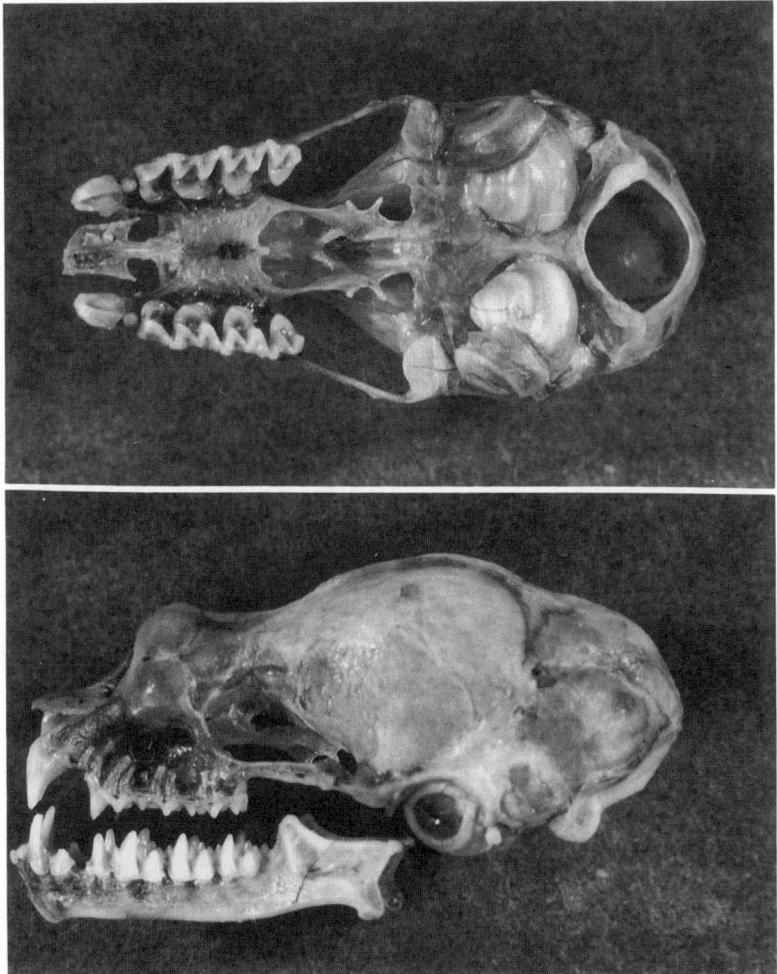

Fig. 22. Skull of a typical rhinolophid bat, *Rhinolophus blasii* from Jordan. Ventral view of skull (upper panel) and lateral view of skull and mandible (lower panel).

3. Larger (forearm 53–60 mm). Ears attenuated near their distal end. First upper premolar is displaced externally from the tooth row. In Mediterranean climatic zone only. . . . *R. ferrumequinum*

Smaller (forearm less than 52 mm). Ears not attenuated near their distal end. First upper premolar lies in tooth row. In desert and subdesert climatic zones. . . . . *R. clivosus*

4. The base of the sella narrows abruptly dorsally (Fig. 21E). The connecting process of the sella with upper projection pointing dorsally. . . . . *R. blasii*

The base of the sella almost with parallel sides. The connecting process of the sella with upper projection pointing forward (Fig. 21C). 5

5. The tip of the lancet tapering evenly and essentially triangular (Fig. 21F). Zygomatic width 9.0–9.8 mm. . . . . *R. euryale*

The sides of lancet more concave (lancet acuminate) with distal tip showing almost parallel sides (Fig. 21G). Zygomatic width 10.0–10.6 mm. . . . . *R. mehelyi*

Rhinolophus blasii   Peter's horseshoe bat   V

Rhinolophus clivosus Blasius, 1857. Fauna Deutchlands, Säugeth., p. 33 (not Cretzschmar, 1828). Type restricted by Ellerman and Morrison-Scott (1951) to Dalmatia, Istria, Sicily. Name not available.

Rhinolophus blasii Peters, 1857. Monatsber. Akad. Wiss. Berl., p. 17 (also, 1866, Monatsber. K. Preuss. Akad. Wiss. Berl., p. 17). New name for Blasius's *R. clivosus*.

Diagnosis.—Another of the medium-sized *Rhinolophus* in the region with a forearm length of 44–48 mm and a greatest length of the skull of 19–20 mm. The connecting process of the sella is pointed upwards. The base of the sella is very distinctive, being wide ventrally and narrow dorsally (Fig. 21E). The wing membrane attaches to the ankles.

Range of species.—In the Mediterranean woodlands from Italy and Morocco east through Arabia to Pakistan and north to Russia. Also in Africa south of the Sahara if *R. andreinii* and *R. empusa* are conspecific with *R. blasii*.

Local status (Fig. 23).—Tristram sent specimens of this species to the British Museum that were collected at the Cave of Adullam (Dobson, 1878). Tristram later (1884) reported that this species was found in the areas of Jerusalem, Bethlehem, and Hebron. Bodenheimer (1958) reported it from Herzelia and Sheikh Abreik. The species is known from Herzelia, Solomon's Quarries, Aqua Bella, Cave of Adullam, Jenin and 2 mi S Jenin (Harrison, 1964a; Atallah, 1977), Abu Ghosh, Bet She'arim, and Jerusalem (HUJ).

East of the Jordan River, this species was collected from Mogharet el Roum (= The Roman Cave) in Jerash Refugee Camp (Qumsiyeh, 1980), Tabqat Fahl, Mogharet Al Mata (5 km NNE Fidan, Wadi Fidan) (Qumsiyeh et al., 1986), and Shobek (Zuhair Amr Collection). A specimen I

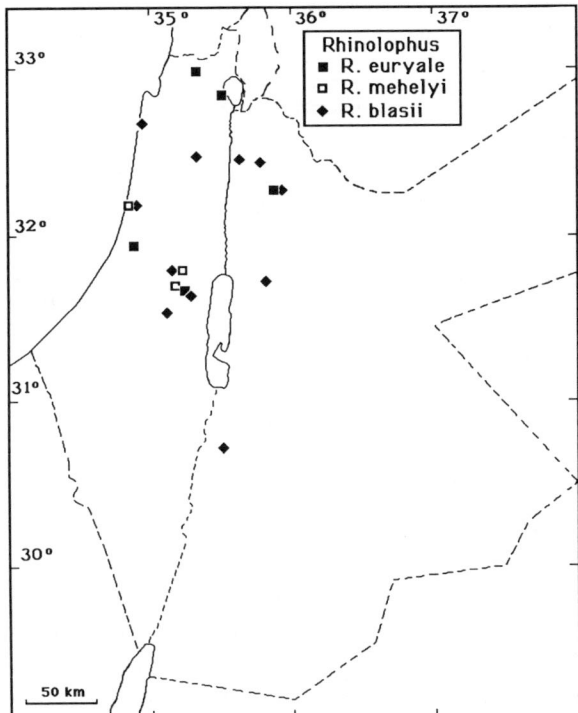

Fig. 23. Locality records of the Mediterranean horseshoe bat (*Rhinolophus euryale*), Mehely's horseshoe bat (*Rhinolophus mehelyi*), and Peter's horseshoe bat (*Rhinolophus blasii*).

deposited at the JUNHM from Mogharet Al Mata was listed erroneously as an additional record of the species from "Wadi Finan" by Amr and Disi (1988). A specimen at JUNHM is from Madaba. It was also collected from Zubiya Forest, northern Jordan by mist nets (Qumsiyeh et al., 1992).

Biology.—I found approximately 30–40 individuals in a visit to Mogharet el Roum (= The Roman Cave) near the Jerash Refugee Camp on 8 October 1976. All observed or collected individuals were females. On 15 October, only three bats were observed and on 1 May 1977 no bats were in the cave. An active colony of 25 individuals (adult males and females) was found in a small cavern in Wadi Finan near the Araba Valley. This is a very dry, desert habitat, much different from that in Dibbine Forest (Mediterranean climate), where this species was also collected. It is thus possible that both a desert form and a montane form of *R. blasii* occur in the Holy Land.

Reproductive data are needed for this species. Dor (1947) reported a specimen found in the pellets of the owl *Tyto alba* in Palestine (no specific locality was given).

Genetics.—2N = 58, AA = 60 for specimens from Jordan (Qumsiyeh et al., 1986), which is similar to that reported for European specimens (Zima, 1982a; Zima and Král, 1984). The X chromosome is metacentric and the Y is a very small acrocentric.

Rhinolophus clivosus   Cretzschmar's or Arabian horseshoe bat   V

Rhinolophus clivosus Cretzschmar, 1828. *In* Rüppell, Atlas Reise Nördl. Afr., Säugeth., p. 47. Type from Mohila (= Muweileh), western Saudi Arabia.

Rhinolophus acrotis Heuglin, 1861. Nova Acta Leopold. Carol., 29, 8:4, 10. Type from Keren, Eritrea. Perhaps a valid subspecies or a synonym of *R. c. clivosus.*

Rhinolophus andersoni Thomas, 1904. Ann. Mag. Nat. Hist., ser. 7, 14:156. Type from Wadi Alagi, eastern Egyptian Desert, Egypt. Synonym of *R. c. clivosus.*

Rhinolophus antinorii Dobson, 1885. p. 156. Type from Shoa, Ethiopia. Synonym of *R. c. brachygnathus.*

Rhinolophus acrotis brachygnathus Anderson, 1905. Ann. Mag. Nat. Hist., 15:53, pl. 37. Type from Giza, Egypt. Valid subspecies: *R. clivosus brachygnathus.* extralimital.

Diagnosis.—This is a medium-sized horseshoe bat with the forearm averaging 48 mm (range 45–52 mm) and a greatest length of skull of 18.5–22 mm. It can be distinguished from other *Rhinolophus* by the shape of the sella (Fig. 21B). Qumsiyeh (1985) discussed the status of the many named forms in this species. Clearly, the Holy Land specimens belong to the subspecies *R. c. clivosus.*

Range of species.—Western Saudi Arabia, Yemen, Palestine, Egypt, Sudan, and most of the savanna areas of Africa (except western Africa).

Local status.—Tristram (1866*b*) reported this species from near Tiberias but later removed it from the list of the mammals of Palestine (Tristram, 1884). This record probably belonged to *R. ferrumequinum.* Authenticated records are available only from the southern, more arid regions of the country (Fig. 24). One is tempted to speculate that the allopatric distributions of *R. ferrumequinum* and *R. clivosus* (Fig. 24) are due to competitive exclusion between these two similarly sized and closely related rhinolophids. Specific records are from Revivim, Abde (Avdat) (Harrison, 1964*a*), Eilat (Makin, 1977), Petra (P. Boye, 1983, unpubl.), Ain Gedi, Neot Hakikar (Yom-Tov et al., 1992*b*), and near Greigra (Wadi Finan) and Disi (Wadi Rum) (Qumsiyeh et al., 1992). The species was also recorded from northern Sinai at Al Arish (Wassif, 1954*b*). The Holy Land specimens are referable to the nominate subspecies, *R. c. clivosus,* originally described from the western coast of Saudi Arabia. Specimens of *R. c. brachygnathus* of Egypt are smaller than those of *R. c. clivosus.*

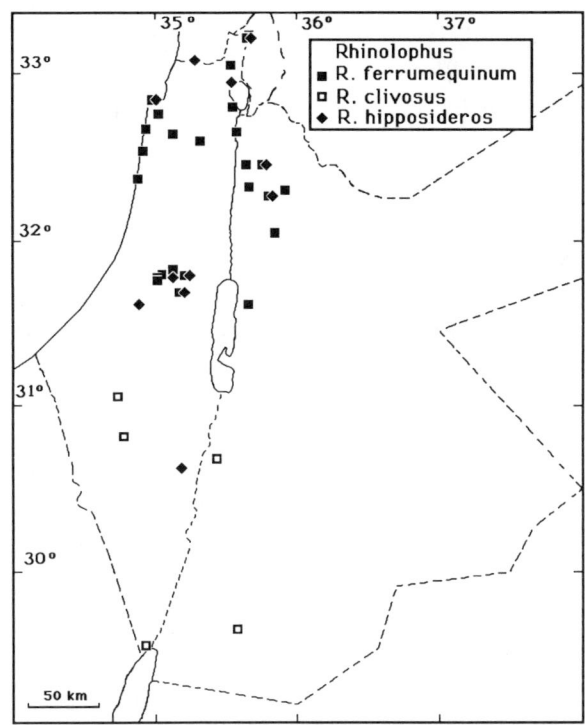

Fig. 24. Locality records of the greater horseshoe bat (*Rhinolophus ferrumequinum*), the Arabian horseshoe bat (*Rhinolophus clivosus*), and the lesser horseshoe bat (*Rhinolophus hipposideros*).

Biology.—In Sinai, the species roosts in old storehouses, stone huts, and hillside caves (Hoogstraal, 1962).

Genetics.—2N = 58, AA = 62 for specimens from eastern Africa (Dulíc and Mutere, 1974).

Rhinolophus euryale   Mediterranean horseshoe bat   E

Rhinolophus euryale Blasius, 1853. Archiv. Naturgesch. 19(1):49. Type from Milan, Italy.

Euryalus judaicus Andersen and Matschie, 1904. Sitzungsber. Ges. Naturf. Fr. Berl., 1904:80. From Cave of Adullam, near Jerusalem (= Mogharet Khureitun, 4 km SE Beit Sahur, *fide* Atallah, 1977). Perhaps a valid subspecies.

Diagnosis.—This is a medium-sized bat with forearm 45–50 mm long and greatest length of skull 18–19.8 mm. The connecting process of the sella is pointed in side view. The sides of the lancet slope gradually creating a triangular lancet (Fig. 21). In the morphologically similar *R. mehelyi*, the sides slope abruptly upwards and create almost parallel sides at the dorsal region. Also, the zygomatic width in *R. euryale* is 9.0–9.7 mm whereas in *R. mehelyi* it is 10 10.6 mm (DeBlase, 1972). The small upper premolar separates the canine and large premolar with a distinct gap. The wing membrane attaches to the tibia above the ankles.

Range of species.—Circum-Mediterranean and east into Iran, Transcaucas, and Turkestan.

Local status. (Fig. 23).—This species was first collected by Tristram from the Lake of Galilee (specimens reported by Dobson, 1878). The type of *judaicus* (Andersen and Matschi, 1904) came from the Cave of Adullam. The species was reported from Ramleh, Ras en Nakura, Herzelia, Solomon's Quarries, Jerusalem (Harrison, 1964*a*), Jerusalem, Ramleh, and Jerash (Atallah, 1977). However, the BM specimens from Jerusalem and from Solomon's Quarries listed by these authors were found to belong to *R. mehelyi* by both lancet shape and cranial measurements (DeBlase, 1972). The specimen from Herzelia is also *R. mehelyi* based on cranial measurements. *Rhinolophus euryale* was reported from the Galilee (Felten et al., 1977), and also from Dibbine National Park (Qumsiyeh, 1985; Harrison and Bates, 1991).

Biology.—In Iran, these bats were found in hibernation caves from September to November (DeBlase, 1980).

Genetics.—2N = 58, AA = 60 for specimens from Europe (Manfreddi-Romanini et al., 1975; Zima and Král, 1984). The X chromosome is metacentric and the Y is a very small acrocentric.

Rhinolophus ferrumequinum
Larger horseshoe bat, greater horseshoe bat   V

Vespertilio ferrum-equinum Schreber, 1774. Die Säugethiere, 1:174, pl. 62 (published 1775). Type from France.

Diagnosis.—This is the largest and most heavily built of the horseshoe bats in the region. Forearm length averages 53–60 mm. A rounded and short connecting process of the sella is a distinguishing character (in side view).

Range of species.—Europe, northwestern Africa, the Near East, and along a belt in the southern Palearctic region from Afghanistan to Japan.

Local status.—The larger horseshoe bat is not uncommon in the northern parts of the Holy Land, especially in the mountainous and forested regions (Fig. 24). Earlier authors reported this species as "common" in Palestine without giving specific localities (Tristram, 1866*b*, 1884; Bodenheimer, 1935, 1937, 1958). At least two of Tristram's specimens came from near the Sea of Galilee (Lake Tiberias) and are at the British Museum (Dobson,

1878). Harrison (1964*a*) reported it from Beit Meir, Tiberias, Haifa, Buzieh (Tel el Qadi), Me'areth Hateamim, Rehavia Cave, Aqua Bella, and Solomon's Quarries. Also noted from Carmel Caves, Sheikh Bureik, Alma Cave, Yarmouk Pumping Station, Merhavya, Hazorea, Beit Hananyah, Bitan Aharon (Makin, 1977), Haifa, and Beit Shemesh (Felten et al., 1977). East of the Jordan River, the species was recorded from the Jerash and Suweilih areas (Harrison, 1959; Atallah, 1977), Zarqa River (Nader and Kock, 1983), Tabqat Fahl (Harrison and Bates, 1991), and Quayliba near Irbid (JUNHM). It has also been reported from Khirbet Sa'ad (near Nadira), Dibbine Forest (Qumsiyeh et al., 1986), and caves near Bitan Aharon (fumigated caves with many dead bats). Recently, this species was also collected from Zubiya Forest, northern Jordan, using mist nets (Qumsiyeh et al., 1992).

Biology.—This species is well studied in Europe. The greater horseshoe bat is a gregarious species in many parts of its range. Only when males separate from the female colonies can they be found to roost singly or in small groups. They prefer to roost in the deep parts of caves or other areas where humidity is high. Horseshoe bats differ from vespertilionid bats in the way they roost. Horseshoe bats prefer to hang freely from horizontal roofs, whereas vespertilionid bats prefer more angular or vertical surfaces (with their bellies against the wall). When hibernating, horseshoe bats acquire a characteristic posture with their wings and tail membranes covering most of their body and the head tucked inwards.

When active, these bats are very alert with their ears moving constantly to pick the reflected sound that is emitted from their nose. This bat locates prey and other objects using sonar emitted from the nose. Horseshoe bats in general fly a peculiar wavy flight by fluttering their wings for short intervals. In Europe, fertilization takes place in October or November but the egg implants in the wall of the uterus only in the spring at the end of the hibernation season (a phenomenon known as delayed implantation). Young are born in late spring or early summer. This chronology also seems to apply to the Holy Land horseshoe bats, as I observed lactating females in the late spring. I obtained hibernating individuals in Dibbine Forest on 20 November and active ones in August. Similar reports of hibernation in November and December are reported from Iran (DeBlase, 1980).

This species may share caves with *R. euryale* (Harrison, 1964*a*). In Dibbine Forest, I noticed that the small *R. hipposideros* may occur in the same small caves as *R. ferrumequinum* but *R. hipposideros* usually roosts closer to the exit.

Genetics.—2N = 58, AA = 60 for specimens from Dibbine Forest in Jordan (Qumsiyeh et al., 1986), which is similar to specimens from Europe (Zima, 1982*a*), and Tunisia (Baker et al., 1975). The X chromosome is metacentric and the Y is a small acrocentric. Qumsiyeh et al. (1988) examined chromosomal and protein data for this and other species of *Rhinolophus* (including *R. blasii* and *R. hipposideros*) and found a rather rapid protein evolution compared to chromosomal evolution. The latter report also shows very close genetic similarity between *R. ferrumequinum* from Jordan and those from Japan.

Human interactions.—Large colonies of this species were present in the 1950s and early 1960s in Jerash (photograph presented in Harrison, 1964*a*), but were absent in the late 1970s (Qumsiyeh, 1980). The reason for their disappearance is not clear although increased tourist activity may be a factor. There are areas in Jordan where this species is still common. At Zubiya Forest, many were caught in one night of mist-netting using a 10-m mist net. This area was well protected from development as a national forest. A new national park is to be established nearby, which should help protect this and other species of bats. The name *ferrumequinum* is from *ferrum*, L., iron, and *equinum*, L., relating to horse (= a horseshoe).

Rhinolophus hipposideros   Lesser horseshoe bat   V

Vespertilio hipposideros Bechstein, 1800. *In* Thomas Pennant, Allgemeine Ueber. Vierfüss. Thiere, 2:629 (also, 1801, Naturg. Deutschl., p. 1194). Type from France.

Rhinolophus minimus Heuglin, 1861. Nova Acta Leopold. Carol., 29, 8:6. Type from Keren, Eritrea. Valid subspecies: *R. h. minimus*.

Rhinolophus hipposideros var. *pallidus* Koch, 1863. Jahrb. Nassau Ver. Naturk., 18:531. Type from the Mediterranean region. Synonym.

Diagnosis.—This is the smallest *Rhinolophus* in the region. The forearm rarely exceeds 39 mm and the skull length is less than 16 mm. The connecting process of the sella is low and blunt in side view. The lancet is thick and almost perfectly triangular in anterior view (Fig. 21D). Identification of subspecies of the lesser horseshoe bat is fraught with taxonomic difficulties. DeBlase (1980) summarized the problem. I reviewed the subspecies designation of the Palestinian specimens and the single specimen from Sinai earlier (Qumsiyeh, 1985) and agreed with other authors that minor size variations distinguish *R. h. minimus* from the nominate subspecies. Because clinal variation exists, there probably is no justification for the subspecific designation *R. h. minimus*. The status of *R. h. midas* from southwestern Asia (with a much lighter color than other forms) is still in need of further reevaluation (DeBlase, 1980).

Range of species.—Southern Europe, northern Africa south to Eritrea, and southwestern Asia east to Kashmir.

Local status.—The lesser horseshoe bat is more common in the northern, Mediterranean climatic zone than in the more arid regions of the south (Fig. 24). The southernmost record for the species in the Near East is from a single specimen from Feiran Oasis in Sinai (Qumsiyeh, 1985). That specimen and another from Ain Yahav in the Negev Desert may represent remnant populations from the retreat of the moist Mediterranean climate due to the expansion of the Sahara Desert. Bodenheimer (1958) reported this species from Herzelia, Beit Araba, and Sheikh Abreik. This species is known from Solomon's Quarries, Aqua Bella, Buzieh (Dan), Beit Guvrin, Ain Yahav (Harrison, 1964a), Shetula, Haifa, Tel Maresha (Makin, 1977), Hartuv (Felten et al., 1977), Berekhot Navit, Neot Hakikar (Yom-Tov et al., 1992b), and Rosh Pinna (HUJ). I reported on specimens from the same localities and added Dibbine National Forest (Qumsiyeh, 1980, 1985; Qumsiyeh et al., 1986). Recently, we also captured this species in Zubiya Forest, northern Jordan, by using mist nets (Qumsiyeh et al., 1992).

Biology.—These bats are found as solitary animals in caves, ruins, and other dark dwellings. Because individuals are solitary, it is not known how abundant this species is. This species is rarely encountered but may be more common and widespread than the meager collection reports indicate.

Usually one or two young are born in June or July following a gestation period of about 7 weeks (Asdell, 1964; Atallah, 1977). I obtained hibernating individuals in Dibbine Forest on 20 November and active individuals on 9 August. Similar hibernation in October and November is known for Iranian specimens (DeBlase, 1980). For other habits of horseshoe bats in general see *R. ferrumequinum.*

Genetics.—2N = 58, AA = 60 for specimens from Dibbine Forest in Jordan (Qumsiyeh et al., 1986), which is different than that reported for European specimens (2N = 54–56, Zima, 1982a; Zima and Král, 1984) and central Asian specimens (2N = 62, Zima et al., 1992a). The X chromosome is metacentric and the Y is a very small acrocentric.

Human interactions.—The name *hipposideros* is derived from *hippos,* Gr., a horse, and *sideros,* Gr., iron (= horseshoe). Notice that *ferrumequinum* and *hipposideros* mean the same thing (a horseshoe) but are derived from Latin and Greek, respectively.

Rhinolophus mehelyi   Mehely's horseshoe bat   E

Rhinolophus mehelyi Matschie, 1901. Sitzungsber. Ges. Naturf. Fr. Berl., 1901:225. Type from Bucharest, Rumania.

Rhinolophus euryale judaicus Allen, 1939. Bull. Mus. Comp. Zool. Harvard Univ. 83:74. (Not *Euryalus judaicus*, Andersen and Matschie, 1904, p. 80, a junior synonym and subspecies of *R. euryale* [Blasius].)

Diagnosis.—This is a medium-sized horseshoe bat with greatest length of skull 19.1–20.5 mm. It is distinguished from *R. clivosus* and *R. hipposideros* by the almost parallel sides of the sella and the relatively pointed connecting process of the sella. From *R. euryale*, it differs in overall larger size and in the shape of the sella and lancet. The connecting process of the sella is shorter and more blunt in *R. mehelyi* and the lancet is more triangular in *R. mehelyi* (Fig. 21F). Females appear to have longer forearm measurements whereas males have larger skull measurements (Baker et al., 1975).

Range of species.—Not well known due to frequent misidentification and confusion with *R. euryale*, but perhaps similar in range.

Local status. (Fig. 23).—Very poorly known due to confusion with *R. euryale*. The following specimens variously noted as *R. euryale* are clearly *R. mehelyi*: Jerusalem (BM), Solomon's Quarries (BM), Herzelia (HZM), and "Palestine" (Field Museum of Natural History) (DeBlase, 1972). Also, a specimen from Jerusalem is at the Senckenberg Museum (Felten et al., 1977). We recently obtained a specimen from Al Naqah in Wadi Araba (Zuhair Amr Collection).

Biology.—See *R. euryale*.

Genetics.—2N = 58, AA = 60 for specimens from Tunisia (Baker et al., 1975) and Europe (Zima and Král, 1984). The X chromosome is metacentric and the Y is a very small acrocentric.

FAMILY HIPPOSIDERIDAE   Leaf-nosed bats

The Hipposideridae (from *hippos*, Gr., a horse, and *sideros*, Gr., iron, meaning horseshoe) are closely related to the Rhinolophidae, as evidenced by numerous morphological similarities including the elaborate nose-leaf structure of these bats. The nose leaf in the Hipposideridae lacks a sella or a connecting process. A transverse lobe is present and corresponds to the lancet in *Rhinolophus*. In the single species of this family known from the area, *Asellia tridens*, this lobe is divided into three dorsally projecting lobes. The dental formula for *Asellia* is i 1/2, c 1/1, p 1/2, m 3/3 = 28.

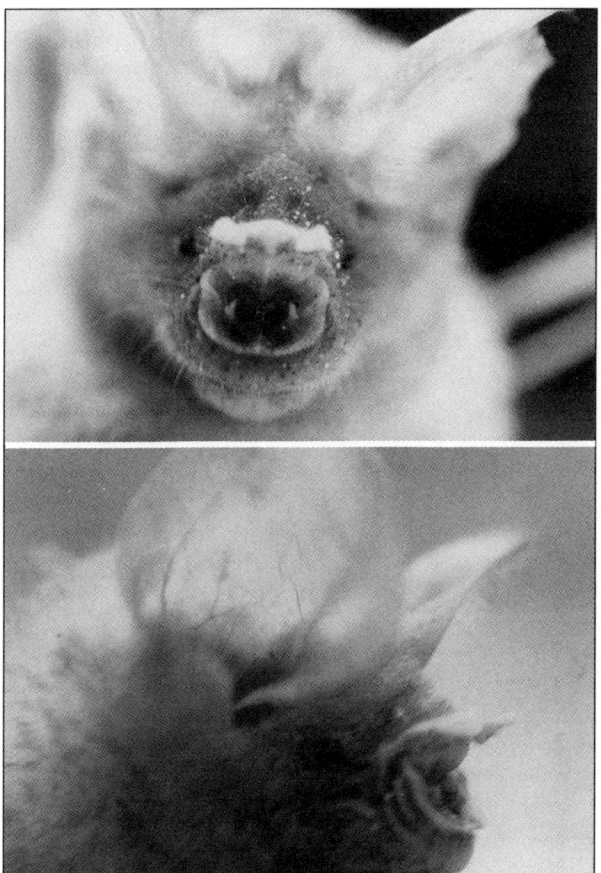

Fig. 25. The trident leaf-nosed bat (*Asellia tridens*).

Genus *ASELLIA*

Asellia tridens   Trident leaf-nosed bat   V

Rhinolophus tridens É. Geoffroy St.-Hilaire, 1812. Description de L'Egypte Hist. Nat., 2:130 (also 1813, Ann. Mus. Hist. Nat. Paris, 20:265). Type from Egypt and restricted to Thebes (= Luxor) by Kock (1969).

Phyllorhina tridens var. *murriana* J. Anderson, 1881. Catalogue of mammals in the Indian Museum, p. 113. Type from Karachi, Sind, Pakistan. Valid subspecies: *A. t. murriana.*

Diagnosis.—*Asellia* (Fig. 25) is distinguished easily from other leaf-nosed bats of the region (*Rhinolophus*) by the presence of three projections of the dorsal nose leaf and by a relatively larger ear size. Many other aspects of the anatomy are also peculiar (Wassif, 1949). There is a reddish individual variant seen in some localities. There is considerable geographic and sexual variation. The specimens in Palestine, Iraq, Iran, and Afghanistan are more

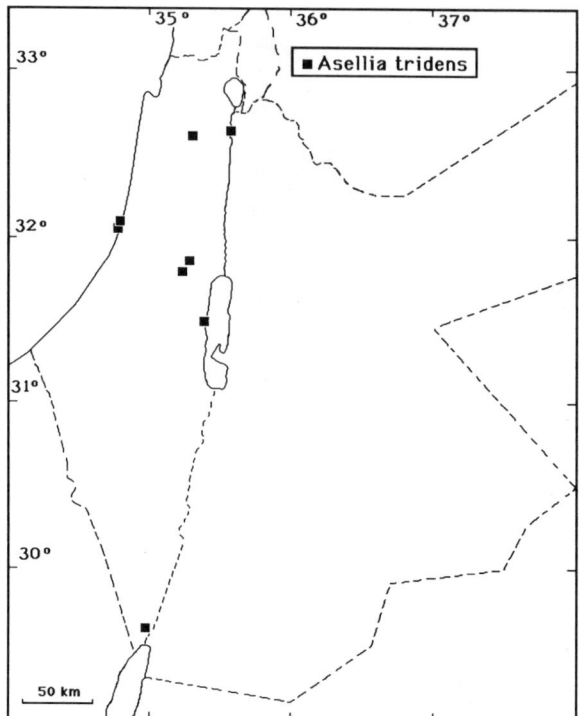

Fig. 26. Locality records of the trident leaf-nosed bat (*Asellia tridens*).

similar in morphology to those from Morocco than to those from Egypt (Owen and Qumsiyeh, 1987). Because of this, Palestinian specimens may be referred to the subspecies *A. t. murriana.*

Range of species.—Northern Africa south to Gambia, east through Arabia and southwestern Asia to Afghanistan, Pakistan, and northern India.

Local status.—This bat is adapted to arid conditions of the Sahara, Arabian, Persian, and Mongolian deserts. It has found its way to the areas around the Rift Valley with a marginal penetration near the coastal regions and in the mountains (Fig. 26). The trident leaf-nosed bat was reported in the Dead Sea Basin (Tristram, 1884; Aharoni, 1930; Bodenheimer, 1935, 1958), Jerusalem, Hezme, Jaffa, Tel Aviv, Nahal Amram (10 km N Eilat) (Bodenheimer, 1958; Harrison, 1968*a*), Yarmouk Pumping Station, Jaffa, Tel Aviv, Jerusalem, Nahal Amram (Qumsiyeh, 1985), Ain Gedi, and Neot Hakikar (Yom-Tov et al., 1992*b*). Specimens at TAUM are available from near Mohila, Yarmouk Pumping Station, and Merhavya and at HUJ from the Yarmouk Dam area, Ashdot-Yaaqov, Jaffa, and Ain Gedi. In Jordan proper, only one specimen is available from Al Naqah near Ghor Safi (Zuhair

Amr Collection). The Jordan Bridge and Meholla records (Owen and Qumsiyeh, 1987) are from west of the Jordan River and not from "Jordan" as stated by Harrison and Bates (1991).

Biology.—This is a desert-adapted colonial species. I have found it in regions of the Sahara Desert where no other species of bat was seen. It has a high temperature tolerance (Weber, 1956). Some roosting sites in Iraq are abandoned in winter, which suggested to Harrison (1957) that they migrate. However, Al-Robaae (1966) showed that actually they have winter quarters that they occupy until April, whereupon they return to their summer quarters. The gestation period is assumed to be 9–10 weeks with a single young born in early June in Iraq (Al-Robaae, 1966).

FAMILY VESPERTILIONIDAE   Vespertilionid or common bats

The Vespertilionidae (from *vesper,* L., evening and *vespertilio,* L., genitive *vespertilionis,* a bat, "an animal of the evening") is the largest of bat families with 33 genera and about 313 species. Vespertilionids are small to medium-sized bats without specializations or much variation in external appearance. They generally lack nose-leaf structures. The ear structure is variably specialized and usually with a well-developed tragus (Figs. 27, 33). They possess a long tail enclosed completely or almost completely by the uropatagium. Eight genera are known from the Holy Land.

Key to the Genera of Vespertilionidae of the Holy Land

1. Sternum with median lobe larger than body of bone. Second
   phalanx of third finger very long (about three times length
   of first phalanx). Skull profile from side view shows abrupt
   elevation of braincase (slopes upward abruptly) with a short
   rostrum. . . . . *Miniopterus*

   Not as above. 2

2. Upper cheekteeth consist of three premolars and three molars on
   each side (six postcanine teeth). . . . . *Myotis*

   Upper cheek with four or five postcanine teeth. 3

3. Ears much longer than head. 4

   Ears shorter or slightly longer than head. 5

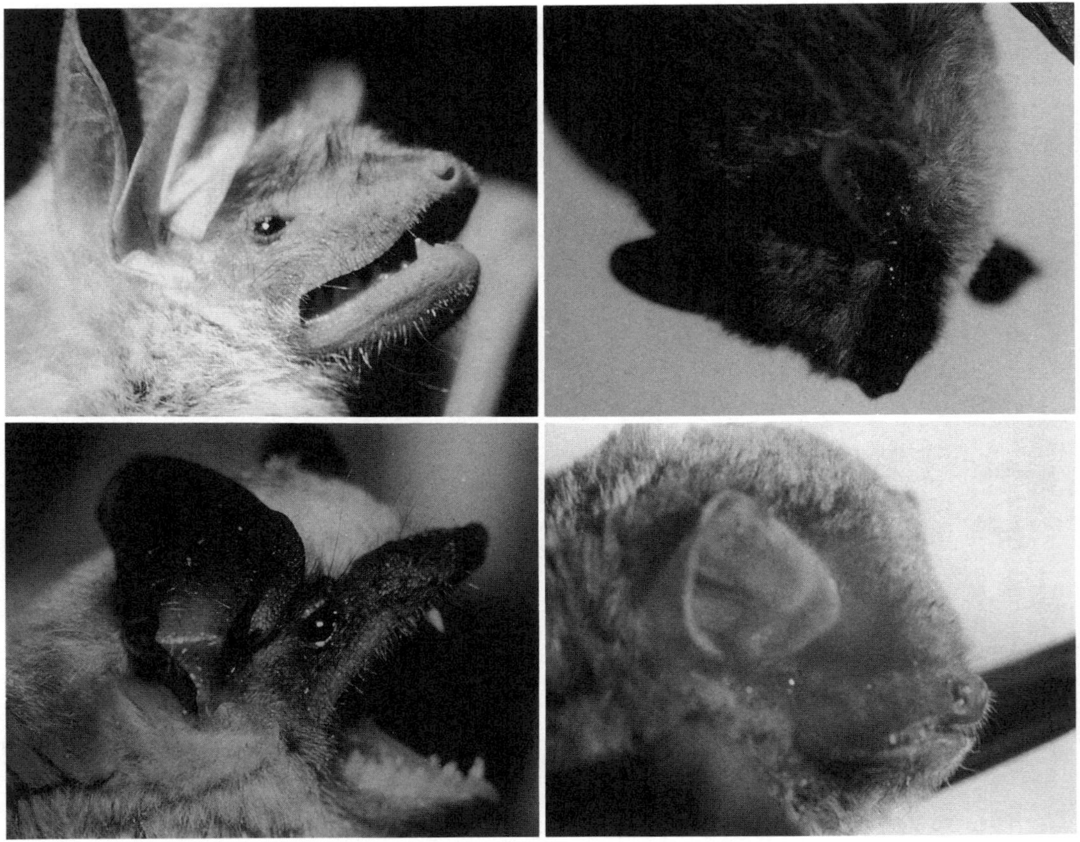

Fig. 27. Facial profiles of some vespertilionid bats. (Upper left) *Myotis blythii*, (Upper right) *Pipistrellus pipistrellus*, (Lower left) *Eptesicus bottae*, (Lower right) *Miniopterus schreibersi*.

4. Ears connected and with a prominent notch at the base of
   anterior margin of ear. Forearm less than 42 mm. Two upper
   incisors on each side. . . . . *Plecotus*

   Ears not connected and without a notch at the base of anterior
   margin of ear. Forearm more than 55 mm. One upper incisor
   on each side. . . . . *Otonycteris*

5. Ears face forward and are joined together across forehead.
   Upper surface of rostrum concave. . . . . *Barbastella*

   Not as above. 6

6. Tragus expanded distally into club shape. Dorsal profile of
   skull almost straight. The fifth digit is shortened with its tip
   not exceeding the level of the first phalanx of the third or fourth
   digits. . . . . *Nyctalus*

   Tragus not expanded distally into club shape. Dorsal profile
   of skull distinctly slopes down anteriorly. The fifth digit is long
   with its tip exceeding the level of the first phalanx of the third
   or fourth digits. 7

7. Larger species, forearm more than 40 mm, skull length more
   than 16 mm. Small upper premolar lacking. . . . . *Eptesicus*

   Smaller species, forearm less than 40 mm, skull length less
   than 16 mm. Small upper premolar usually present (absent
   in some specimens of *Pipistrellus savii*). . . . . *Pipistrellus*

Genus *BARBASTELLA*   Barbastelles

Barbastelles are a distinct and little-known group of bats. The muzzle is
short. The upper lip is divided on each side by a deep groove connecting to
the nostrils. The ears are unique in being "pulled forward." The margins of
the ears lie anterior to the eyes. The inner margins of the ear unite on the
forehead. The skull is very characteristic, with an elevated forehead and a
short obtuse muzzle. The dental formula is i 2/3, c 1/1, p 2/2, m 3/3 = 34.

Barbastella barbastellus   Barbastelle   R

Vespertilio barbastellus Schreber, 1774. Die Säugethiere, 1:168, pl. 55 (published 1775).

Vespertilio leucomelas Cretzschmar, 1826. *In* Rüppell, Atlas Reise Nördl. Afr., Säugeth., p. 73, pl. 28b. Type from "Arabia Petraea" = Sinai. Valid subspecies: *B. b. leucomelas.*

Diagnosis.—A medium-sized vespertilionid bat with a forearm length of 37–40 mm and a skull length of 14–14.5 mm. This bat is distinguished from others in the region by its broad forward-facing ears that are joined across the forehead. The lateral sides of the ear lobes show fleshy protrusions. There is a long, triangular, pointed tragus. The color is blackish with silver tips, giving a distinctive mottled appearance on the back. The ventral fur is whitish. The first upper incisor is strongly bicuspid. Although Harrison treated *leucomelas* as a species (Harrison, 1964*a;* Harrison and Makin, 1988), other investigators (Kock, 1969; Qumsiyeh, 1985) placed it as a subspecies of *B. barbastellus.* Evidence for specific status of *leucomelas* must come only from finding the two species in sympatry with no hybridization (perhaps in southern Russia) or by genetic studies.

Range of species.—Most of the temperate areas of the Palearctic region.

Local status.—This species is poorly known from the Holy Land. The type of *leucomelas* most likely came from Sinai. In the Holy Land, this species was reported only once from the northern shores of the Gulf of 'Aqaba at Elot, N of Eilat (Qumsiyeh, 1985; Harrison and Makin, 1988) (Fig. 28).

Biology.—Little is known of the biology of the barbastelle in the Holy Land. The bat mist-netted at Eilat in April 1970 was a pregnant female. In Europe and Iran these bats were found to roost in trees (under bark or in holes), in tunnels, and in small caves (Ognev, 1928; Lay, 1967). Two young are usually born in the spring.

Genetics.—2N = 32, AA = 50 was reported for specimens from Europe (Manfreddi-Romanini et al., 1975; Zima, 1978).

Human interactions.— Barbastelle is French meaning little beard; these bats were also known as "hairy-lipped bats."

Genus *EPTESICUS*   Serotines

A difficulty exists in the taxonomic separation of *Eptesicus* from *Pipistrellus.* It is generally suggested that *Eptesicus* lacks the upper small premolar that exists in *Pipistrellus.* However, this character is variable in *Pipistrellus*

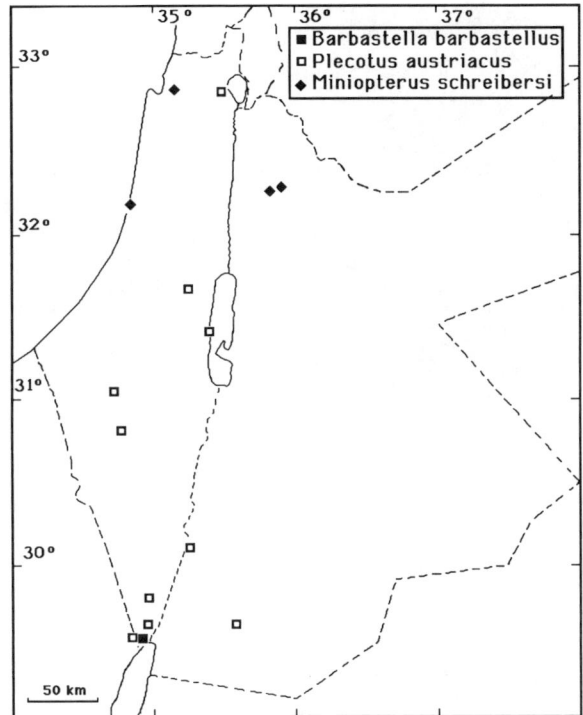

Fig. 28. Locality records of the barbastelle (*Barbastella barbastellus*), gray long-eared bat (*Plecotus austriacus*), and the long-winged bat (*Miniopterus schreibersi*).

and no additional characters were reported to resolve this issue. Most scientists continue the use of separate genera because of historical reasons. More research is now being performed at the molecular level to resolve these questions. The name *Eptesicus* is probably derived from the Greek *epten* (I fly) and *oikos* (house) meaning "house flyer."

Key to the Species of *Eptesicus* of the Holy Land

1. Large species. Forearm 47–58 mm, skull length 19–23 mm.
    In Mediterranean habitats. . . . . *E. serotinus*

    Small species. Forearm 41–47 mm, skull length 16–18.5 mm.
    In desert and subdesert habitats. . . . . *E. bottae*

Eptesicus bottae   Botta's serotine   V

Vesperus bottae Peters, 1869. Monatsber. K. Preuss. Akad. Wiss. Berl., p. 406. Type from Yemen, Arabia.

Vesperugo (Vesperus) innesi Lataste, 1887. Ann. Mus. Stor. Nat. Genova, 4:625. Type from Cairo, Egypt. Valid subspecies: *E. b. innesi*.

Fig. 29. Botta's serotine (*Eptesicus bottae*). Photo by M. B. Qumsiyeh.

Diagnosis.—Smaller than *E. serotinus* with the forearm less than 47 mm. The dorsal fur color is pale brown (lighter than *E. serotinus*) but the membranes, the ears, and the facial profile are dark (blackish) (Fig. 29). The skull is less robust (skull length 16–18.5 mm) than the larger serotine and is not as prominently flat in dorsal profile. The inner (or first) upper incisor is bicuspid and the small second premolar, present in pipistrelles of similar size, is always absent.

Range of species.—Arabia, Egypt, Palestine, Turkey, Iraq, western Iran, the Caucasus, Turkestan, and east to Uzbekistan and Afghanistan.

Local status.—Botta's serotine is found in the arid regions in the south including Wadi Araba, the southeastern Jordan Desert, and probably the Negev Desert (Fig. 30). Records are available from Yotvata (Ain Ghidyan) (Harrison, 1963*a*), Ain Gedi (Makin, 1977; Qumsiyeh, 1985), Neot Hakikar and other uspecified localities southwest of the Dead Sea (Yom-Tov et al., 1992*b*), "Lawrence's pool" (near Rum) (Bates and Harrison, 1989), and Disi (Wadi Rum) (Qumsiyeh et al., 1992).

Biology.—Little is known of the biology of this species in the region. Harrison (1964*a*) caught this bat near cultivated fields at Yotvata, an area surrounded by sandy desert. The specimen collected there on 22 April was a pregnant female with two embryos.

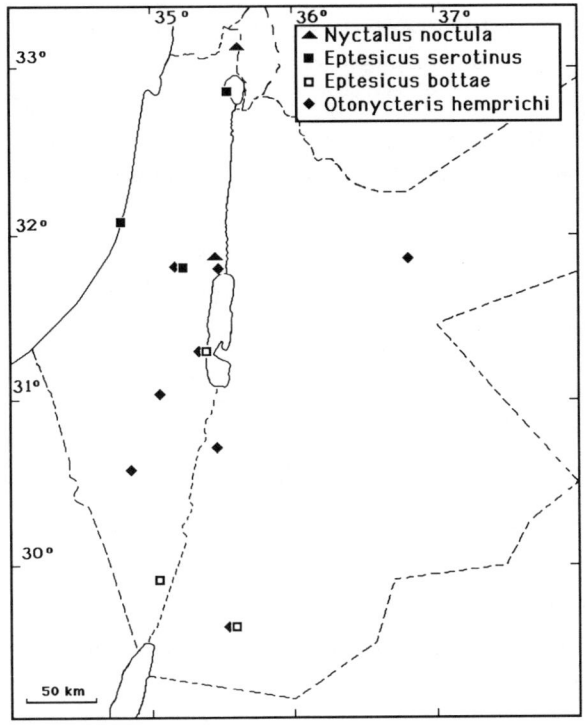

Fig. 30. Locality records of the common noctule (*Nyctalus noctula*), the serotine (*Eptesicus serotinus*), Botta's serotine (*Eptesicus bottae*), and Hemprich's long-eared bat (*Otonycteris hemprichi*).

Genetics.—2N = 50, AA = 46 is found in specimens from Jordan (M. B. Qumsiyeh, unpubl. data), similar to Eurasian specimens (Zima et al., 1991). The X is submetacentric and the Y is a small acrocentric.

Human interactions.—This bat was named after Botta, who collected the first specimens from Yemen.

Eptesicus serotinus   Serotine   T

Vespertilio serotinus Schreber, 1774. Die Säugethiere, 1:167, pl. 53 (published 1775). Type locality: France.

Vespertilio isabellinus Temminck, 1840. Monographies de Mammalogie, 2:205, pl. 52, figs. 1, 2. Type from Tripoli, Libya. Valid subspecies: *E. s. isabellinus*.

Diagnosis.—Larger and more heavily built than *E. bottae* with forearm 47–58 mm in length in adults. The color is dark brown (darker than *E. bottae*). The skull is also massive, measuring 19–23 mm in length. The dorsal profile of the skull is almost straight. The first upper incisor is bicuspid when unworn.

Range of species.—Europe, northern Africa, southwestern Asia, northern India, and into Mongolia, China, and Korea.

Local status (Fig. 30).—This species was recorded from Jerusalem, Tel Aviv, and Wadi Amud (Bodenheimer, 1935, 1958; Harrison, 1964*a*). Further studies in the mountain regions in the north will certainly show that this is a more widespread species than is presently known.

Biology.—This is a species found in well-vegetated areas. It roosts in buildings, tree holes, crevices, and similar places. Summer breeding colonies in Europe are known with up to 50 females. Food consists of mostly insects, especially beetles, which these bats can easily manipulate with their powerful teeth. Serotines may migrate up to 300 km between summer and winter quarters (Corbet and Ovenden, 1980).

Harrison (1964*a*) reported collecting two females (one collected on 18 April and one on 7 May) each with a single embryo. Specimens of the serotine were obtained by Harrison (1964*a*) near the rocky ravines at Wadi Amud.

Genetics.—2N = 50, AA = 46 was reported for specimens from Tunisia (Baker et al., 1975). The X is submetacentric and the Y is a very small acrocentric or submetacentric.

Human interactions.—The name *serotinus* is Latin for late or backward, an allusion to the bats' habit of coming out at night.

Genus *MINIOPTERUS*

The genus *Miniopterus* is so distinct from other vespertilionid bats that some authors place it in its own subfamily, Miniopterinae. The cranium is abruptly elevated above the muzzle. The dental formula is i 2/3, c 1/1, p 2/3, m 3/3 = 36.

Miniopterus schreibersi   Schreibers' bat, long-winged bat   V

Vespertilio schreibersi Natterer *in* Kuhl, 1819. Ann. Wetterau. Ges. Gesammte Naturk., 4(2):185. Type from Kulmbozer Cave, on the left bank of the Danube River, now in Romania.

Miniopterus schreibersi pallidus Thomas, 1907. Ann. Mag. Nat. Hist. ser. 7, 20:197. Type from the southern shore of the Caspian Sea, Iran. Valid subspecies.

Miniopterus schreibersi pulcher Harrison, 1956. J. Mammal., 37:257. Type from Ser'Amadia, Kurdistan, northern Iraq. Synonym.

Diagnosis.—The genus is very unique among vespertilionid bats. The second and third metacarpals are almost equal but the third finger is unique in having an elongated second phalanx (almost three times as long as the third phalanx). The skull is delicate with a very elevated frontal region of the braincase. The muzzle is short and the head appears to be domed. The two upper incisors are almost equal in size.

Range of species.—Widespread in the Ethiopian, Oriental, and southern Palearctic regions.

Local status.—The long-winged bat is a European species adapted to more mesic habitats. It is thus only found in the northern parts of the Holy Land (Fig. 28). It was reported from caves in the Jordan Valley (Tristram, 1884; Aharoni, 1930). Bodenheimer (1935, 1937, 1958) also reported this species as "common" in Palestine but gave no specific localities. The species is also reported from Tel Al Kurdani (near Acco), Herzelia (Harrison, 1964*a*), and Mugharet el Wardani in Dibbine Forest (Harrison and Bates, 1991). Harrison (1956) reported a specimen from Jerash, which he referred to *M. s. pulcher* that he described from northern Iraq. This subspecies was based on slight color differences. Because there is much individual variation in color with age and locality in specimens of this species from the Middle East and northern Africa, Harrison later (1964*a*) considered *M. s. pulcher* a synonym of *M. s. pallidus.* Specimens are also available at HUJ from the Jordan Valley, Wadi Tal'ah, and Herzelia.

Biology.—This is a highly colonial species. In caves in Algeria, I found female colonies numbering in the hundreds or thousands with young in dense clusters. There are even suggestions of communal feeding. Dor (1947) reported this species from owl (*Tyto alba*) pellets in Palestine.

Genetics.—2N = 46, AA = 50 was reported for specimens from Europe (Manfreddi-Romanini et al., 1975; Zima, 1978) and Tunisia (Baker et al., 1975). The X is metacentric and the Y is a very small acrocentric.

Genus *MYOTIS*

In *Myotis* the muzzle is long and simple with small nostrils. The ears are long with a well-developed tragus. The tail is usually less than the length of the head and body and is completely or almost completely enclosed by membranes. The dental formula is i 2/3, c 1/1, p 3/3, m 3/3 = 38. This is the largest number of teeth reported for bats and is higher than the other vespertilionid genera known in the Holy Land. The name *Myotis* derives from the Greek *mys* (mouse) and *otos* (ear) because of the similarity to mouse ears of many members of this group. The ear also has a long slender tragus, which is characteristic for this group.

Key to the Species of *Myotis* of the Holy Land

1. Large species with forearm more than 55 mm in length in adults. 2

   Smaller species with forearm less than 50 mm in length. 3

2. Greatest length of the skull more than 24 mm. Forearm more than 64 mm. If ear is pulled forward it extends more than 5 mm beyond nostril (Panous, 1951). . . . . *M. myotis*

   Greatest length of the skull less than 24 mm. Forearm less than 64 mm. If ear is pulled forward it extends much less than 5 mm beyond nostril. . . . . *M. blythii*

3. Feet very large, exceeding three-quarters of the length of the tibia. Outer edge of tibia hairy. Wing membrane attaches to the leg at or above the ankle. . . . . *M. capaccinii*

   Feet not large, not exceeding one-half of the length of the tibia. Outer edge of tibia not hairy. Wing membrane attaches to the outer digit. 4

4. Ear edge with marked indentation on the lateral aspect. No dense hair on border of interfemoral membrane. . *M. emarginatus*

   Ear edge without indentation on the lateral aspect. Dense hair extends on border of interfemoral membrane between tip of calcar and tail. . . . . *M. nattereri*

Note: *Myotis mystacinus* is reported from "Syria" (old name includes parts of the Holy Land) (Dobson, 1878:315). Although Harrison (1964*a*) does not include this bat in the fauna of the region, it perhaps exists in the Near East.

Myotis blythii   Lesser mouse-eared bat   V

Vespertilio blythii Tomes, 1857. Proc. Zool. Soc. Lond., 1857:53. Type from Nasirabad, Rajputana, India.

Myotis myotis omari Thomas, 1906. Proc. Zool. Soc. Lond., 1906:521. Type from Derbent, 50 mi W Isfahan, Iran. Perhaps a valid subspecies in the Holy Land.

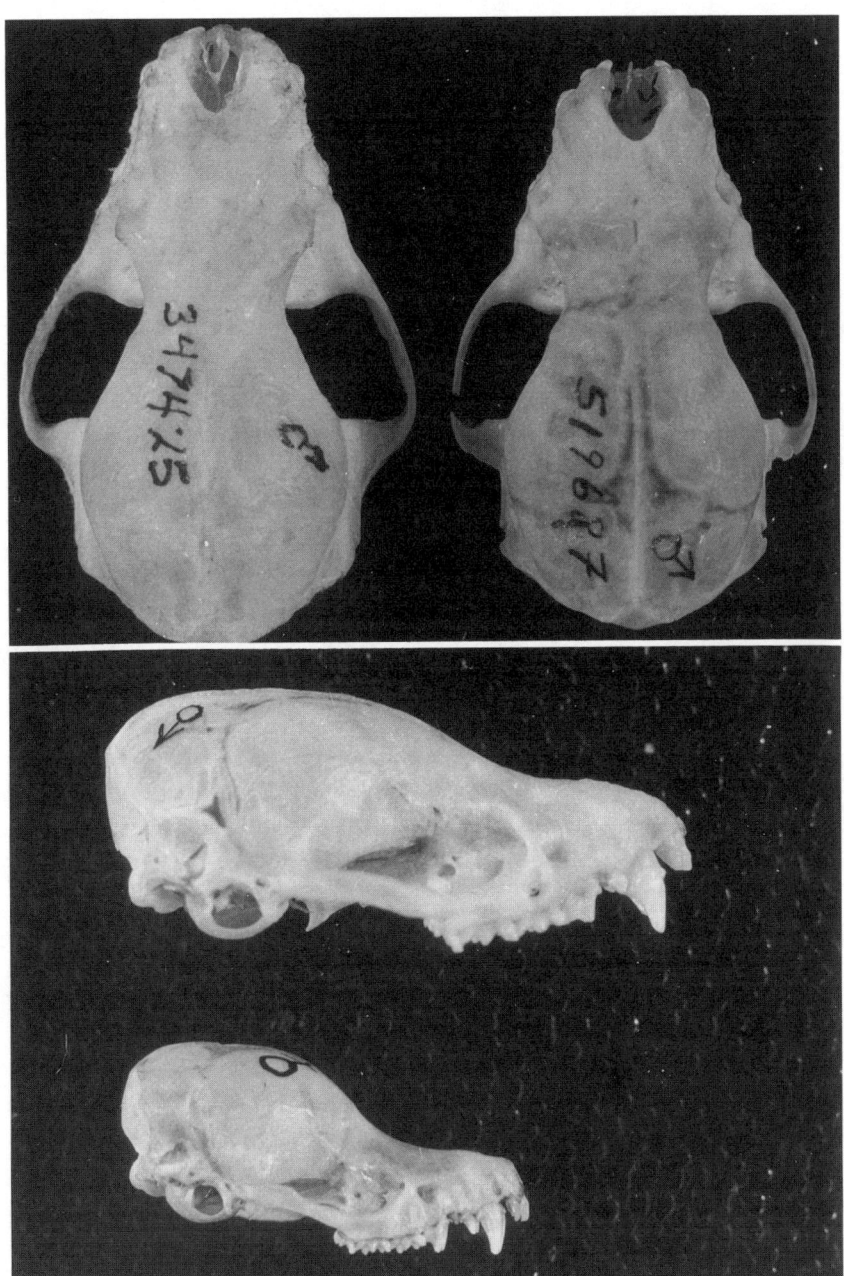

Fig. 31. Skulls of *Myotis*. (Upper panel) Dorsal views of skulls of *Myotis myotis* (left) and
M. blythii (right). (Lower panel) Lateral views of skulls of *M. blythii* (larger skull) and *M.
emarginatus*.

Diagnosis.—This species is slightly smaller and more lightly built than *M. myotis* (fig 31). The forearm measures 56–64 mm in length and the greatest length of the skull is 21–24 mm. The ears are relatively smaller than those of *M. myotis*. The subspecies *M. b. omari* is perhaps distinct as a light-colored variety. However, color variations exist and the status must await genetic data.

Range of species.—Southern Europe and northwestern Africa and east through Turkey and southwestern Asia to western China.

Local status.—Theodor and Moscona (1954) list this species as a host for parasites collected in northern Galilee. Harrison (1964a) recorded a specimen at HZM from "Israel" (no exact locality). Because it occurs in Lebanon (Harrison and Lewis, 1961), it probably enters northern Palestine.

Biology.—Little is known of the biology of this species in the Holy Land but it is presumed to be similar to *M. myotis*. In Lebanon, specimens were taken in caves and in crevices in natural bridges, sometimes roosting with *M. myotis* (Atallah, 1977). In Iran, there are data to suggest that sexes segregate for the breeding season (DeBlase, 1980).

Genetics.—2N = 44, AA = 50 reported for specimens from Tunisia (Baker et al., 1975) and Europe (Manfreddi-Romanini et al., 1975; Zima and Král, 1984). The X chromosome is metacentric and the Y is a very small acrocentric.

Myotis capaccinii   Long-fringed bat   V

Vespertilio capaccinii Bonaparte, 1837. Fauna Ital., 1, fasc. 20. Type from Sicily, Italy.
Leuconoe capaccinii bureschi Heinrich, 1936. Mitt. Naturwiss. Inst. Sofia, 9:38. Type from Karamlek, Strandjabalkan, 800 feet, Bulgaria. Valid subspecies: *M. c. bureschi.*

Diagnosis.—This is a small bat with the forearm measuring 39–41 mm in length and greatest length of the skull 15–16 mm. It is immediately recognized by its relatively large feet, almost three-quarters of the length of the tibia. In this species the outer edge of the leg (tibia) and the dorsal portion of the proximal surface of the interfemoral membrane near the feet are both covered with hair. Unlike *M. nattereri*, the posterior edge of the interfemoral membrane does not have dense hair. Also, the posterior border of the wing membrane attaches to the tibia just above the ankle. The color is light brown on the back. Harrison (1964a) referred specimens from Arabia (including the Holy Land) to *M. c. bureschi* on the basis of the whiter color of the belly and paler back relative to the nominate subspecies.

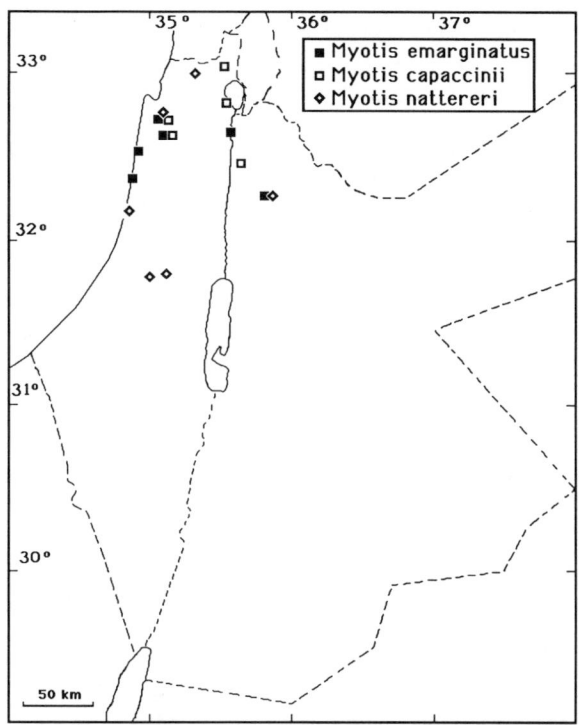

Fig. 32. Locality records of the notch-eared bat (*Myotis emarginatus*), the long-fringed bat (*M. capaccinii*), and Natterer's bat (*M. nattereri*).

Range of species.—Northwestern Africa, southern Europe, and the eastern Mediterranean region east to Uzbekistan, with further records from China, Korea, and Japan.

Local status.—The long-fringed bat occurs only in northern, mesic habitats (Fig. 32). This species was reported from "Palestine" based on an owl pellet specimen (Dor, 1947). Harrison (1964*a*) reported specimens of this species from Carmel Caves and Tiberias. Other specimens are available from Alma Cave, Hazorea (TAUM, HUJ), and Tabqat Fahl (TTU) (Qumsiyeh et al., 1986).

Biology.—Harrison (1964*a*) reported on a maternity colony in Carmel Caves south of Haifa where each female had a single, large embryo in April. Atallah (1977) suggested that the mating season may be in the fall. Following the mating season, males and females segregate to separate quarters.

It is speculated that the large feet of these animals are used to capture insects on surfaces of water where these bats can be frequently seen (Helversen, 1989).

Genetics.—2N = 44, AA = 50 was reported for specimens from Europe (Manfreddi-Romanini et al., 1975; Zima and Král, 1984). The X chromosome is metacentric and the Y is a very small acrocentric.

Myotis emarginatus   Geoffroy's bat, notch-eared bat   **V**

Vespertilio emarginatus Geoffroy, 1806. Ann. Mus. Hist. Nat. Paris, 8:198. Type from Charlemont, Givet, Ardennes, France.

Vespertilio desertorum Dobson, 1875. *In* Blanford, Ann. Mag. Nat. Hist., ser. 1, 16:309. Type from Jalk, Baluchistan, Iran. Valid subspecies, extralimital.

Diagnosis.—*Myotis emarginatus* is a small *Myotis* with the forearm measuring 36–43 mm in length and the greatest length of the skull 16–16.6 mm. This species is best characterized by a conspicuous notch on the lateral border of the ear, a character lacking in any other species of *Myotis* in the region. The posterior border of the wing membrane attaches to the base of the outer toe. It is further distinguished from *M. nattereri* by the lack of the fringe of hairs on the interfemoral membrane. The color is a distinctive reddish brown, unlike other *Myotis* of the Holy Land. Other species of *Myotis* in the region are various shades of gray, never with an orange tinge. The skull is slender and has a characteristically sloped forehead (Fig. 31).

Range of species.—Northwestern Africa, southern Europe, and east through Turkey and the eastern Mediterranean region to Afghanistan.

Local status.—The notch-eared bat has been reported only from northern, mesic habitats (Fig. 32). Specific records of this species are from Mount Carmel (Tristram, 1884), "Palestine" (Dor, 1947), and Nahal Oren, Mount Carmel, and "Jebal" (Harrison, 1964*a*). Specimens at HUJ and TAUM are from Nahal Oren, Rehobot, Yarmouk Pumping Station, Hazorea, and Beit Hananya (Makin, 1977). I collected specimens from Dibbine Forest (Qumsiyeh et al., 1986) and from caves in Beit Hayariyi (= Bitan Aharon).

Biology.—Harrison (1964*a*) recorded a single embryo found in each of 21 females (out of 26 collected) from a large maternity colony ("several hundred individuals") at Nahal Oren, Mount Carmel. The single male specimen I found at Dibbine Forest on 9 August was in a torpid state. These observations agree with studies in other gregarious bats in suggesting that maternity colonies form and single males can be found at most times of the year except the periods of copulation.

Genetics.—2N = 44, AA = 50 reported for specimens from Europe (Zima and Král, 1984). The X chromosome is metacentric and the Y is a very small acrocentric.

Myotis myotis   Greater mouse-eared bat   V

Vespertilio myotis Borkhausen, 1797. Deutsche Fauna, 1:80. Type from Thuringia, Germany.
Myotis myotis macrocephalis Harrison and Lewis, 1961. J. Mammal., 42:373. Type from Mogharet Saleh, 2 km E Amchite, Lebanon. Valid subspecies.

Diagnosis.—This is the largest *Myotis* in the Holy Land with the forearm measuring 60–71 mm in length and the greatest length of skull 24–28 mm. The ears are relatively longer and the skull is longer, narrower and more robust than in *Myotis blythii* (Fig. 31).

Range of species.—England, central and southern Europe, Turkey, and south to northern Palestine.

Local status.—Theodor and Moscona (1954) list this species as a host for parasites collected in Palestine (no exact locality). This record is dubious and may also be *M. blythii*. However, *M. myotis* is known from Lebanon (Harrison and Lewis, 1961), and Harrison and Bates (1991) report that D. Makin collected a specimen from El Haja Cave near Mizpe Yodefat (Lower Galilee). More data are needed to confirm the presence of this species in northern Palestine.

Biology.—This species roosts in large colonies in caves in wooded areas with a Mediterranean climate. Lewis and Harrison (1962) and Atallah (1977) found it to roost along with *M. blythii* in Lebanon. Usually a single young is born around June (Asdell, 1964). The young are born in maternity colonies with hundreds of females roosting together.

Genetics.—2N = 44, AA = 52 was reported for specimens from Europe (Manfreddi-Romanini et al., 1975; Zima and Král, 1984). The X chromosome is metacentric and the Y is a very small acrocentric.

Myotis nattereri   Natterer's bat   V

Vespertilio nattereri Kuhl, 1818. Ann. Wetterau Ges. Gesammte Naturk., 4(1):33. Type from Hessen, Hanau, Germany.
Myotis nattereri hoveli Harrison, 1964. Z. Säugetierkd., 29:58. Type from Aqua Bella, near Jerusalem, Israel. Synonym of *M. nattereri*.

Diagnosis.—This is a small bat with a forearm of 38–43 mm in length and greatest length of the skull 16–16.6 mm. A fringe of hair is present on the edge of the membrane between the tail and the feet. The ears are relatively larger than those in other species of *Myotis* in the region. The subspecies *M. n. hoveli* (Harrison, 1964b) is considered here a synonym of *M. nattereri* because the characters used to distinguish it (for example, lighter color) vary in a geographic cline. The color of specimens from the Holy Land is light brown. The tragus is rather long, about three-fourths the length of the ear. The wings attach to the outer toe.

Range of species.—Northwestern Africa, Europe, and east through southwestern Asia to Japan.

Local status.—Natterer's bat was reported only from the northern, mesic habitats of the Holy Land as far south as near Jerusalem (Fig. 32). It was reported as "common" in the Galilee, Aqua Bella, and Herzelia (Bodenheimer, 1958). Harrison (1964*a*, 1964*b*) reported on specimens from Aqua Bella, Hartuv, Mount Carmel, and Herzelia. It was also reported from Dibbine Forest (Qumsiyeh, 1980).

Biology.—Harrison (1964*a*, 1964*b*) reported a maternity colony at Aqua Bella near Jerusalem, where each female carried a single young on 30 April. Thus it is estimated that delivery occurs in late May or early June. This species is found to share the roost with *M. capaccinii* in Lebanon (Atallah, 1970*b*). I found a colony of about 250 (mostly nonpregnant females and subadults) on 7 July 1977 in Dibbine Forest. Fewer animals were found on a return visit on 20 November and only one was found on 27 February 1981.

Genetics.—2N = 44, AA = 50 was reported for specimens from Europe (review in Zima and Král, 1984). The X chromosome is metacentric and the Y is a very small acrocentric.

Genus *NYCTALUS*   Noctules

The general appearance is like a large *Pipistrellus*. The head is large with swollen nasal regions. Ears are thick and wide and the tragus is short and club-shaped (unlike that of most *Pipistrellus* in the Holy Land). In *Nyctalus*, the fifth finger is shortened and scarcely longer than the metacarpal of the third or fourth fingers. The calcar has a large keel stretched by a cartilaginous spur at a right angle to the calcar. One species is reported from the Holy Land but it is possible that the two other European species (*Nyctalus lasiopterus* and *N. leisleri*) also occur in this region. The dental formula is i 2/3, c 1/1, p 2/2, m 3/3 = 34. The name *Nyctalus* is derived from *nux*, Gr., genitive *nuktos*, night, and *nuktalos (nustalos)*, Gr., drowsy; an allusion to the bats' habit of coming out at dusk.

Nyctalus noctula   Common noctule   V

Nyctalus noctula Schreber, 1774. Die Säugethiere, 1:166, pl. 52 (published 1775). Type locality in France.

Nyctalus noctula lebanoticus Harrison, 1962. Proc. Zool. Soc. Lond., 139(2):337, pl. 1. Type from Natural Bridge, Faraya, Lebanon. Valid subspecies.

Diagnosis.—*Nyctalus noctula* is a rather large vespertilionid bat with the forearm well over 50 mm in length and the greatest length of skull more than 18.3 mm. The wing is long and narrow mainly because of the shortened fifth finger. The tragus of noctules differ from other vespertilionids of the region in being broad and expanded distally (club-shaped). *Nyctalus n. lebanoticus* was named as a distinct subspecies from Lebanon that is darker than the nominate subspecies. The overall color is brownish with a golden tinge. This color is uniform along the length of the hair, a character that distinguishes noctules from other, similar-sized bats. Harrison (1962) described a new subspecies from Lebanon and suggested that the Palestinian specimens belong to this subspecies. Considering the widespread distribution and few specimens examined, I think it is unwise to separate the eastern Mediterranean forms into a separate subspecies.

Range of species.—Northwestern Africa, Europe, the Near East (south to Palestine) and east to the Himalayas, mainland China, Japan, and Taiwan. This species is a migrant and single specimens obtained at distant localities such as Mozambique and Oman might be migrant or vagrant and not endemic (Harrison, 1980).

Local status.—This species was reported by Festa in 1894 from Jebel Qarantal (*fide* Bodenheimer, 1958) and from the Hula Nature Reserve (Harrison and Makin, 1988; Fig. 30). Further studies in the northern parts of the Holy Land will doubtless show a wider distribution.

Biology.—This species migrates in Europe, with one individual recovered 755 km from the place it was marked (Strelkov, 1969; Nowak and Paradiso, 1983). It would be interesting to see if similar migrations occur in the Near East. Other studies in Europe suggest that the noctule is colonial and roosts in hollow trees. They are noisy animals and produce various audible sounds, especially males in defending a territory (Ognev, 1928). One to two young are born after a gestation period of 50–70 days (Nowak and Paradiso, 1983). In Lebanon, the species was reported in crevices or fissures in a natural bridge (Harrison, 1962).

Fig. 33. Long-eared bat (*Otonycteris hemprichi*).

Genetics.—2N = 42, AA = 50 was reported for specimens from Europe (Zima, 1978). The X chromosome is metacentric and the Y is a very small acrocentric or submetacentric.

Human interactions.—The name *noctula* is derived from *nuktalos (nustalos)*, Gr., drowsy; an allusion to the bats' habit of coming out at dusk.

Genus *OTONYCTERIS*

The ears in this genus are very large. Based on genetics and appearance, this genus is closely related to the long-eared bats of the genera *Plecotus* and *Barbastella* (Horáček, 1991; Qumsiyeh and Bickham, 1993). However, unlike these genera, the ears are not connected across the forehead. The dental formula is i 1/3, c 1/1, p 1/2, m 3/3 = 30. A single species is known in this genus. Further characters for the species are described under the diagnosis. The name *Otonycteris* derives from *otos*, Gr., ear, and *nukteris*, Gr., a bat (the eared bat).

Otonycteris hemprichi   Hemprich's long-eared bat   V

Otonycteris hemprichi Peters, 1859. Monatsber. K. Preuss. Akad. Wiss. Berl., p. 223. Type from northeastern Africa. Kock (1969) restricted the type locality to the Nile Valley between Dandara, Egypt and Chondek, Wadi Halfa, Sudan.

Otonycteris jin Cheesman and Hinton, 1924. Ann. Mag. Nat. Hist. ser. 9, 14:549. Type from Hufuf, Hasa, Saudi Arabia. A valid subspecies in the Holy Land and Arabia: *O. h. jin*.

Diagnosis.—This is a large insectivourous bat with a forearm of 62–69 mm in length and a greatest length of skull of 24–26 mm. Two pairs of nipples occur in this species. The ears are very large (Fig. 33), as in *Plecotus*. *Plecotus* ears are, however, joined in the midline. *Otonycteris hemprichi jin* is larger (forearm 64–69 mm) and paler than *O. h. hemprichi*. Clinal variation may, however, preclude the separation of subspecies in *Otonycteris* (Kock, 1969). If subspecific names are used, *O. h. jin* should be used for specimens from Saudi Arabia and the Holy Land and *O. h. hemprichi* for specimens from Egypt and Sudan (Harrison, 1964*a;* Koopman, 1975; Atallah, 1977; Qumsiyeh, 1985).

Range of species.—The desert regions of northern Africa, Arabia, and southwestern Asia (to Turkestan and Afghanistan).

Local status.—Hemprich's long-eared bat probably occurs throughout the arid regions of the Holy Land (Fig. 30). The first specimen from the Holy Land was probably that collected by Pastor Schmitz at the turn of the century from Wadi Al Mukallik, near Nebi Musa (Bodenheimer, 1958). This species was also reported from Shen Ramon (Harrison, 1964*a*), near Jerusalem (Nader and Kock, 1983), Jerusalem, Ain Gedi, Mamshit, Shen Ramon (Qumsiyeh, 1985), 4 km NW Azraq Shishan (near Azraq Druz) (Atallah, 1966, 1967*b*, 1977), and Karyatein (specimens at HUJ). The species was collected from "Lawrence's pool" (near Rum) (Bates and Harrison, 1989), near Quraiqira (Wadi Al Fidan) and Disi (Wadi Rum) (Qumsiyeh et al., 1992). The species is also known from northern Sinai at El Arish (Wassif, 1954*b*).

Biology.—The very unique morphology of this genus with its limited distribution restricted to desert habitats suggests a very specialized biology. Unfortunately, little is known of the habits and habitats of this species. The wing morphology (large and broad) and the smooth flight suggest an ability to forage close to the ground. The overall morphology is reminiscent of carnivorous bats (Norberg and Fenton, 1988) and studies on this species are much needed to elucidate its feeding habits. One specimen from Egypt was caught in a mouse trap, also indicating foraging close to the ground. The specimens I obtained from Jordan were collected in mist nets set over water pools.

DeBlase (1980) examined four females collected in mid-June from northeast Iran and reported no embryos. Gromov et al. (1928) reported a female with two embryos on 12 June from central Asia. Three pregnant females were obtained on 2 May from Azraq Oasis in northeastern Jordan where early June was estimated as the date of birth (Atallah, 1977). Two of three females collected on 24 June 1974 in the Air Mountains (Niger) were lactating (Fairon, 1980). Six females from central Asia (no date given;

Horáček, 1991) as well as the three from Jordan (Atallah, 1977) were found carrying two embryos each. Two males collected from Saudi Arabia, one in August and one in October, had testes 4.2 and 4.8 mm wide, respectively (Harrison and Bates, 1991).

Otonycteris roosts in the fissures of rocks or in human constructions. It was reported to be mostly solitary, but occasional clusters of up to 18 females were found (Bogdanov, 1953). This species is found in xeric and usually rocky habitats that have little vegetation. This bat seems to be well adapted to arid climates. Flight in *Otonycteris* has been observed by few investigators. This bat seems to hover close to the ground (M. B. Qumsiyeh, pers. obs. in southern Jordan), probably because it hunts nonflying prey. Norberg and Fenton (1988) showed that the wing parameters in this species are similar to those of carnivorous bats such as *Antrozous*. *Otonycteris* has been caught in nets set close to the ground at heights of 10 cm to 3 m (Horáček, 1991). Stomach contents indicated diets consisting of Tenebrionidae, Blattoidea, and Orthoptera (Horáček, 1991). Captive individuals have also been fed geckos (Gorelov, 1977, cited in Horáček, 1991).

Hemprich's long-eared bats start their activity just before dusk, flying low along rocks. Later in the evening, they are observed flying 4–9 m above the ground. They have been reported to fly in circles 20–60 m in diameter and in straight flight without fluttering or quick maneuvers (Gromov et al., 1928; Horáček, 1991). *Otonycteris* emits short series of low frequency clicks with a regular low repetition rate that increases when approaching a prey item and terminates with feeding. Calls ranged from 18 to 40 kHz with a maximum intensity at 30–32 kHz. When flying low this bat seems to stop echolocating, which led to the conclusion that when feeding on a nonflying prey item, *Otonycteris* is a facultative echolocator. According to Horáček (1991) the second harmonic has half the intensity of the first and a third harmonic has a low intensity of about 75 kHz. Heim de Balsac (1965) reported this bat from pellets of the owl *Tyto alba* at Djanet, Algeria.

Genetics.—In Turkestan, this species has a 2N = 30, AA = 48 (Zima et al., 1991, 1992*b*). Data from a single Jordanian specimen revealed a 2N = 28, AA = 46 (Qumsiyeh and Bickham, 1993). In both areas, the X is submetacentric and the Y is a small acrocentric. G-banding studes indicate that only two rearrangements distinguish Holy Land *Otonycteris* from *Plecotus* (Qumsiyeh and Bickham, 1993) and only one rearrangement distinguishes between Turkestan *Otonycteris* and *Plecotus* (M. B. Qumsiyeh, personal comparison of karyotypes published by Zima et al., 1992*b*, with karyotypes from Jordan).

Human interactions.—Peters named this species after Otto von Hemprich. If this species indeed forages on scorpions and other desert arthropods, it is probably an important biological link in the food chain.

Genus *PIPISTRELLUS*

Pipistrelles are small bats (forearm less than 40 mm). The ears are longer than broad and are separated from each other. The tragus is long and its maximum width is usually below the midpoint. The area between the eyes and the tip of the nose is covered with fur. The dental formula is i 2/3, c 1/1, p 2/2, m 3/3 = 34. However, *Pipistrellus savii* may be missing the small upper premolars and thus have only 32 teeth. This genus is one that presents difficulty in identification even to the experienced mammalogist. The following key is only provisional and reference should be made to the diagnosis section of each species for a more accurate identification. The name *Pipistrellus* is latinized from the Italian *pipistrello,* meaning bat.

Key to the Species of *Pipistrellus* of the Holy Land

1. Belly white, presenting a clear contrast to the back, which is grayish brown to sandy. First upper incisor large and deeply cleft, the second incisor small and barely exceeds the cingulum of the first (Fig. 34E). *P. rueppelli*

   Color not as above with the belly never white nor sharply contrasted with the back. First upper incisor unicuspid or bicuspid but never deeply cleft. When the first upper incisor is bicuspid, the second is always large, exceeding half the length of the first. . . . 2

2. First upper incisor unicuspid. 3

   First upper incisor bicuspid. 4

3. The second upper incisor small and short, just exceeding the cingulum of the first (Fig. 34A). Wing membrane usually with a whitish border. . . . . *P. kuhlii*

   The second upper incisor larger and taller, exceeding half the length of the first (Fig. 34F, but see text). Wing membrane without a whitish border. . . . . *P. ariel*

4. First upper premolar visible from the lateral side (not displaced internally). Second upper incisor about one-half as long as first upper one (Fig. 34B). . . . . *P. pipistrellus*

First upper premolar displaced internally or missing. Not visible from the lateral side. Second upper incisor very tall, attaining height of second cusp on first incisor (Fig. 34C and 34D). 5

5. Large species. Forearm length 31.5–33.5 mm. Skull length 13–14 mm. Membranes blackish throughout. . . . . *P. savii*

Small species. Forearm length 28–32 mm. Skull length 10.5–12 mm. Membranes blackish except interfemoral and ears are light colored. . . . . *P. bodenheimeri*

Note: *Pipistrellus nathusii* could possibly occur in the Holy Land. It is essentially similar to *P. pipistrellus* but with a longer fifth finger (greater than 43 mm), a long thumb (longer than width of wrist), the calcar usually branched, and the second and third lower incisors separated by a distinct gap.

Pipistrellus ariel   Pygmy pipistrelle   R

Pipistrellus ariel Thomas, 1904. Ann. Mag. Nat. Hist. ser. 7, 14:157. Type from Wadi Alagi, eastern Egyptian Desert.

Diagnosis.—This is a small pipistrelle with the forearm between 28 and 31 mm in length. It is distinguished from *P. savii* and *P. kuhlii* by its smaller size. From *P. kuhlii*, *P. ariel* is distinguished by the long outer (or second) upper incisor, which reaches almost two-thirds the length of the inner upper incisor (Fig. 34F). It is distinguished from *P. savii*, *P. pipistrellus*, and *P. bodenheimeri* by the unicuspid first upper incisor. However, a specimen of *P. bodenheimeri* from Ain Gedi had a unicuspid incisor (Harrison and Bates, 1991). Further confusion may exist with the poorly diagnosed *Pipistrellus arabicus* (see Harrison and Bates, 1991). Thus, further material and studies on the relationships of these three species (*P. ariel*, *P. bodenheimeri*, and *P. arabicus*) are neccessary.

Range of species.—Egypt, Sudan, and the Holy Land.

Local status (Fig. 35).—Reported from two localities only: Nahal Zeelim (Wadi Siyal) (Makin and Harrison, 1988) and a possible specimen from Disi in Wadi Rum (Qumsiyeh et al., 1992).

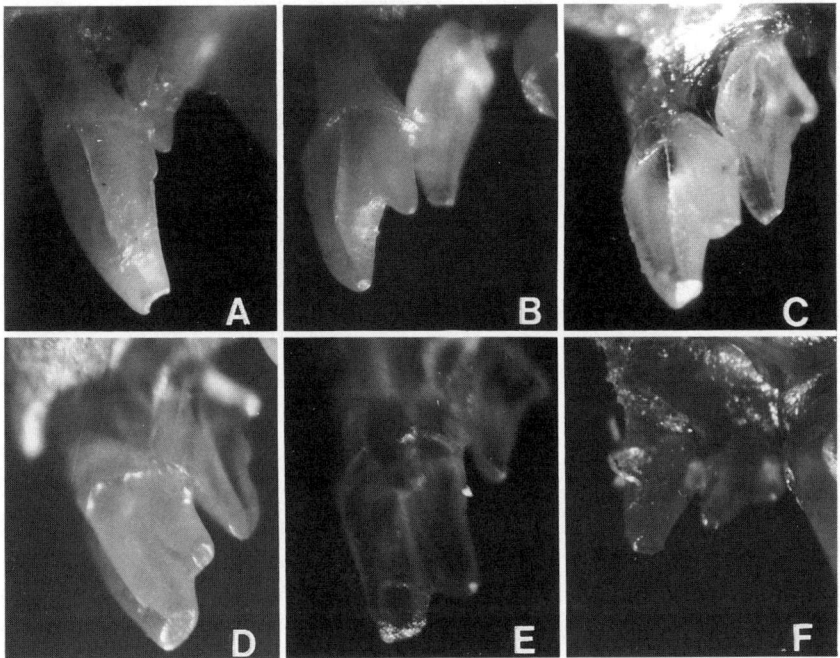

Fig. 34. Upper incisors of the *Pipistrellus* of the Holy Land as viewed from inside the mouth region. (A) *Pipistrellus kuhlii*, (B) *P. pipistrellus*, (C) *P. bodenheimeri*, (D) *P. savii*, (E) *P. rueppelli*, (F) *P.* cf. *ariel* (see text).

Pipistrellus bodenheimeri   Bodenheimer's pipistrelle   V

Pipistrellus bodenheimeri Harrison, 1960. Durban Mus. Novit., 5(19):261. Type from Yotvata, 40 km north of Eilat, Wadi Araba, Israel.

Diagnosis.—This is a very small pipistrelle similar in size to *P. ariel* with a forearm 26–31 mm in length. The ears are relatively large a with well-developed tragus and antitragus. The color is very distinctive with a pale buff color dorsally, whitish ventrally. This pale fur color contrasts with the dark wing membranes and uropatagium color. It is distinguished from *P. ariel* by the rather narrower muzzle and by having a bicuspid upper first incisor (Harrison, 1960; Makin and Harrison, 1988). The second upper incisor is high and subequal with the secondary cusp on the first incisor (Fig. 34C). The upper canine and large premolar are in contact due to the reduction and displacement of the small premolar.

Range of species.—Aden, Sinai, and the Holy Land.

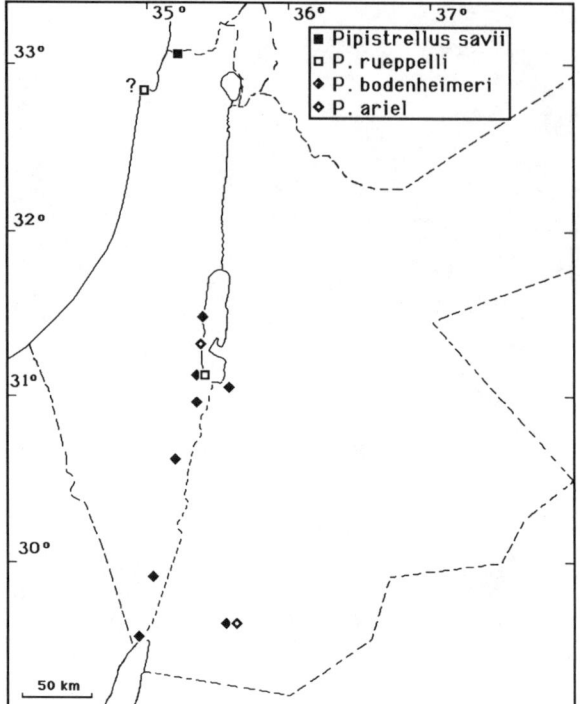

Fig. 35. Locality
records of Savi's
pipistrelle (*Pipistrellus
savii*), Rüppell's bat
(*Pipistrellus rueppelli*),
Bodenheimer's
pipistrelle (*Pipistrellus
bodenheimeri*), and the
pygmy pipistrelle
(*Pipistrellus ariel*).

Local status.—This species is not uncommon in Wadi Araba, around the Dead Sea, and the wadis of southern Jordan (Fig. 35). Specimens were collected from Yotvata (Ain Ghidyan), Ain Gedi (Harrison, 1960), Ain Yahav, Eilat (Makin and Harrison, 1988), Ain Boqeq (Ain Um Baghak), Neot Hakikar (HUJ), 2 km E Ghor As Safi, and from Disi in Wadi Rum (Qumsiyeh et al., 1992). Yom-Tov et al. (1992) found this to be the most common bat netted in areas west and southwest of the Dead Sea.

Biology.—Yom-Tov et al. (1992*a*) reported on some aspects of the biology of this species. Where it occurs in the Holy Land it appears to be common. These pipistrelles are rare between October and April near water sites. This was explained as indicative of hibernation. Another explanation maybe that they can acquire their water from other sources during these more mild and occasionally rainy seasons. These bats feed intensively in the evening hours and may gain up to 15% of their weight during that time. A female collected in April had a single embryo and the few specimens examined were usually shot near *Acacia* trees in a wadi (Harrison, 1960). I obtained pregnant females in May in Wadi Rum and As Safi.

Human interactions.—David L. Harrison named this bat after Bodenheimer, who collected many species of mammals in the Holy Land in the first half of this century.

Pipistrellus kuhlii   Kuhl's pipistrelle   O

Vespertilio kuhlii Natterer *in* Kuhl, 1819. Ann. Wetterau. Ges. Gesammte Naturk., 4(2):199. Type from Trieste, Italian–Yugoslavian border.

Vespertilio marginatus Cretzschmar, 1830 vel 1831. *In* Rüppell, Atlas Reise Nördl. Afr., Säugeth., p. 74, fig. 29a. Type from "Arabia Petraea" (= Sinai) and "Nubia." According to Koopman (1975) from the Nile Valley or eastern desert of Egypt. Perhaps a valid subspecies: *P. k. marginatus* (see Qumsiyeh, 1985).

Pipistrellus kuhlii ikhwanius Cheesman and Hinton, 1924. Ann. Mag. Nat. Hist. ser 9, 14:549. Type from Hufuf, Hasa, Saudi Arabia. Synonym of *P. k. kuhlii.*

Diagnosis.—Kuhl's pipistrelle is a small vespertilionid with a forearm of 32–38 mm in length and a greatest length of skull of 13.3–14.7 mm. This pipistrelle is distinguished from other pipistrelles in this region by the following features: a white posterior margin is usually noted on the wing membrane (the plagiopatagium), especially between the foot and the nearest finger; and the first upper incisor is unicuspid and the second upper incisor is very small, not exceeding the cingulum of the first (Fig. 34A). There is extensive geographic and individual variation in color and size in these bats, which indicate that *P. k. ikhwanius* (Cheesman and Hinton, 1924) and *P. k. marginatus* should be synonymized with *P. k. kuhlii* (Lewis and Harrison, 1962; Qumsiyeh, 1985).

Range of species.—Most of Africa (except some areas in western Africa and the Sahara Desert), southern Europe, and the Near East to Pakistan.

Local status.—Kuhl's pipistrelle is very common throughout the Holy Land because it adapted well to living in towns and cities. However, it seems to avoid the extreme desert conditions in the Negev, the Araba, and south and southeast Jordan (Fig. 36). The following is a list of localities reported for the species west of the Jordan River: Arbel, Ayelot Hashakhar, near Beersheba, Beit Oren, Beit Sahur, Bethlehem, Cave of Adullam, Dafna, Dan, near Haifa, Hulata, Givat Yeshayahu, Jericho, Kfar Menachem, Kfar Ruppin, Kishon River, Lake of Galilee, Menahamya, Moledot, Mount Carmel, Plain of Acre, Rehovot, Sanour, Tantura, Tel Aviv, Tiberias, near Tivon, Wadi Amud, Wadi Tira, and the Yarmouk Valley (Tristram, 1866*b*, 1884; Aharoni, 1930; Bodenheimer, 1958; Harrison, 1964*a;* Atallah, 1977; Makin, 1977; Qumsiyeh, 1985; M. B. Qumsiyeh, pers. obs.). Many other localities west of the Rift Valley are represented by museum specimens but

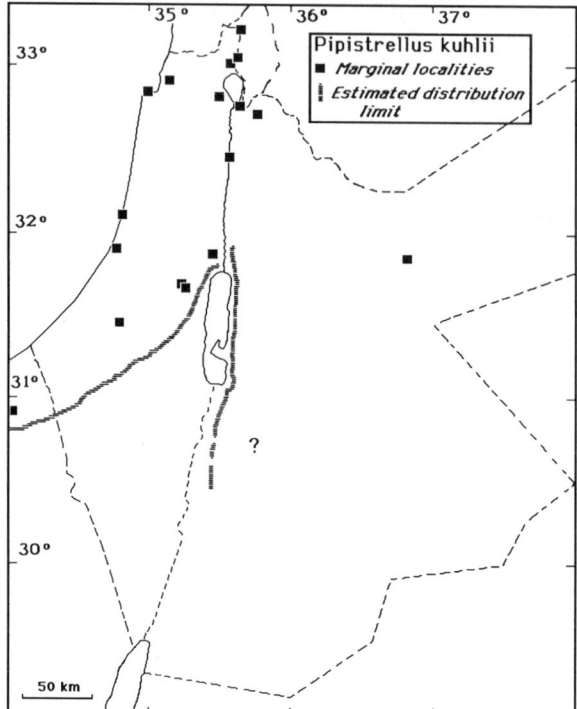

Fig. 36. Estimated
distributional limits of
the common
vespertilionid bat,
Kuhl's pipistrelle
(*Pipistrellus kuhlii*) with
some marginal localities
shown.

are not listed here because they do not add much to our knowledge of this
widespread species.

East of the Jordan River, this species is known from Azraq ed Druz, Azraq
Shishan (Atallah, 1966, 1967*b*), Ramtha, and 3 km W Suweilih (Zuhair
Amr Collection). A specimen at JNHM is from Aqraba. The species was
also recorded from northern Sinai from Magdabah (Wassif, 1954*b*) and from
several localities in Lebanon and Syria including Beirut, Damascus, Deir az
Zur, and between Homs and Latakia (Harrison, 1964*a;* Nader and Kock, 1983).

Biology.—Stencel (1961) studied the distribution of this species in Lebanon and commented on its relationship to man. Its abundance may be due
to its adaptability to roost in almost any shelter including inhabited human
dwellings. In Beit Sahur, a colony of about 50 was found in a crevice in a
stone house. Active bats were seen from January through late September.
Bats emerged individually shortly after dark and returned shortly before
dawn. Males with enlarged testes were noted in August and September.
Reproducing males also have enlarged buccal pads (Harrison, 1964*a*). The
function of these glands during the reproductive season is not well documented. Similar observations were noted for this species from the Holy Land

(Barak and Yom-Tov, 1991). It appears that dominant reproductive males roost singly and court females along specific routes, whereas other males roost in groups.

Parturition probably occurs around late April and May in many areas in the Near East including the Holy Land (Anderson and de Winton, 1902; Lewis and Harrison, 1962; Harrison, 1964a). Lactating females were observed on 24 June (M. B. Qumsiyeh, pers. obs.). The suggestion of migratory behavior in Egypt (Gaisler et al., 1972) is disputed (Qumsiyeh, 1985). Dor (1947) obtained 24 specimens of this bat from some 6000 owl (*Tyto alba*) pellets in Palestine.

Genetics.—2N = 44, AA = 50 was reported for specimens from Tunisia (Baker et al., 1975) and Libya (Zima, 1982b). I also obtained similar data on two specimens from Jordan (M. B. Qumsiyeh, unpubl. data).

Human interactions.—Kuhl's pipistrelle is the most common bat in the towns and villages of the Holy Land and is thus the familiar "bat" *(khaffash, watwat, atlaphon)* of the locals. The species was named after Kuhl, who collected and described many species of mammals, including bats, around the turn of the 19th century.

Pipistrellus pipistrellus   Common pipistrelle   V

Vespertilio pipistrellus Schreber, 1774. Die Säugethiere, 1:167, pl. 54 (published 1775). Type from France.

Diagnosis.—This is a small species with a forearm less than 33 mm. The ears are short with a curved blunt tragus. The color is dark cinnamon-brown throughout the back and slightly lighter ventrally. The skull is small and more delicate than the similar species *P. kuhlii*. The teeth are, however, distinctive. The inner (first) upper incisor is bicuspid and the second is about one-half as high as the first (Fig. 34B). The first upper premolar is not displaced internally and is thus visible from the lateral side.

Range of species.—Europe, northern Africa, Turkey, Lebanon, and east to Afghanistan and China (questionable records from Korea, Taiwan, and Japan).

Local status.—Because the species is common in Lebanon (Lewis and Harrison, 1962) and many Mediterranean habitats from northern Africa (Qumsiyeh and Schlitter, 1982), it is expected to occur commonly in the northern regions of the Holy Land. It is thus surprising that only a single record from Mount Meiron is available (Mendelssohn and Yom-Tov, 1987). A photograph taken by Al Shah from northern Jordan clearly shows this bat.

Biology.—Common pipistrelles prefer to roost in tree holes, small crevices, between building blocks, or in similar areas. In Europe, summer colonies of breeding females may number in the hundreds and roost in caves. This is not observed in the Holy Land, perhaps because the species is rare here. A single male collected in Lebanon on 20 August had enlarged testes indicating a possible breeding season (Harrison, 1964*a*). In Europe, young are born in June and July. Nursery colonies were found in Iran in July and September (DeBlase, 1980). This species migrates in Europe (Strelkov, 1969) and it would be interesting to see if similar migrations occur in the Near East.

Genetics.—2N = 42–44, AA = 50 was reported for specimens from Europe (Fedyk and Ruprecht, 1976; Zima, 1978; Zima and Král, 1984). The X chromosome is metacentric and the Y is a very small acrocentric.

Pipistrellus rueppelli   Rüppell's bat   V

Vespertilio temminckii Cretzschmar, 1826. *In* Rüppell, Atlas Reise Nördl. Afr., Säugeth., p. 17, fig. 6. (Preoccupied, not of Horsfield.)

Vespertilio rüppellii Fischer, 1829. Synops. Mamm., 109. Type from Dongola, Sudan.

Diagnosis.—This bat is distinguished from all other pipistrelles in the region by having a pure white fur underneath. It is a robust bat similar in size to *P. kuhlii* with a forearm of 32–35 mm. The skull is slightly smaller than that of *P. kuhlii*. The first upper incisor is large and strongly bicuspid (deeply cleft); the second upper incisor is very small, barely exceeding the cingulum of the first (Fig. 34E). The upper canines and the large premolar are clearly separated by the small premolar.

Range of species.—Africa south of the Sahara, Algeria, and Egypt to Iraq.

Local status (Fig. 35).—Reported from Navit Pools in Ain Boqeq (Ain Um Baghak) (Harrison and Makin, 1988), Neot Hakikar and other unspecified localities near the Dead Sea (Yom-Tov et al., 1992*b*), and a questionable record from Haifa (H. Hovel *fide* Harrison and Bates, 1991).

Biology.—Little is known of the biology of this species. In Egypt the species was collected from under rocks in arid regions (Qumsiyeh, 1985). This combined with the color (white underneath) suggests that this animal may be foraging on or close to the ground. Other specimens are occasionally caught in mist nets near water.

Human interactions.— The species was named after the 19th century mammalogist Rüppell, who collected and described many species of mammals, including bats, around the turn of the 19th century.

Pipistrellus savii   Savi's pipistrelle   V

Vespertilio savii Bonaparte, 1837. Fauna Ital., 1, fasc. 20. Type from Pisa, Italy.

Vesperugo (Vesperus) caucasicus Satunin, 1901. Zool. Anz., 24:462. Type from Tiplisi, Georgia. Valid subspecies.

Diagnosis.—This is a rather large pipistrelle (forearm 30–38 mm) with very fine silky hair. There is a pronounced difference between the dark dorsal and lighter ventral color. The tip of the tail projects about 3 mm from the interfemoral membrane. The first upper incisor is bicuspid, the second is large and reaches the height of the second cusp of the first incisor (Fig. 34D). The small upper premolar is reduced and displaced internally or absent. The subspecies *P. s. caucasicus* is distinguished from the nominate form by its larger size, paler color, and more frequent loss of the small upper premolar (Harrison, 1961).

Range of species.—Northwestern Africa, southern Europe, and the Near East east into central Asia, south to India, and with isolated records as far east as Japan.

Local status (Fig. 35).—Occurs in Lebanon (Harrison, 1961) and in the northern regions of the Holy Land. A single specimen from near Kibbutz Eilon is known (Makin and Harrison, 1988).

Biology.—Nothing is known of the biology of this species in the Holy Land.

Genetics.—2N = 44, AA = 50 was reported for specimens from Italy (Capanna et al., 1967) and Bulgaria (Zima, 1982*b*).

Human interactions.— The species was named after the 19th century mammalogist Savi, who collected and described many species of mammals, including bats, around the turn of the 19th century.

Genus *PLECOTUS*

This genus is distinguished externally by the very large ears that are connected across the forehead. The dental formula is i 2/3, c 1/1, p 2/3, m 3/3 = 36.

Plecotus austriacus   Gray long-eared bat   V

Vespertilio auritus var. B *austriacus* Fischer, 1829. Synopsis Mamm., p. 117. Type from Vienna, Austria.

Vespertilio auritus var. A *aegyptius* Fischer, 1829. Synopsis Mamm., p. 105. A junior homonym of *Vespertilio pipistrellus* var. *aegyptius,* ibid., p. 105.

Plecotus christiei Gray, 1838. Mag. Zool. Bot., 2:495. Type from northern Africa. The type locality was restricted to southern Egypt between Qena and Aswan (Qumsiyeh, 1985). Valid subspecies: *P. austriacus christiei*

Diagnosis.—This is a medium-sized bat with the forearm measuring 38–42 mm in length. Ears are large (more than 35 mm) and connected across the forehead. At rest the ears are folded back, whereas the slender tragus remains erect and pointing forward. The external nares are unique in having a posterior fissure. The skull is delicate with an expanded braincase and a narrow rostrum. The specimens from some parts of the Holy Land (north and west) are slightly larger than the Egyptian ones and may represent *P. a. austriacus* (Qumsiyeh, 1985). Those from the Syrian Desert in the southeast of the Holy Land are closer to *P. a. christiei* (Qumsiyeh et al., 1992). *Plecotus austriacus christiei* differs from the nominate subspecies in its relatively smaller size and a smaller baculum, with a more angular shape as illustrated by Qumsiyeh (1985).

Range of species.—Southern Europe and northern Africa south to Senegal and Ethiopia. Palestine, Arabia, and southwestern Asia east to Mongolia.

Local status.—The gray long-eared bat is commonly caught in mist nets in the Holy Land and most probably occurs in all habitats (Fig. 28). Tristram (1884:27) found this bat to be common in Palestine "especially in caves and tombs about Bethlehem and Jerusalem, and by the sea of Galilee." One of Tristram's specimens, an adult male collected at the Cave of Adullam is at the British Museum (Dobson, 1878). Aharoni (1930) reported this species from the Dead Sea Basin. Bodenheimer (1958) did not give localities but referred specimens from Palestine to the form *christiei*. Records exist from the following localities: Nahal Amram, Eilat, Cave of Adullam, Wadi Meneya (Timna), and Nahal Hever (Wadi Khabra) (Harrison, 1964a), Cave of Adullam (10 km S Jerusalem) (Atallah, 1977), Revivim, Mogharet Khureitun, Avdat (Abde), Eilat, Timna, Wadi Khabra (Makin, 1977; Qumsiyeh, 1985), Neot Hakikar and Arad (Yom-Tov et al., 1992b). Recently, the species was reported from near Gharandal (Wadi Araba), Disi (Wadi Rum) (Qumsiyeh et al., 1992), and Ras En Naqab (Zuhair Amr Collection).

Biology.—This bat is found roosting in caves, abandoned mines, ruins, and underground tunnels in the Holy Land. It is usually found to be solitary and thus an estimate of its abundance is difficult. Some of the prey of this species is gleaned from foliage, trees, rocks, cliffs, and so forth. This bat is well adapted to hovering flight for a gleaning method of foraging. The long ears are held back close to the body when resting. In open areas, these bats fly with the ears erect. This bat is an expert climber, as I have observed one in captivity easily climb very small branches. Harrison (1964a) reported a female with a single small fetus in March at Nahal Amram.

Genetics.—The karyotype of 2N = 32, AA = 50 of a specimen from the Holy Land (Qumsiyeh and Bickham, 1993) is indistinguishable from those reported from Tunisia (Baker et al., 1975) and Europe (Zima, 1978). G-bands also appear to suggest a conserved karyotype for three studied species of *Plecotus* (see Qumsiyeh and Bickham, 1993).

Human interactions.—The name *Plecotus* is derived from the Greek words *plekos* (braided or twisted) and *ous*, genitive *otos* (ear). The name *austriacus* refers to the locality of Austria where this species was first described.

FAMILY MOLOSSIDAE   Free-tailed bats, mastiff bats

Mastiff bats are easily distinguished by their external appearance. These bats have wrinkled, "bull-dog" faces. About half the length of the tail extends beyond the uropatagium (Fig. 37). The skull has no postorbital processes. There are 12 genera with about 84 species reported. *Tadarida teniotis* is the only molossid so far reported from the Holy Land. However, it is possible that *T. aegyptiaca* also occurs in this region. The dental formula for *Tadarida* is i 1/2-3, c 1/1, p 2/2, m 3/3 = 30 or 32.

Genus *TADARIDA*

Tadarida aegyptiaca   Egyptian free-tailed bat   **V**

Nyctinomus aegyptiacus Geoffroy, 1818. Description de L'Egypte Hist. Nat., 2:128, pl. 2, no. 2. Type from Egypt.

Diagnosis.—See under *Tadarida teniotis*. A medium-sized bat with a forearm length of 47–55 mm and a greatest length of skull of 20–22 mm.

Range of species.—Most of Africa, and into Arabia and India.

Local status.—Not yet recorded from the Holy Land but doubtless occurs in the Negev and Arabian deserts.

Biology.—Probably similar to *T. teniotis* but little is known of the biology.

Human interactions.—The name *aegyptiaca* refers to Egypt, where the first specimens of this species were caught and described.

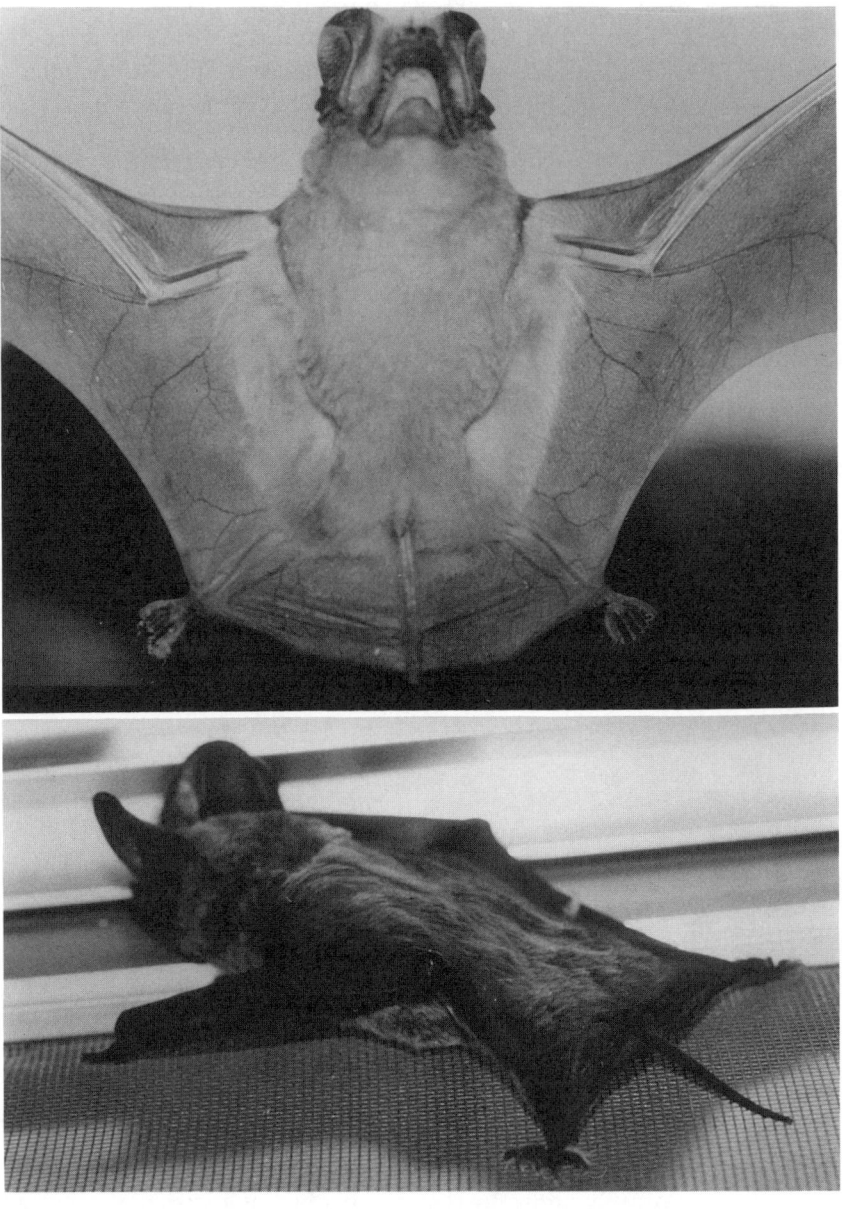

Fig. 37. The European free-tailed bat (*Tadarida teniotis*).

Tadarida teniotis   European free-tailed bat   V

Cephalotes teniotis Rafinesque, 1814. Precis des Découvertes et travaux somiologiques de zool. Bot. Palermo, p. 12, 55. Type from Sicily, Italy.

Dinops cestoni Savi, 1825. N. Giorn. Lett. Pisa, Sci. 10:235. Type from Pisa, Italy.

Dysopes rüppellii Temminck, 1826. Monographies de Mammalogie, 1:224, pl. 18. Type from Egypt; restricted to Cairo, Egypt by Qumsiyeh (1985). Valid subspecies: *T. t. rueppellii*.

Diagnosis.—See under the family for diagnosis. *Tadarida teniotis* is a large free-tailed bat with forearm measuring 54–64 mm in length and a greatest length of skull of 23.5–24.5 mm.

Range of species.—This species occurs in the Mediterranean coastal zone of southern Europe, northern Africa, and the eastern Mediterranean region. The range further extends into central Asia, China, Korea, and Japan. A lectotype for *rüppellii* was designated by Kock and Nader (1984), who also discussed this species in the western Palearctic region.

Local status.—Known locality records and personal collecting indicate that the European free-tailed bat is present throughout the Holy Land (Fig. 38). Its abilities to fly at high altitudes and for long distances are perhaps responsible for this wide distribution (see biology). This species was first reported (as *Xantharpyia aegyptiaca*) from Wadi Kern and caves along the Jordan Valley by Tristram (1866*b*, 1884). Anciaux de Faveaux (1953) reported this species from Wadi Nahr. Specimens were later reported from Eilat, Ain Gedi, Yotvata (Ain Ghidyan), Jericho, Yagur (Yadjour), Herzelia, Jerusalem, Jaffa, and near Beit Guvrin (Schmiedeknecht, 1906; Bodenheimer, 1935, 1958; Harrison, 1964*a*). East of the Jordan River, this species was found at Faidhat ed Dhahikiya (Atallah, 1967*b*, 1977), and Jordan University Campus, Jubeiha (Qumsiyeh, 1980). A specimen from Jerash is at JUNHM. Recently, we collected this species from 2 km E Ghor as Safi near the Dead Sea, Disi in Wadi Rum (Qumsiyeh et al., 1992), and Amman (Zahair Amr Collection).

Biology.—In places where this species was collected in Europe, Lebanon, and the Holy Land, individuals live in fissures in the sides of cliffs, in crevices in natural rock formations, or in crevices in the roofs of caves. The bats fly high and can travel long distances. They descend to roost, drink, and rest. Their flight is fast and direct due to unique adaptations of their wings. In Jordan, the species was found in extreme desert conditions in the north end of Wadi Sirhan in eastern Jordan (Atallah, 1967*b*, 1977), as well as in very mild Mediterranean forests of the north. In Lebanon, pregnant females with

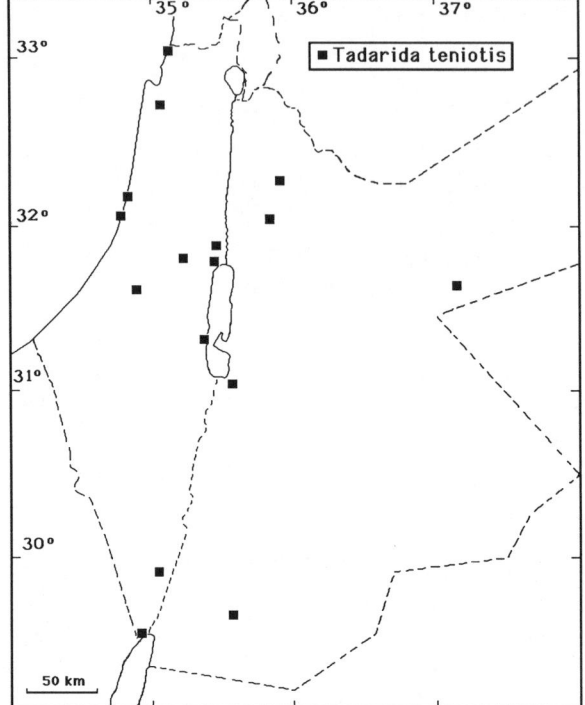

Fig. 38. Locality records of the European free-tailed bat (*Tadarida teniotis*).

a single embryo each were collected on 31 May with an estimated parturition in mid-June (Lewis and Harrison, 1962).

Genetics.—2N = 48, AA = 76 was reported for specimens from Yugoslavia (Dulíc and Mrakovcíc, 1980). The X is submetacentric and the Y is a very small acrocentric or submetacentric.

Human interactions.—The name *Tadarida* is a dubious coinage by C. S. Rafinesque (1783–1840), and *teniotis* is from *taenia*, L., a head-band, and *otos*, Gr., ear (the one with earbands).

ORDER CARNIVORA   Carnivores

Members of the order Carnivora are characterized mainly by specialized dentition. Most carnivores have a modified and large upper premolar (P4) and lower molar (m1), forming a pair of teeth called the carnassials, which have a shearing function for cutting meat. Evolution of omnivory in certain groups (for example, bears) has resulted in a reduction in the carnassials, which have become less specialized. In all living carnivores, some bones of the wrist are fused giving a more rigid structure. There are 17 families, 109 genera, and about 270 species in this order. The name Carnivora is derived from *caro*, L., genitive; *carnis*, L., flesh; and *voro*, L., devour (flesh-eating).

Key to the families of Carnivora of the Holy Land

1. Doglike animals. Upper toothrow longer than one-half skull length. 2

   Cat- or weasellike animals. Upper toothrow shorter than one-half skull length. 3

2. Strong forequarters and neck giving the appearance of weak hind legs. Striped coat color. Skull has broad massive rostrum. . . . . Hyaenidae

   Morphology not as above. Coat color not striped (more uniform coloration). Skull has narrow rostrum. . . . . Canidae

3. Three lower cheekteeth. Skull appears rounded because of a short rostrum and a short rounded cranium. Catlike animals. . . . . Felidae

   Four or more lower cheekteeth. Skull appears elongated. Weasellike animals. 4

4. Rostrum relatively long. Bulla constricted and divided by a septum. . . . . Viverridae

   Rostrum relatively short. Bulla not constricted or divided by a septum. . . . . Mustelidae

FAMILY CANIDAE   Dogs, wolves, foxes, jackals

The canids have specialized dentition for carnivory including large, powerful canines and strong carnassial teeth. The dental formula is usually i 3/3, c 1/1, p 4/4, m 2/3 = 42. Twelve genera with 34 recent species are known from throughout the world. I have adopted the classification of the family Canidae proposed by Clutton-Brock et al. (1976).

Key to the Canidae of the Holy Land

1. Size larger, skull length in adults greater than 145 mm. Doglike animals. The frontal region of the skull is swollen. 2

    Size smaller, skull length in adults less than 145 mm. Foxes. The frontal region of the skull is not swollen. 3

2. Size larger with skull length greater than 180 mm. Cingulum on outer edge of the first molar small. . . . . *Canis lupus*

    Size smaller with skull length between 145 and 175 mm. Cingulum on the first molar well developed. . . . . . *Canis aureus*

3. Back of ears black or dark brown. . . . . *Vulpes vulpes*

    Back of ears same color as head and body. 4

4. Color grayish, footpads not covered in hair. . . . . . . *Vulpes cana*

    Color sandy brown, footpads covered in hair. 5

5. Small fox with large ears. Tympanic bulla swollen. . . *Vulpes zerda*

    Larger fox but with relatively small ears. Tympanic bulla not significantly swollen. . . . . *Vulpes rueppelli*

Genus *CANIS*   Dogs, wolves, jackals

The genus *Canis* includes eight species. *Canis* is the Latin word for "dog."

Canis aureus   Jackal, golden jackal   E

Canis aureus Linnaeus, 1758. Syst. Nat., 10th ed., 1:40. Type from Laristan (now Fars), Iran.

Canis syriacus Hemprich and Ehrenberg, 1833. Symb. Phys. Mamm., 2:sig. z, pl. 16. Type from the coast of Lebanon between Beirut and Tripoli. Valid subspecies: *C. a. syriacus*. In the eastern Mediterranean region.

Canis lupaster Hemprich and Ehrenberg, 1833. Symb. Phys. Mamm., 2, folio ff. Type from El Faiyum, Egypt. Valid subspecies: *C. a. lupaster*. In Egypt and perhaps Sinai and the Negev. However, see Ferguson (1981*a*).

Diagnosis.—The jackal (Fig. 39) is similar to a small dog in appearance. Jackals weigh 5–12 kg and have a head and body length of 60–90 cm, a tail length of 20–30 cm, and a skull length of 14.8–18 cm. The fur is rather short and coarse. The dorsal color is variable black or brown-yellowish to mottled

Fig. 39. The jackal (*Canis aureus*). Photo: H. Mendelssohn.

gray. A dark band runs along the back from the nose to the tip of the tail. This mane becomes wider on the back, extending into the lateral surfaces. The tail is relatively short, usually with a black tip.

The skull is similar to that of a dog or a small wolf. The teeth are high crowned. The sagittal crest is present. The cingulum on M1 is usually well developed. Some authors have suggested (although with no direct evidence) that hybridization occurs between dogs and jackals. The jackal is distinguished from the dog by having larger canines and more rounded bullae. Osborn and Helmy (1980) summarized the literature on differences between dogs and jackals. They listed 18 characters that help distinguish a jackal skull from that of a dog, including: the jackal skull generally has smaller inflation of the frontal region, less steep forehead, less inflated and elongated postorbital regions, smaller upward curvature of the zygomatic arches, longer and thinner lower jaw, more inflated auditory bulla with smooth surface, and more straight posterior margin of ramus.

Nehring reported *lupaster* as the name for a specimen collected by E. Schmitz in 1908 and this designation was repeated later (Flower, 1932; Bodenheimer, 1958). It is possible that both the smaller *syriacus* and the larger *lupaster* occur in the eastern Mediterranean region (Flower, 1932). Alternatively, the size variation could be clinal and these two names may be considered synonyms (Lewis et al., 1968). The status of the jackal is difficult to ascertain because of the possibility of its interbreeding with wild and semiwild pariah dogs in the region, a problem commented on by early workers (Aharoni, 1930; Bodenheimer, 1958). The reports of locals and some earlier workers of wolves probably refer to jackals. Some locals cannot distinguish between wolves and jackals and use *Deeb* for both. Scientists have also confused the two. According to Ferguson (1981*a*), *lupaster* of Egypt and Sinai is actually a small subspecies of the wolf rather than a large subspecies of the jackal. This hypothesis is based on its conformity with Bergmann's and Allen's rules rather than on any genetic or clear morphological evidence. This issue is not further discussed here because such evidence is not yet available and *lupaster* does not appear to occur in Palestine.

Range of species.—Throughout northern Africa south to Senegal in the west and Kenya in the east. Also ranges in Arabia, southwestern Asia, and the Caucasus, east through India, Assam, Burma, and Thailand.

Local status.—Judging from the diversity in the climates in the many localities reported (Fig. 40), the golden jackal probably occurs throughout the Holy Land. Linnaeus (1762) recorded this species from Jaffa. Tristram (1866*b*) reported that the jackal "swarms in incredible numbers in every part of the country." The jackal was very common in Palestine until early in this century (Allen, 1915) and was reported from the Ghor and at Gaza (Hart, 1891). Specimens were reported by Harrison (1968*a*) from 15 mi W Jerusalem, Bir Salem, and Nahr Rubin. It is known from near Azraq, Wadi Rum (Mountford, 1965), and Lubban (specimen at SANHC). Specimens at the JHMN and JUNHM are from Wadi Rum and Azraq (Amr and Disi, 1988) where animals were seen earlier (Mountford, 1965; Atallah, 1967*b*; Nelson, 1973). Specimens at HUJ are from Qumran, Jerusalem, near Bet Yosef, Ashdot-Yaaqov, near Naharia, Beersheba, Ain Fashkhah, Huldah, Lod, and Tel Aviv.

Biology.—Jackals are found in open and wooded areas of variable climates. They are omnivorous and opportunistic, feeding on dates, mulberries, other fruits, snails, insects, rodents, carcasses of large animals, fish, and other foods (Anderson and de Winton, 1902; Lay, 1967; Osborn and Helmy, 1980). The jackal seems to remain in well-protected areas during the day and forage in more open areas at night. Occasionally, they may be active during daytime, especially in the evenings or mornings.

Behavioral studies on a pair of jackals in the Holy Land were made by Golani and Keller (1975). Like other canids, jackals will scent mark their territory. In Egypt, jackals appear to give birth to a maximum of eight young (usually three to five) between March and April (Flower, 1932). The gestation period is 9–10 weeks.

Genetics.—2N = 78, AA = 76 was reported for specimens from India (Ranjini, 1966) and Yugoslavia (Soldatovíc et al., 1970).

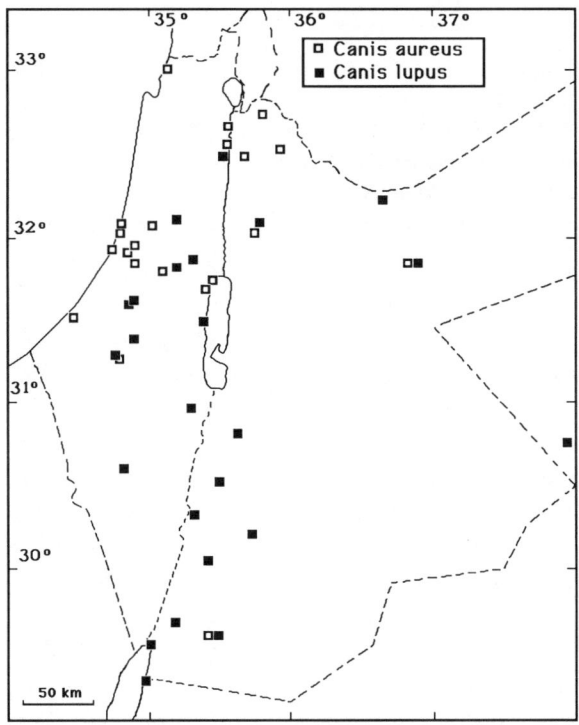

Fig. 40. Distributions of the jackal (*Canis aureus*) and the wolf (*Canis lupus*).

Human interactions.—The jackal is well known to the locals who have given it many names *(husseini, wawi, deeb, ibn awaten)*. The Latin name *aureus* means golden (hence, the golden jackal). Howells (1956) mentioned that the jackal enters villages in Palestine to prey on domestic animals such as chickens, young goats, and sheep. No scientific evidence for such predation on domestic animals was found in Lebanon (Lewis et al., 1968). In both Lebanon and the Holy Land, jackals were more common 40–50 years ago than they are today (Lewis et al., 1968, and M. B. Qumsiyeh, personal interviews). Their decline was most likely caused by humans but competition with the red fox could be an additional factor (Lewis et al., 1968). The assumed danger of rabies and the supposed threat to livestock caused massive official poisoning campaigns, which were very effective in causing a decline of jackals in Israel (Yoffe, 1968). Urgent conservation measures may be needed to prevent further declines in their population.

Canis lupus   Wolf   E

Canis lupus Linnaeus, 1758. Syst. Nat., 10th ed., 1:39. Type from Sweden. The nominate subspecies is extralimital.

Canis pallipes Sykes, 1831. Proc. Zool. Soc. Lond., 1831:101. Type from Deccan, India (Sykes, 1831). Valid subspecies, northern Palestine.

Canis lupus arabs Pocock, 1934. Ann. Mag. Nat. Hist. ser. 10, 14:636. Type from Ain, Qara Mountains, Oman. Valid subspecies in the southern deserts of the Holy Land.

Diagnosis.—The wolf (Fig. 41) is the largest member of the family Canidae and closely resembles its descendant, the domestic dog (*Canis familiaris*), with which it can readily interbreed. The body is heavily built (weight 14–30 kg and head and body length 80–115 cm) with a large head and strong long legs. The tail is long (30–40 cm) and bushy. The skull is large (19–24 cm long) with a well-developed sagittal crest. The jaws are massive with well-developed teeth that are larger than those of the domestic dog. The most consistent diagnostic feature distinguishing wolves from dogs appears to be the angle of the orbital bones relative to the midline (Iljin, 1941). In wolves, this angle is 40–45° as opposed to 53–60° in dogs. Additionally, the tympanic bulla is relatively larger and more spherical in wolves than in dogs (Harrison, 1973). *Canis lupus arabs* is relatively smaller than *C. l. pallipes*. The wolves of northern Arabia and those from northern areas of the Holy Land may be intermediate in size (but closer to *C. l. pallipes*); thus, the different subspecies may be clinal variants (Harrison, 1973).

Range of species.—Formerly found throughout the Northern Hemisphere north of 20°N latitude in almost all habitats. Hunting and habitat destruction have reduced the range considerably in recent years (Mech, 1974). The status of the wolf in the Arabian regions has not been well studied but reviews are available for Saudi Arabia (Nader and Buttiker, 1980) and Israel (Mendelssohn, 1982).

Local status (Fig. 40).—In the late 19th century, the wolf was still found in many parts of the Holy Land (east and west of the Jordan River) as indicated by numerous reports. Tristram (1884) stated that the wolf is "still common in Palestine [and] is found in every part of the country." Hart (1885) reported on verbal records from Wadi Lebweh.

Aharoni (1930) stated that the wolf was limited to the mountains. Reports exist of wolves shot from Salfit and 'Aqaba (Anonymous, 1946*a*). Bodenheimer (1958) claimed that the wolf was an intruder from the east and north rather than a true inhabitant of Palestine in the 1950s. Reports of wolf exist from Wadi Araba, 48 km N 'Aqaba, Wadi Sirhan, Umm el Quetain (Bromage, 1954), Mukhmas (Wadi Suweinit), Lachish (Harrison,

1968*a*), Beersheba, Bashan, Gilead, Galilee (Tristram, 1866*b*), Nahal Zin in 1982 (Anonymous, 1982), and in Makhtesh Ramon in the Negev in 1983 (Anonymous, 1983*b*).

There are two subspecies in the region, *C. l. pallipes* occurs in the northern regions as far south as the 400 mm isohyet and *C. l. arabs* occurs in the southern regions (Mendelssohn, 1982). Mendelssohn (1982) discussed available records east of the Rift Valley and showed the earlier decline of this species and its recent resurgence following conservation efforts and greater food availability (such as at garbage dumps). Bromage (1954) reported wolves from Wadi Araba and eastern deserts of Jordan. Nelson (1973) reported several seen and or shot near Azraq. Specimens at the JUNHM were obtained from Wadi Rum in 1979, Wadi Finan in 1981, and Al Rishah in 1986 (Amr and Disi, 1988). Recent (1985–1987) verbal records are from Ma'an and Abu Anseer (Amr and Disi, 1988). Specimens at HUJ are from Lahav, Jerusalem, Beit Guvrin, and Ain Gedi. Sight records from northern Saudi Arabia near the border with Jordan are available from Wadi Sirhan (Nader and Buttiker, 1980) and Haql (Gasperetti et al., 1985).

Biology.—The wolf is an opportunistic feeder and has been reported to feed on domestic animals, garbage, jirds, voles, hares, birds and occasionally gazelles. Packs of two to eight individuals may form. Wolves are mainly

Fig. 41. The wolf (*Canis lupus*). Photo: A. Shoob.

nocturnal but may occasionally hunt during the day. Wolves are very social animals with hierarchies formed within the pack. In North America, a male and female may form a bond lasting a lifetime. Courting begins with elaborate rituals by the males that include sniffing, rubbing heads, wagging tails, and wrestling. Copulation occurs during an estrus period of 5–7 days with the female lifting her tail and exposing her genitals when receptive. The male mounts the female from behind but soon dismounts by rotating his body 180°, lifting one leg over the female's back. The pair then remains locked by their genitals in a rump to rump positions for about 30 minutes, during which time several ejaculations may occur (Mech, 1974). Up to six young are born in April or May following a gestation period of about 2–3 months. Wolves appear to have an uneven sex ratio with a preponderance of male pups (Mech, 1974; Mendelssohn, 1982). Wolves in the wild live for about 10 years. In captivity, animals were kept for more than 16 years. Further observations on the biology of wolves in the region can be found in the review by Mendelssohn (1982).

Genetics.—2N = 78, AA = 76 was reported for both the wolf and its descendant, the dog, from many parts of the world (review in Zima and Král, 1984).

Human interactions.—The wolf is known in almost every culture in the Northern Hemisphere. In the Holy Land, the wolf is known from cave drawings and was mentioned in the earliest writings of the Near East. Examples of early writings are: "A wolf of the deserts shall destroy them" (Jeremiah, 5:6) and "like wolves tearing the prey, by shedding blood . . ." (Ezekiel, 22:27).

Wolf population size in the Holy Land decreased considerably between 1964 and 1980, due mainly to extermination by farmers and others. Protection now is much better in both Jordan and Israel. However, the main threat now is destruction of habitat and human overpopulation, which is reaching into areas that are the last refuges of the wolf. The name *lupus* is Latin, meaning wolf.

Genus *VULPES*   Foxes

There are 11 species of foxes in this genus, of which 4 are known from the Holy Land. The dental formula is i 3/3, c 1/1, p 4/4, m 2/3 = 42. *Vulpes* is the Latin word for fox.

Fig. 42. Blanford's fox (*Vulpes cana*). Animal from Ain Gedi. Photo: A. Shoob

Vulpes cana   Afghan fox, Blanford's fox   V

Vulpes canus Blanford, 1877. J. Asiat. Soc. Bengal, 46(2):321. Type from Gwadar, West Pakistan.

Diagnosis.—Blanford's fox is small with a head and body length of 38–45 cm, tail length of 30–36 cm, and skull length of 90–100 mm. *Vulpes cana* is intermediate in size between *V. rueppelli* and *V. zerda*. *Vulpes cana* has a relatively longer and more bushy tail than *V. rueppelli* (compare Figure 42 with 45). The tail measures 77% of head and body length (as opposed to 70% in *V. rueppelli*; Mendelssohn et al., 1987). *Vulpes cana* differs from *V. zerda* (which is its closest relative phenetically, Clutton-Brock et al., 1976) in its darker color, naked footpads, and slightly larger skull dimensions (Peters and Rödel, 1994). The rostrum of the skull is more delicate than in *V. rueppelli*.

Range of species.—Uzbek and Turkman SSR, Pakistan, Iran, Afghanistan, Oman, the Holy Land, Sinai, and Egypt.

Local status (Fig. 43).—Blanford's fox was recorded from Ain Gedi, Nahal Ze'elim (Wadi Siyal), Makhtesh Ramon, and near Eilat (Mendelssohn et al., 1987). More recently, Geffen et al. (1993) reported the following additional localities: Wadi Qelt, near Qumran, Qidron, Bir Mashash, Nahal Hever (Khabra), and Nahal Boqeq (Wadi Um Baghkak), among others.

Biology.—This species prefers rocky mountain habitats throughout its known range, which is very limited (only a few localities have been reported).

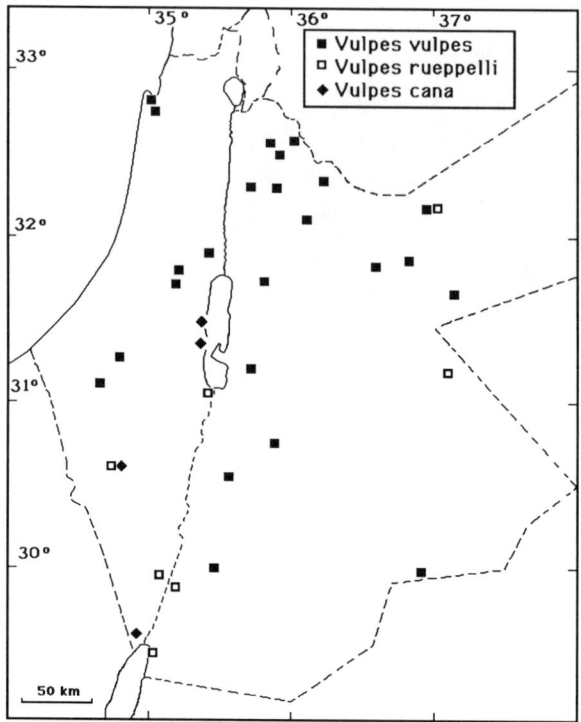

Fig. 43. Records of the red fox (*Vulpes vulpes*), Blanford's fox (*Vulpes cana*), and Rüppell's fox (*Vulpes rueppelli*). For *V. cana* some approximate localities were taken from a map by Geffen et al. (1993).

Doubtless it is more common than recorded in the literature. One to three pups were born in April in captivity and in February under field conditions (Mendelssohn et al., 1987). The same authors report that these animals feed on insects (beetles, grasshoppers), scorpions, and small mammals. A more detailed account of diet and foraging behavior can be found in Geffen et al. (1992), who report that food consists of plants, arthropods, reptiles, birds, and mammals.

Human interactions.—See the red fox, *V. vulpes.* In Pakistan and Afghanistan, the pelts of this fox are valued and the species was hunted extensively which threatened its survival.

Vulpes (*Fennecus*) *zerda*   Fennec fox   V

Canis zerda Zimmerman, 1780. Geogr. Gesch. Mensch. Vierf. Thiere, 2:247. Type from the Sahara Desert.

Diagnosis.—This is a small buff-colored fox. Weighing about 1 kg, this is the smallest of the foxes. The head and body length is 36–42 cm, the tail length is 16–23 cm, and the skull length is 83–89 mm. The ears are excessively large relative to head size and the tail has a dark tip (Fig. 44).

There is a small dark patch dorsal to the tail glands. The remainder of the fur is light colored and well camouflaged in a sandy desert environment. The skull is distinguished by being small and delicate and having extremely enlarged smooth bullae. See under other species for further identification.

Range of species.—From Kuwait across Arabia, Sinai, Egypt, and most of the Sahara to Mauritania.

Local status.—This species is almost certain to be found in sandy desert areas in the Negev and in eastern Jordan, because it was reported in similar habitats in Kuwait, Egypt, and western Sinai (Harrison, 1968*a;* Harrison and Bates, 1991). Hatough-Bouran and Disi (1991) cite a paper by Garrard (1983–not seen) that records an Epipaleolithic *Fennecus* from Qasr Al Kharana in Jordan.

Biology.—These animals live in sandy areas with little vegetation. They seem to acquire their water entirely from their diet, which consists of insects, small birds, rodents, reptiles, and some plants (Schmidt-Nielson, 1964). Although most of their activity is nocturnal, I observed many individuals in the Libyan Sahara at midday. As in other foxes, males are territorial and aggressive during the breeding season. Algerian fennecs kept in captivity had a gestation period of 50–52 days and gave birth to one to three young in May and June (Saint-Girons, 1962).

Fig. 44. The fennec fox (*Vulpes zerda*). From western Sinai. Photo: A. Shoob.

Human interactions.—*Fennecus* and *zerda* are derived the Arabic words for this fox *(fenek* and *zerdawi).* See also comments on the red fox, *V. vulpes.*

Vulpes rueppelli   Rüppell's fox, sand fox   **E**

   Canis rüppellii Schinz, 1825. *In* G. Cuvier, Das Thierreich, 4:508. Type from Dongola, Northern Prov., Sudan.

   Canis niloticus É. Geoffroy Stèt. Mamm. Mus. Nat 1803:134. Type probably from near the pyramids. Synonym.

   Canis famelicus Cretzschmar, 1826. *In* Rüppell, Atlas Reise Nördl. Afr., Säugeth., p. 15. Type from Nubian Desert and Kordofan. Synonym.

   Vulpes rüppellii sabaea Pocock, 1934. Ann. Mag. Nat. Hist., ser. ? 14:636. Type from Rub el Khali, Saudi Arabia. Synonym.

   Diagnosis.—A small fox with prominent ears (Fig. 45). Head and body length 40–51 cm, tail length 29–37 cm, ear length 8–11 cm, skull length, 99–117 mm. The color is reddish gray. The face has a dark patch between the eyes and the nose. The tail has dark guard hairs. See under *V. vulpes* for diagnosis. The lighter color (light tan or sandy brown) of the back of the ear is the best external character that distinguishes *V. rueppelli* from desert forms of the red fox (*V. vulpes*), which has a black back of ear. The skull of *V. rueppelli* is distinguished from that of *V. vulpes* in its smaller size and more inflated bulla. From *V. zerda*, *V. rueppelli* is distinguished by its larger size, but relatively smaller ears and bulla. From *V. cana*, *V. rueppelli* is distinguished by its slightly larger size, relatively smaller ears and bulla, and, most

Fig. 45. Rüppell's fox (*Vulpes rueppelli*). Animal from Wadi Araba. Photo: H. Mendelssohn.

prominently, by the presence of hair covering the footpads. *Vulpes rueppelli* is also darker and has a white tail tip (tip not distinct in *V. cana* and black tip in *V. zerda*). Harrison and Bates (1991) recognize *V. r. sebaea* as a valid subspecies for Arabian (including Holy Land) specimens. The difference in color between *V. r. sebaea* and the nominate subspecies is minimal, with intermediate specimens found in the Sinai.

Range of species.—Semiarid and arid regions of Afghanistan, Iran, Iraq, the Arabian Peninsula, and northern Africa west to Algeria and south to Somalia.

Local status.—This is a desert-adapted species and is known from the Negev, Wadi Araba, and the Syrian Desert (Fig. 43). Tristram (1884) and Aharoni (1930) used the name *niloticus* for Palestinian foxes. Setzer (1961) used this name as a subspecies of *Vulpes vulpes*. Osborn and Helmy (1980) did not deal with this name in their monograph. If this name belongs to the red fox, it should be a senior synonym for the Egyptian red fox. For now, I consider this name a synonym of *rueppelli* and until reevaluations of subspecific status in these animals, all are referred to the nominate subspecies. This fox is mainly found in central and southern Palestine (Aharoni, 1930). Harrison (1968*a*) reported seeing animals at Tel Aviv Zoo from the Negev and from the southern tip of the Dead Sea. A specimen at the JUNHM is from Hadraj (Amr and Disi, 1988). I also obtained specimens from H5 in eastern Jordan (collected by George Saad) and 2 km SE 'Aqaba. Other specimens at HUJ are from Grofit and Makhtesh Ramon.

Biology.—This species lives in sandy areas in wadis and plains in semiarid conditions. Osborn and Helmy (1980) reported that in Egypt this fox drinks water and feeds on insects, rodents, small birds, lizards, and fruits. Three fetal scars were reported by Osborn and Helmy (1980) in Egypt. Little additional information is available on the biology of this interesting fox.

Genetics.—Matthey (1954) reported a 2N = 40 for a specimen from northern Africa.

Human interactions.—See the red fox, *V. vulpes.*

Vulpes vulpes   Tawny fox, common fox, red fox   O

Canis vulpes Linnaeus, 1758. Syst. Nat., 10th ed., 1:40. Type from Upsala, Sweden.

Canis aegyptiacus Sonnini, 1816. Nouv. Dict. Sci. Nat., Vol. VI. Type from Giza Pyramids, Giza Prov., Egypt. Valid subspecies: *V. v. aegyptiacus.*

Vulpes vulpes arabica Thomas, 1902. Ann. Mag. Nat. Hist., ser. ? 10:489. Type from Muscat, Oman. Valid subspecies.

Vulpes vulpes palaestina Thomas, 1920. Ann. Mag. Nat. Hist. ser. 9, 5:122. Type from Ramleh, near Jaffa, Palestine. Synonym of *V. v. aegyptiacus.*

Diagnosis.—Although size is variable in different areas of the range of the species, the red fox is the largest fox in the genus. In the Holy Land, the red fox is not very variable in size (Davis, 1977) and is the largest fox, especially in skull measurements. Red foxes weigh 2–4 kg and have a head and body length of 52–66 cm. The tail is bushy and long (29–45 cm). The skull is massive at 11–14.5 cm in length. *Vulpes vulpes* further differs from other *Vulpes* in the region in having darker color, black back of the ears, dark underside (instead of whitish), and less inflated bulla. The red fox has a conspicuous middorsal stripe of fulvous dark color. The basic color is slate gray with some reddish and brown tinges due to the presence of guard hairs. The tail is large, very bushy, and is reddish brown on a portion of its dorsal surface. Most animals have a white tip on the tail.

Four species of foxes occur in the Holy Land. *Vulpes zerda* is the smallest fox, and is easily distinguished by its light color and large ears. *Vulpes vulpes* and *V. rueppelli* rarely occur in the same locality, with the latter being more desert adapted. Finally, *V. cana* is a rock dweller, with a long bushy tail that usually has a black tip and footpads that are not covered by hair. Thomas described *V. v. palaestina* (type locality, Ramleh, near Jaffa) as characterized by its more grayish coloration. However, Bodenheimer (1958) suggested that clinal variation exists between Egypt, Palestine, and Arabia, and thus the oldest available name, *V. v. aegyptiacus* Sonnini, 1816, should be applied to the Palestinian fox. Tristram (1866*b*, 1884) and Aharoni (1930) stated that the Persian subspecies, *V. v. flavescens* (originally described from Iran), occurs in the Galilee in northern Palestine, but this is unconfirmed by Bodenheimer (1958). Lewis et al. (1968) added *V. v. flavescens* to the subspecies that are considered as forming a cline with and thus are synonyms of *V. v. aegyptiacus.*

Range of species.—*Vulpes vulpes* is the most widespread canid species. It occurs throughout temperate (including subdesert) regions of the Palearctic region and North America.

Local Status.—*Vulpes vulpes* is very common and occurs in most habitats except the extreme arid regions (Fig. 43). It is less common in areas where *V. rueppelli* and *V. zerda* occur (more arid regions), perhaps due to interspecific competition. The red fox was reported from near Haifa, Mount Carmel, Jerusalem, Beersheba (Harrison, 1968*a*) as well as from these areas and the Jerusalem and Bethlehem areas, Sinai, the Dead Sea Basin, Haluza Sands, and Negev (Davis, 1977; M. B. Qumziyeh, pers. obs.). Atallah (1967*b*) collected specimens from 3 km N Azraq Shishan and observed others around Qasr Amrah, between Ma'an and Al Hasa, and Faidhat ed Dhahikiya.

Searight (1987*b*) observed it at Jawa. Specimens are available from Madaba, Zarqa, Shobek, Karak, Kufranjah, Azraq, Irbid, Al Inab (Saliba and Amr, 1985; Amr and Disi, 1988; JUNHM), Beit Sahur, and 6 km NNE Beit Sahur (SANHC). Specimens at the JNHM are from Wadi Araba, Al Husn, Azraq, Ramtha, Nu'aymah, 25 km W H5, Jerash, Mafraq, and Wadi Aqraba. Specimens at HUJ are from Ramleh, Abu Ghosh, Beit Meir, Eilat, Ain Kerem, Lahav, Meholla, Qiryat Saide, and along Jerusalem–Jericho road.

Biology.—The red fox is the most common carnivore in the Holy Land, especially in mountainous regions. Although mostly nocturnal, animals are active during many daylight and nighttime hours. I have watched activities of the red fox many times, especially during evening or morning hours. Like the jackal, the red fox appears to be opportunistic and omnivorous. Atallah (1967*b*) reported them feeding on jerboas. They feed on fruits, insects, small rodents, fish, reptiles, and other food items (Lewis et al., 1968; M. B. Qumsiyeh, pers. obs.). I observed foxes near Bethlehem turning rocks to feed on insects, scorpions, and centipedes. In Iran, food consisted primarily of rodents and tenebrionid beetles (Lay, 1967). The pelage in foxes is thicker and fuller in the winter months than in the summer months. This change in fur structure is common for carnivores.

Dens are usually dug on the sides of hills in protected areas such as those between boulders or in thick plant cover. Data on reproduction in the Holy Land are scanty. In Lebanon, foxes give birth to two to five young in March or April (Lewis et al., 1968). In European forms, the gestation period is 50–53 days. In the wild in Europe, the red fox is though to have a mean life expectancy of about 4.5 years (Southern, 1964).

Genetics.—A polymorphic diploid number ranging from 34 to 42 was reported for specimens from Europe (Renzoni and Omodeo, 1972; Zima and Král, 1984).

Human interactions.—Many Arab tales include stories of the *tha'lab* (pl. *tha'aleb*). However, no distinction is recorded between the different species. The fox is believed to be a sly animal with the bad habit of "stealing" chickens and other domestic animals.

FAMILY FELIDAE   Cats

Morphologically and behaviorally, cats are a uniform group of carnivores. The diversification into unique habitats created such superficially distinct forms as the lions and the domestic cats. Cats have a broad skull with large orbital spaces and enlarged auditory bullae. Claws are retractable (except in the cheetah). The small incisors are chisellike, the canines are large, and the carnassials are well developed. The dental formula is i 3/3, c 1/1, p 2–3/2, m 1/1 = 28–30. There are 37 species of felids, of which seven are found in the Holy Land or were known to occur there in recent times.

Key to the Felidae of the Holy Land

1. Size very large, shoulder height about 50 cm, skull greater
   than 16 cm in length. 2

   Size smaller, skull length less than 16 cm in length. 3

2. Spots on body are all solid; black stripe on face running from
   eyes to mouth. . . . . *Acinonyx jubatus*

   Most spots on back are rosettes; no black stripe on face.
   . . . . *Felis pardus*

3. Ears with distinct long tuft. A black facial stripe extends
   from the nose to the forehead on either side. Nasal branch
   of premaxilla long. . . . . *Felis caracal*

   Ears with short or no tuft. Facial markings not as above.
   Nasal branch of premaxilla broad. 4

4. Color pale. Ears large. Elbows have distinct dark marks.
   Furry paws. Tympanic bulla greatly inflated. . . . *Felis margarita*

   Color dark. Ears small. Elbows have no distinct dark marks.
   Paws not very furry. Tympanic bulla not inflated. 5

5. Tail longer than 60% of head and body length, size small,
   no ear tufts, a distinct cheek stripe present. . . . . . *Felis silvestris*

   Tail short, less than 50% of head and body length, size larger,
   ear tufts present, no cheek stripe present. . . . . *Felis chaus*

Genus *ACINONYX*   Cheetah

Acinonyx jubatus   Cheetah, hunting leopard   X?

Felis jubata Schreber, 1776. Die Säugethiere, 3(1777):392, 586, pl. 105. Type from Cape of Good Hope, South Africa.

Felis venatica Griffith, 1821. The general and particular descriptions of the vertebrate animals. Carnivora, p. 93. Type from India. Valid subspecies: *A. j. venaticus.*

Diagnosis.—This is a large cat (head and body length 110–150 cm, tail length 55–80 cm, weight 35–65 kg) with a pattern of solid black spots on a yellowish gray background. The cheetah can be distinguished from the leopard by the solid spots and by the more slender body of the cheetah (Fig. 46). The underside is whitish. The ears are small, rounded, and with a dark base. There is a black stripe running from the inner corner of the eye to the corner of the mouth.

Range of species.—African savanna and steppe regions north to Morocco, east through Arabia, Iran, and Baluchistan, and south to India. Extinct in much of its previous range except in Africa south of the Sahara.

Local status.—The cheetah must have been common in the Middle Ages because it is mentioned frequently in the writings of Arabian travelers and by European explorers. The name "fahd" also appears on many towns and as a surname for many in Jordan and Palestine. By the time of Tristram (1866*b*, 1876, 1884), the cheetah was scarce and restricted to wooded hills of Galilee, near Tabor, and in the Ajlun Mountains. The following statements by Tristram (1866*b*) are worth repeating: "The Cheetah is scarce . . . few still haunt the neighborhood of Tabor and the Hills of Galilee. In Gilead it is more common, and a sheikh there presented me with three skins of cheetah shot by his people." The cheetah probably occurred in Palestine until the last half of the 19th century with remnant populations in Jordan left around Moab until 1900–1912, where E. Schmitz obtained a specimen killed by a Bedouin east of Wadi Zarqa Ma'in and discussed the presence of the leopard in the area (Fig. 46). Aharoni (1930) mentioned that Bedouin reported seeing animals in the southern region of Palestine early in this century. A specimen was collected at Deir El Zor in Syria in 1931 (now at HUJ). Carruthers (1909) reported tracks near Taymah oasis and near Jebal El Tubaiq, both in Saudi Arabia near the Jordanian border. The cheetah is probably no longer to be found in the Holy Land. However, cheetahs were killed at Turaif in the 1950s (Morrison-Scott, 1951; Hatt, 1959) and other observations in northern Saudi Arabia earlier (Raswan, 1935 observation and photograph) suggest that there may be a remnant population surviving.

In 1962, a female was shot about 60 km E 'Aqaba and her cub was raised in captivity. The desert in northern Arabia shared by Iraq, Jordan, and Saudi Arabia is little explored or inhabited and may be the last holdout for these animals in the Arabian Peninsula.

Biology.—There is no information on the biology of cheetahs in the Holy Land. The following observations are based on African cheetahs from Haltenorth and Diller (1980). This species is diurnal and rarely hunts at night (only when the moon shines bright). This may be the reason that it was more severely affected by humans than the leopard, which hunts easily at night. This species is known to occur in grassland and subdesert regions. Food includes hares, gazelles, jackals, porcupines, addax, and ibex. Cheetahs rely on sight and spurts of high speed chasing (up to 110 km per hour) to catch prey. In desert areas, animals may go without drinking water for many days, obtaining their water from desert melons and from their prey.

After a gestation period of about 3 months, one to six young are born. Young are weaned before 2 months and reach sexual maturity at 10–14 months. Cheetahs have lived up to 16 years in captivity and are known to live 10–12 years in the wild.

Fig. 46. The cheetah *(Acinonyx jubatus)*, collected by Ernst Schmitz in the mountains of Zarqa Ma'in.

Genetics.—2N = 38, AA = 70 was reported for zoo specimens (Hsu and Benirschke, 1967–1971). The X is submetacentric and the Y is a small acrocentric.

Human interactions.—The scientific name *Acinonyx* is probably derived from *akaina,* Gr., a thorn, a goad, and *onux,* Gr., a claw. However, a different derivation can be postulated from *a-,* Gr., not, and *kineo,* Gr., I move + claw. Both refer to the nonretractable claws of the cheetah (more similar to dogs than other cats). The word *jubatus* (= *iubatus*) is Latin meaning maned. The word cheetah is from *cital* (Hind) from *chitraka* (Sanskrit) meaning having a speckled body.

The leopard and the cheetah were the two most common big cats in the Holy Land in the 19th century. The cheetah was more vulnerable to humans because of its habits. It is diurnal and not shy like the leopard. The cheetah was found to be easily tamed and both ancient Egyptians and Assyrians kept tame cheetahs and used them for hunting. Bedouins used tamed animals to hunt gazelles (Metaxas, 1891).

Genus *FELIS*   Cats

Felis caracal   Caracal, red lynx   E

Felis caracal Schreber, 1776. Die Säugethiere, 3(1777):413, 587, pl. 10. Type from Table Mountain, Cape Town, South Africa.

Felis (Caracal) schmitzi Matschi, 1912. Schriften Berl. Ges. Naturf. Fr. Berl., 1912:64. Type from "Wadis opening to the Dead Sea." Type specimen at the BZM is from Ain ed Dachubeijir, Jordan. Valid subspecies.

Felis (Caracal) caracal aharonii Matschi, 1912. Schriften Berl. Ges. Naturf. Fr. Berl., 1912:66. Type from mouth of Chabur River, Upper Euphrates, Syria. Synonym.

Diagnosis.—The caracal (Fig. 47) is a medium-sized felid. The head and body length is 55–105 cm and the tail measures 23–30 cm. Males are significantly larger than females (for example, males weigh as much as 13 kg, females weigh up to 9 kg). The caracal has a long black tuft on the ear. The back of the ear is also black. The fur color is a rather uniform reddish brown becoming light grayish with age. The ventral color is lighter, almost white. A thin indistinct black stripe runs from the nose to the eyes.

Range of species.—South Africa north through most of eastern and northern Africa, Arabia, the eastern Mediterranean region, and east to India and Russian Turkistan.

Local status.—The caracal is reported from many areas, especially in the south and east of the Holy Land (Fig. 48). Its exact distribution and habitat preference remain to be elucidated. However, the available studies indicate that caracals are equally content in desert as well as Mediterranean climates. Caracals were thought to be rare in Palestine at the time of Tristram (1884). The species was not very rare in the Arabian and Syrian deserts at the time. Schmitz obtained a specimen from Wadi Kelt. This and other specimens (mainly from around the Dead Sea: Dcheier, Khareitun, El Messra, Safje, Wadi Kelt) allowed Matschi (1912) to describe the form *schmitzi*. A skull of a specimen of this series from Wadi Kelt is pictured in Harrison (1968*a*). Harrison (1968*a*) added Khareitun and Eilat and provided a picture of a female specimen with a kitten collected near Amman that were presented to the London Zoo. The species was recorded from Mount Carmel (Aharoni, 1930), Mount Herodes near Bethlehem (Schmitz, 1911), near Eilat (Boden-heimer, 1958), Sede Boqer (Skinner, 1979), and the Hai-Bar Nature Reserve, near Yotvata (Weisbein, 1988). Specimens at HUJ are from Nahal Bosmat (collected 1959), Ramleh, and Beit Guvrin (collected 1960). A specimen collected in 1983 from the Al Hazim area is at the JUNHM (Amr et al., 1987). A specimen was collected at the Shawmari Wildlife Reserve

Fig. 47. The caracal (*Felis caracal*). Photo: A. Shoob.

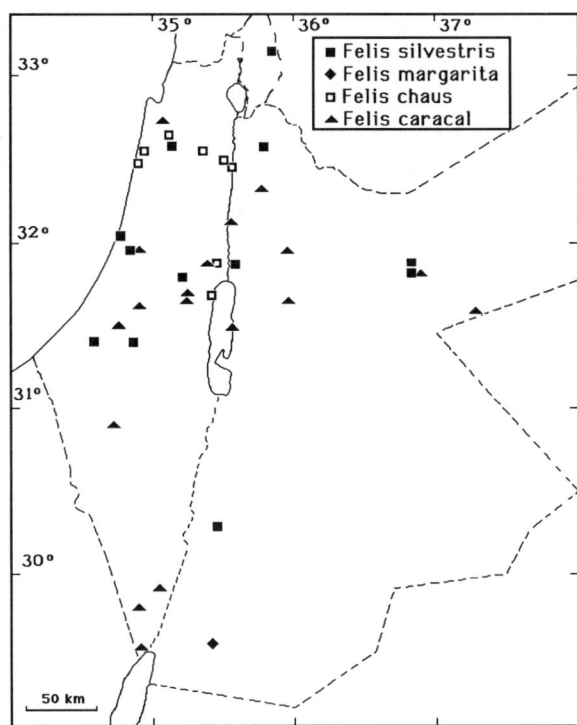

Fig. 48. Locality records of the wild cat (*Felis silvestris*), the jungle cat (*Felis chaus*), the caracal (*Felis caracal*), and the sand cat (*Felis margarita*).

and released in 1985 (Amr et al., 1987). The species also occurs in localities in northern Sinai such as Al Arish (Flower, 1932).

Biology.—The caracal occurs in plains, deserts, and subdesert areas where prey occurs. The principal food items are gazelles, hares, rodents, birds, and reptiles. The caracal is a rare species in the deserts of northern Arabia (western and southern Iraq and eastern Jordan and Syria). Factors affecting its distribution are the availability of large prey species and water (Thalen, 1975). Caracals do not seem to require an extensive shelter, as do other cats in the desert (Prater, 1965).

An introduced female caracal had a home range of 34 km² (Weisbein, 1988). This animal had peaks of activity in the evening and early morning hours. Pregnancy lasts 82 days with two or three (unusually five) young born and hidden in the burrows of other animals or in crevices, hollow trees, thickets, burrows, and caves (Gasperetti et al., 1985; Weisbein, 1988).

Genetics.—2N = 38 was reported for a zoo specimen (Hsu and Arrighi, 1966).

Human interactions.—See also under *F. silvestris*. The name caracal comes from *karkulak* (Turkish) meaning black ears. The Arabic name *'anak el ard*, literally meaning that which is attached to the land, is suitable for an animal that is so well camouflaged that it can barely be distinguished from the background. In fact, some travelers commented that when the animal lays down among sand or sandstone, the only visible parts are the black ear tufts, which are mistaken for grasses and allow the caracal to easily approach its prey. Caracal kittens were raised by locals in certain parts of India and used for hunting.

Weisbein (1988) stated that populations of the caracal increased in the Holy Land after 1964, following the poisoning of canids (especially the jackal) by farmers. Decreased predation by jackals on prey items such as hares and Chukar partridges resulted in a prey population increase. Hare populations also increased following the establishment of more extensive agriculture. According to Weisbein, populations of caracals increased and the range expanded further north and also into the Araba Valley. However, no meaningful census data are available to document these conclusions and the caracal remains endangered in most parts of its range.

Felis chaus   Jungle cat   E

Felis chaus Güldenstädt, 1776. Nova Comm. Acad. Sci. Imp. Petropoli., 20:483. Type from Terek River N of Caucasus, USSR.

Felis chaus furax de Winton, 1898. Ann. Mag. Nat. Hist., ser. 6, 2:293 Type from Jericho, Palestine (based on a specimen collected by Tristram). Valid subspecies.

Lyncus chrysomelanotis Nehring, 1902. Schriften Berl. Ges. Naturf. Fr. Berl., 1902:145. Type from near the Jordan River. Synonym of *F. c. furax*.

Felis rüppelii Brandt, 1832. Bull. Soc. Nat. Moscou, IV:211. Type from Mensaleh, Nile Delta, Egypt (at Senckenberg Museum). Synonym of *F. c. nilotica*.

Diagnosis.—This is a small cat with a grizzled gray color. Adults weigh 6–12 kg with a head and body length of 50–90 cm and a tail length of 23–35 cm. The greatest length of the skull is 115–140 cm. The ear is reddish brown with a black tufted tip. *Felis silvestris* differs from this species in having more distinct body markings including a cheek stripe and in the absence of the black ear tufts.

Range of species.—From Vietnam and Thailand north and west through China, Nepal, India, southwestern Asia, and the eastern Mediterranean to Egypt.

Local status.—During the last century, Tristram (1876) stated that the jungle cat was "not uncommon, especially in jungle and thickets, as by the Jordan." Tristram also collected it in Jericho (1866*b*). However, the localities reported in the Holy Land and habitat studies in Egypt (see below) suggest that this species is limited to marshes, reed beds, and other similar areas in the Holy Land (Fig. 48). The "jungle" cat of the Holy Land was not reported in the thickets of the upper Galilee. Nehring (1902*b*) described *Lychus chrysomelanotis* from Jordan, a synonym of *F. chaus* (type in the BZM), and also reported it from the Jordan River and the Dead Sea areas. Aharoni (1930) claimed that they were "extremely rare" but Bodenheimer (1958) stated that these animals were fairly common and present in all large local collections. However, neither author gave any data. It was also reported from the vicinity of Kfar Ruppin (Harrison, 1968*a*) and Hazorea (Harrison and Bates, 1991). In 1978, I observed this cat at Ain Fashkhah, close to the shore of the northern end of the Dead Sea. Specimens are available at HUJ from Beit Shean (Beisan), Ain Harod, Hephzibah, and Ma'agan Michael (all collected in the 1970s). Kock et al. (1993) reported on a specimen at Yarmouk University collected near Damia Bridge on the east Bank of the Jordan River.

Biology.—In Egypt this animal occurs in areas of low vegetation, especially among reeds and in marshes and feeds on fish and reptiles, especially snakes (Flower, 1932; Osborn and Helmy, 1980). In Iran and Afghanistan, this species preyed on rodents and birds (Lay, 1967; Hassinger, 1973). In the Holy Land, this cat can be seen even during the day, hunting among the tamarisk and other vegetation around water sources. This is especially true in the Jordan Valley. The previously mentioned observation I made at Ain Fashkha was in tall reeds. Three to five young are born in the spring (Flower, 1932; Harrison, 1968*a*).

Genetics.—2N = 38, AA = 68 was reported for specimens from Europe (see Zima and Král, 1984).

Human interactions.—See under *F. silvestris*. The name *chaus* is an ancient name with an obscure origin for the Near Eastern jungle cat.

Felis leo   Lion   X

Felis leo Linnaeus, 1758. Syst. Nat., 10th ed., 1:41. Type from Constantine, Algeria.

The lion formerly was widespread in the central and western Palearctic region and Africa. Presently, there are no lions in the Palearctic region (except for a small pocket in India) and the species has a severely limited distribution in Africa. The lion became extirpated in the Holy Land around the time of the Crusaders according to Tristram (1866*b*, 1884) and perhaps in the 13th century when "the last specimen was hunted at Lejun (near

Megiddo)" (Bodenheimer, 1958). Bodenheimer (1960) mentioned the classical text of Diodorus Siculus as stating that prides of lions were present in the Nabatean plains near 'Aqaba and were so abundant that shepherds constantly had to fight for their flocks' security. The former existence of the lion is also indicated by the frequent (more than 130 times) references to it in the Scriptures; for example: "Therefore a lion from the forest shall slay them" (Jeremiah, 5:6). It is also of interest to note that many Arab families in Palestine, Jordan, Syria, and Lebanon carry the familial name "Asad" or "Saba'," the Arabic names for lion. In Persia, the last specimen was killed in 1942 (Lay, 1967). The name *leo* is Latin for lion.

Felis margarita   Sand cat   E

  Felis margarita Loche, 1858. Rev. Mag. Zool., Paris 10, 2:49, pl. 1. Type from near Negonca, north of Ouargala, Algeria. Extralimital subspecies.

  Felis margarita harrisoni Hemmer, Grubb, and Groves, 1976. Z. Säugetierkd., 41:286–303. Type from Umm as Samim, Oman. Valid subspecies for the Holy Land sand cat.

Diagnosis.—This is a small cat with a distinctive broad face, large ears, green-yellow irises, and overall pale coloration. The head and body length is 40–47 cm and the tail length is 20–30 cm. The greatest length of skull in males is 85–90 mm (smaller in females). Males are relatively larger than females in most measurements. The subspecies *F. m. harrisoni* is supposedly distinct from the nominate subspecies by a relatively broader skull with large bulla, high occiput, and large carnassials, and more sharply marked patterns on the pelage (Hemmer et al., 1976).

Range of species.—The Sahara and Arabian deserts through Iran to Turkestan in the north and Pakistan in the east.

Local status (Fig. 48).—The sand cat was reported observed in Wadi Rum (Mountfort, 1965). A specimen was collected there in March or April of 1977 (Hemmer, 1978). A living specimen from Sinai was at TAUM in the early 1970s (Hemmer et al., 1976). Further study is needed to understand the distribution and status of this cat. The sandy habitat at Wadi Rum is also found in many areas of the southern and eastern regions of the Holy Land.

Biology.—Little is known of the biology of this rare cat in the Holy Land. Anecdotal remarks in Pakistan were reported by Roberts (1977). Sand cats dig burrows for shelter during daytime. They hunt gerbils, geckos, snakes, and other desert animals. Sand cats may breed twice a year in spring and summer.

Genetics.—2N = 38, AA = 68 was reported (Schauenberg, 1974).

Human interactions.—See under *F. silvestris. Felis margarita* was named after Général Margueritte who was on duty in Algeria in the 1850s.

Felis pardus   Leopard, panther   E

Felis pardus Linnaeus, 1758. Syst. Nat., 10th ed., 1:41. Type from Valley of the Nile either in Egypt or the Sudan.

Felis nimr Hemprich and Ehrenberg, 1833. Symb. Phys. Mamm., 2:gg, pl. 17. Type from Syria, Arabia, and Abyssinia, restricted to Arabia by Ellerman and Morrison-Scott (1951) and to mountains in the vicinity of Qunfidah, Asir Province, Saudi Arabia by Harrison (1968a). Valid subspecies in Arabia and the southern Holy Land.

Felis tulliana Valenciennes, 1856. C. R. Acad. Sci., Paris, 421035-1039. Type from Ninfi, 40 km E Smyrna, Turkey. Valid subspecies in the northern Holy Land: *F. p. tulliana.*

Panthera pardus jarvisi Pocock, 1932. Abstr. Proc. Zool. Soc., Lond., no. 347, p. 33. Type from southwestern Sinai. A synonym of *F. p. nimr.*

Diagnosis.—The leopard (Fig. 49) is a large cat with a head and body of about 1 m, and a tail slightly shorter than the head and body length. Adults attain a body weight of 30–40 kg. The head is broad and the ears are short and round. The species is easily diagnosed by the pattern of black and brown spots on a yellowish background. The spots are usually in the form of rosettes of three black spots surrounding a lighter area. These spots fuse on the terminal region of the tail to appear as stripes.

Fig. 49. The leopard (*Felis pardus*), at Ain Gedi. Photo: Y. Odd.

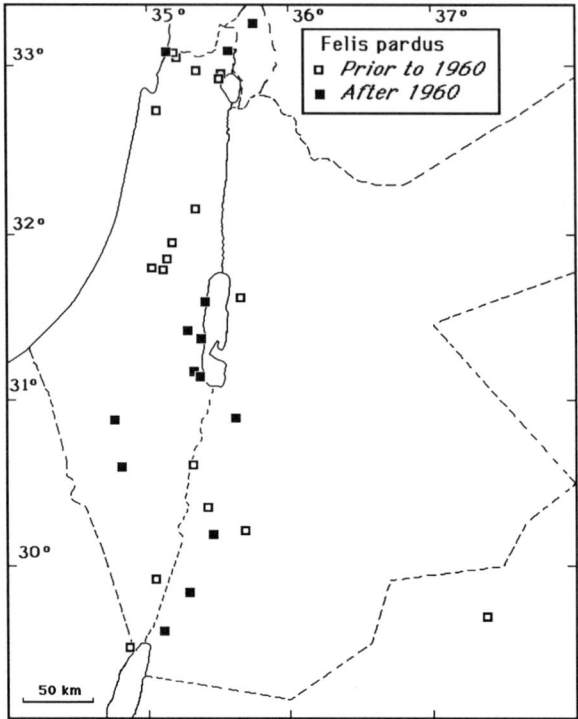

Fig. 50. Known locality records of the leopard (*Felis pardus*). Open squares are records prior to 1960 and solid ones are records after 1960.

Range of species.—From eastern Siberia and Manchuria throughout most of Asia, the Arabian Peninsula, and most of Africa. Range severely reduced by local extinctions.

Local status.—Tristram (1866*b*, 1884) reported leopards as more common than the cheetah, though still scarce. He stated they are: "found all round the Dead Sea, in Gilead and Bashan, and occasionally in the wooded districts of the West. I saw a fine pair which had been killed on Mount Carmel. . . . Its ancient abundance in the Holy Land is testified not only by the numerous allusions in the Scripture, but also by the frequent occurrence of the word Nim'r in the names of places."

The distribution of the leopard decreased significantly due to habitat destruction and hunting. The leopard now is found only in certain areas near the Dead Sea and in the extreme north of the region (Fig. 50). Hart (1885) reported on verbal records from Serbal and Umm Shaumer in Sinai. Tracks were seen by Hart (1891) at Ain el Taba (Sinai) and Ayun Buweirdeh (Wadi Araba). The leopard became rare after the turn of the century but continued to be hunted and was almost exterminated in the first half of the 20th century in the Holy Land.

Ernst Schmitz recorded five leopards shot "near Jerusalem" in the first decade of this century (Hardy, 1947*b*). A specimen I examined at the Schmitz collection in Jerusalem was collected near Emmaus Kubebe on 13 October 1910. A second specimen collected by Schmitz was a female from Khirbet Kasle (20 km W Jerusalem) on 16 November 1911, and is now at the Zoological Museum of Humboldt University (Kumerloeve, 1971). A third Schmitz specimen is at the BZM and was collected on 26 November 1911 from El Ammur, 20 km from Jerusalem (measurements provided in Harrison, 1968*a*).

Aharoni (1930) reported a leopard killed in 1911 in "Zichron Jakob" between Ramallah and Emmaus Kubebe (Qubeiba). Another was killed in 1938 on "the Palestinian–Lebanon frontier" and a cub was secured in 1940 near Safad, west of Lake Galilee (Hardy, 1947*b*). The latter specimen was kept at the Tel Aviv Zoo for several years. Additional specimens mentioned by Hardy (1947*b*) include an individual from near Edon (= Elon?) killed in 1938, a female from the same locality trapped in 1942, a female killed near Safad in 1939, and an individual shot at Ain Ghidyan in 1945. Hardy (1947*b*) stated that the leopard still inhabits the wadi south of Petra as well as occurs in Wadi Zarqa Ma'in. Harrison (1968*a*) reported that C. Seton-Browne shot one in 1934 near "Black Rock," W of Ma'an. Two individuals were reported seen in 1950–1951 near Safad (Rivon, 1957). Observations were made by G. B. Corbet of a leopard on the west coast of the Dead Sea in 1964 (von Lehmann, 1965). Indeed, a leopard was killed by a Bedouin in Wadi Darajah, in the Judean Desert in October 1965 (Blake, 1966). Harrison (1968*a*) cited data he received from W. Ferguson of leopards seen and/or collected at Hanita in 1925 and 1952, Pekin in 1948, Kfar Aruma in 1952, and Ashona in 1956. Clarke (1977, unpub. report) listed the following localities in Jordan based on earlier records: Petra, Wadi Zarqa Ma'in, Ain el Taba, Ain Buweirdeh, and Ma'an. Amr and Disi (1988) reported that one was seen by shepherds in the Tafileh area where it attacked and killed two sheep. The collection at TAUM includes several specimens with the following localities and dates of collection: Safad (1938, died 1955), Akbara (1957), near Wadi Sayad (1964), Wadi Karkara (1965), Masada North (1967), Nahal Ze'elim (W Siyal, 1971—excrement), 10 km ESE Sede Boqer (1973—shot), Ain Boqeq (1986—strychnine poisoning), Nahal Harduf (1988—died of "old age—12 years").

A few individuals are still seen in the mountains of the Judean hills, especially in the areas near Ain Gedi and Nahal Ze'elim (Wadi Siyal) (Ilani, 1988*c*) and recently, an unconfirmed sighting on the road between Kibbutz Yiftah and the Hula Valley (Anonymous, 1988) and another near Makhtesh

Ramon in the Negev (Anonymous, 1983a). Leopards were also reported early in the century from areas in northern Saudi Arabia (for example, Jebel Tubeiq) near the Jordanian border (Raswan, 1935; Harrison, 1968a). Intensive conservation efforts may yet save the few remaining individuals of this magnificent animal.

Biology.—The leopard formerly inhabited all hill and mountain country in the region. It is usually active both day and night but where disturbed by humans, it is nocturnal and secretive (Haltenorth and Diller, 1980). An individual will make a rough noise similar to a hoarse cough and may growl in alarm or threat. The leopard is a solitary animal and males fiercely defend their territories. In the Sinai, it is known to feed on hyraxes and ibex (Murray, 1930) and this is doubtless the case in our region because those species also are present. In Afghanistan, wild goats and sheep are the primary food but occasionally even carnivores such as the wolf are taken (Hassinger, 1973).

The breeding season is probably in the spring. The gestation period is about 100 days. A litter of one to six cubs is born and hidden in crevices between rocks, under thickets, or in small caves. The young suckle for 3 months, become independent after about 2 years, and reach sexual maturity a few months later (Haltenorth and Diller, 1980).

Genetics.—2N = 38, AA = 68 was reported for specimens from zoos (Pathak and Wurster-Hill, 1977).

Human interactions.—Leopards in the region were hunted for millennia for sport and out of fear for man and his domestic animals. Leopards can be seen in the art of Assyrians, Canaanites, Romans, and Umayyed cultures. The subspecies in Sinai, *F. p. jarvisi*, was named after C. S. Jarvis, the British governor of Sinai in the 1930s. The leopard is referred to in the Scriptures: "Like a leopard I will lie in wait by the wayside . . .." (Hosea, 13:7); ". . . a leopard is watching their cities " (Jeremiah, 5:6); "Can the Ethiopian change his skin or the leopard his spots? Then you also can do good who are accustomed to do evil" (Jeremiah, 13:23).

Felis silvestris   Wild cat   V

Felis (Catus) silvestris Schreber, 1777. Die Säugethiere, 3:397. Type from Germany.
Felis libyca tristrami Pocock, 1944. Ann. Mag. Nat. Hist. ser. 11, 11:125. Type from Ghor Seisaban, Moab, Jordan (collected by Tristram). Perhaps a valid subspecies: *F. s. tristrami.*
Felis syriaca Tristram, 1867. Natural history of the Bible, p. 67. Type from Syria. Perhaps an available senior synonym of *F. s. tristrami.*

Fig. 51. The wild cat (*Felis silvestris*). Specimen from Sede Boqer. Photo: H. Mendelssohn.

Diagnosis.—The wild cat (Fig. 51) is a small cat about the size of a house cat. Adults weigh 2.4–4.4 kg and have a head and body length of 45–67 cm. The tail measures 26–36 cm long and the skull length is 89–105 mm. Some authors maintain that the domestic cat and wild cat are the same species (Lay, 1967; Hassinger, 1973). Domestic cats probably originated from the northern African *Felis libyca* (Kratochvil and Kratochvil, 1976; Ragni and Randi, 1986; Randi and Ragni, 1991). However, *libyca* is most likely a subspecies of *F. silvestris* (Haltenorth, 1953). The fur is thick and is a grizzled buff (tawny) to olive-brown with some areas darker than others, giving the appearance of stripes and spots. The ears are rusty reddish brown with indistinct grayish black markings. The soles of the feet are naked. The ears are small and lack a tuft.

Range of species.—British Isles and western Europe south across northern Africa and to Somalia. Eastern Mediterranean region and Arabia to Chinese Turkestan.

Local status.—The status of the wild cat in the Holy Land is uncertain. The few records available suggests a wide distribution (Fig. 48). Tristram (1876) stated that this cat (synonym *F. syriaca*) was very common east of the Jordan River but scarce to the west. Pocock (1944) named Tristram's specimens *Felis libyca tristrami*, with the type collected from Ghor Seisaban in the Moab Mountains. Specimens were collected by Schmitz from Jerusalem (Anonymous, 1946*b*). The species was reported from Tel Abu Hareireh (Hart, 1891), Jaffa, Ramleh (Pocock, 1951), the Galilee (Haltenorth, 1953), 3 km N Azraq Shishan, Shawmari Wildlife Reserve (Atallah, 1966; Amr and Disi, 1988), and Tivon (Harrison, 1968*a*). The species was also recorded from Dareiya (Damascus) (Harrison, 1968*a*) and near Quneitra in the Golan Lahav and Mishmar Ha'Emeq, and at JNHM (collected in 1985) from Kufr Som.

Biology.—This is a nocturnal animal that feeds on small vertebrates, insects, and fruits. In Oman, stomach contents revealed remains of insects, lizards, small mammals, and dates (Harrison, 1968*a*). Wild cats are known to dig their own shelters. In the wild, two to five young are born in a litter. The wild cat is similar in habits to the domestic cat. As in other cats, the female takes care of the young.

Genetics.—2N = 38, AA = 68 was reported for specimens from Europe (see Zima and Král, 1984).

Human interactions.—The derivation of the scientific name is a combination of *feles,* L., genitive *felis,* a cat, and *silvestris,* L., belonging to the woods (from *silva,* a wood), hence meaning the cat of the woods. In formal Arabic, the word for cat is *qit* or *qut.* In colloquial dialects in the Holy Land it is called *hirr* or *hirra.* Wild cats are simply called that, *qit berri* or *senoor* (Hebrew *natol*). The wild cat is the ancestor of our domestic cat. They were domesticated (and worshipped) in ancient Egypt (see discussion under domestic animals).

FAMILY HERPESTIDAE   Mongooses, civets, and genets

These agile animals are ubiquitous and common in the Holy Land. The name Herpestidae is derived from the Greek *herpestes,* a creeper. In the family Herpestidae, the canines are elongate and the second lower incisor is higher than the first and third incisors on each side. These animals are found in Africa, Europe, and southern Asia. There are 36 genera with about 70 species recognized, of which two species occur in the Holy Land.

Records of the European genet (*Genetta genetta*), from the Holy Land are now known to be erroneous. Tristram (1884) reported a genet from the Mount Carmel area. Newmann (1902; Sitzurgsber. Ges. Naturf. Fr. Berl., p. 183) described this specimen as *Genetta terraesanctae.* However, as pointed out by Schlawe (1981), the specimen of Tristram actually originated in Algeria. Aharoni mistakenly stated that he had two specimens of *Genetta genetta terraesanctae* from Sejera (Schedschera) and Wadi Fauar, in the

Fig. 52. The Egyptian mongoose (*Herpestes ichneumon*). Specimen at JNHM. Photo: M. B. Qumsiyeh.

Mount Carmel area (Aharoni, 1930:334). Both records were suggested to be those of *Vormela peregusna* rather than the genet (Kock, 1983). Thus, there are no authenticated specimens of the genet from the Holy Land.

Genus *HERPESTES*   Mongooses

Herpestes ichneumon   Ichneumon, Egyptian mongoose   O

Viverra ichneumon Linnaeus, 1758. Syst. Nat., 10th ed., 1:43. Type from Egypt, "ad ripas Nili."

Ichneumon pharaon Lacépède, 1799. Tabl. Div. Ord. Gen. Mamm., 1799:7. Type from Egypt. Synonym.

Ichneumon aegyptiae Tiederman, 1808. Zoologica, 1:364. Type from Egypt. Synonym.

Ichneumon major É. Geoffroy St.-Hilaire, 1812. Description de L'Egypte Hist. Nat., 2:139 (footnote). Type from Egypt. Synonym.

Diagnosis.—These are medium-sized herpestids weighing 2–4 kg with a head and body length of 45–60 cm, and tail length of 38–48 cm. The skull length is 94–107 mm. The Egyptian mongoose is easily distinguished by its elongated body and by its pelage (Fig. 52). The pelage is rough and the color is grizzled gray or olive brown/black/cream. The tail is long and tapering at the end with a black tip. The ears are short and rounded. As in most members of the family Herpestidae, males are significantly larger than females in most measurements.

Range of species.—In Mediterranean and sub-Mediterranean habitats from Portugal and Spain through northern Africa, the eastern Mediterranean to Turkey, and extending south through most of Africa.

Local status.—This is a common species in the coasts, hills, and the Jordan Valley (Fig. 53). It is also common in disturbed habitats throughout the

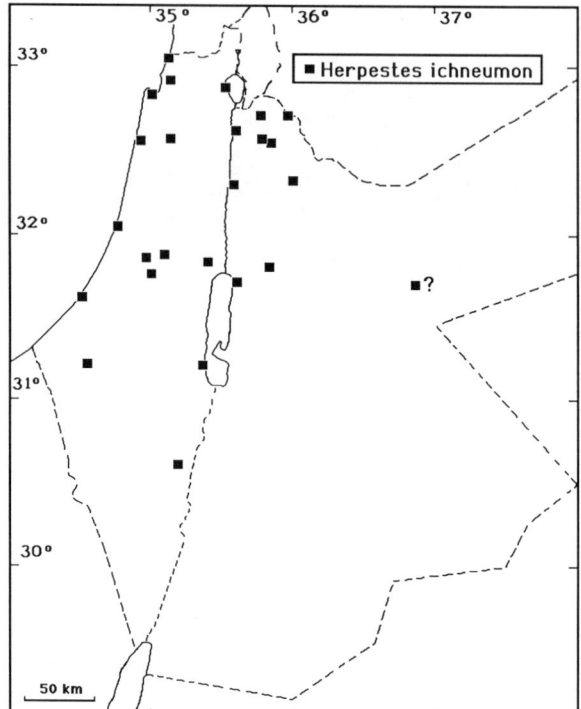

Fig. 53. Locality
records for the
Egyptian mongoose
(*Herpestes ichneumon*).

northern regions of the area. It appears to have adapted to man's agricultural
activities, as mentioned by Tristram (1866*b*, 1876). Tristram reported it
from Wadi Kurn and the Plain of Acre. Bodenheimer (1958) also commented on its widespread distribution and stated that it is known as a
"predator of fowl houses."

This species was reported from Latrun (Anonymous, 1946*b*), Tabgha,
Mishmar Ha'Emeq, Kasr Hadschla, Bir Salem, Beit Anan, Haifa (Harrison,
1968*a*, 1972), Ma'agan Michael (Ben-Yaacov and Yom-Tov, 1983), Maqarim
Dam Station, Shuna Ash Shamaliya (visual records) (Amr et al., 1987),
Jordan Valley, and Ain Ghazal (specimens at JUNHM; Amr and Disi,
1988). Specimens at JNHM are from Kuraimah, Turabah, near Kufr Som,
and Aqraba. Ilani (1988*a*) reported on specimens from Zo'ar, Nahal Zeelim,
and Ain Yahav. I have also seen it frequently on several field trips between
1975 and 1982 in many areas of the western highlands. Many specimens are
killed on the road by vehicles in areas near Jerusalem (for example, I have a
specimen from 1.4 km N Beit Shemesh).

Biology.—This species appears to have become more common around agricultural areas in the Holy Land. It is perhaps the only carnivore that has increased rather than decreased in numbers because of human activities. It occurs in many areas, especially around water. This is a diurnal animal that feeds on rodents, birds, reptiles, amphibians, fish, and various invertebrates (insects, molluscs, crustaceans). Its biology in the Holy Land has been recently reviewed (Ben-Yaacov and Yom-Tov, 1983). In Egypt, litters have been found in February, May, July, September, and October (Flower, 1932). A lactating female killed by a car was obtained near Jerusalem on 25 August. The gestation period is reported to be 60 days with one to four (average 3.3) young in a litter (Ben-Yaacov and Yom-Tov, 1983). Weaning occurs at 1–2 months of age.

Genetics.—2N = 43 in males, 44 in females, AA = 64. This is due to X-autosome translocations resulting in an X1X1X2X2/X1X1Y sex-determining mechanism. This karyotype was reported by Fredga (1972) for individuals of unknown origin (Zima and Král, 1984).

Human interactions.—The scientific name of the species is descriptive of its habits: *Herpestes* is Greek, meaning a creeper and *ikhneumon* (Greek) means the tracker. Respect for the mongoose is well documented in studies of ancient Egypt (Anderson and de Winton, 1902); this respect was most likely gained because mongooses fed on the eggs of the feared Nile crocodiles. The mongoose is well known in Jordan (where it is called *zerdi* or *nims*) and is respected as a snake killer. I have seen many specimens mounted in a combat posture with snakes sold as decorative items in Amman.

FAMILY HYAENIDAE   Hyenas, aardwolves

Genus *HYAENA*   Hyenas

Hyenas are carnivores with unspecialized incisors, powerful canines, well-developed carnassials, and large and strong molars (Fig. 54). The dental formula for the hyena is i 3/3, c 1/1, p 4/3, m 1/1 = 34. The hind limbs are shorter than the forelimbs. There are only four species in the family and they range from Africa through southwestern Asia to India. The striped hyena (*Hyaena hyaena*) is the only contemporary hyena in the Holy Land. The spotted hyena, genus *Crocuta*, is reported to occur in Pleistocene fossils in Palestine.

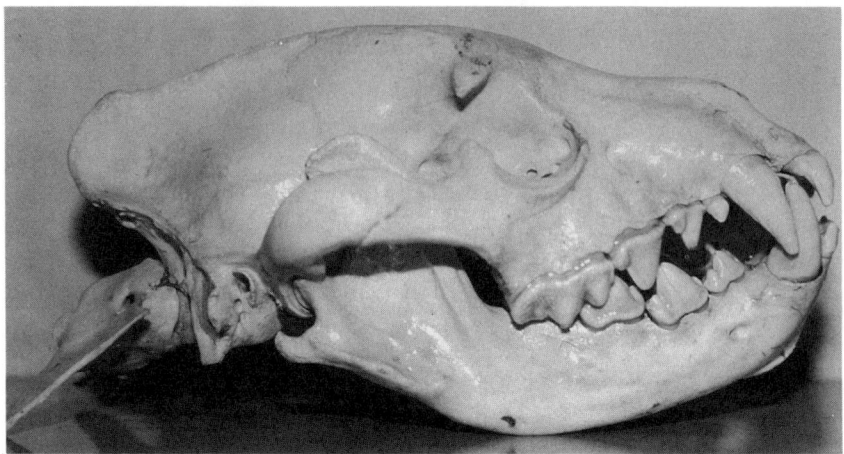

Fig. 54. Skull of the striped hyena (*Hyaena hyaena*).

Hyaena hyaena   Striped hyena   V

Canis hyaena Linnaeus, 1758. Syst. Nat., 10th ed., 1:40. Type from Benna Mountains, Laristan, Iran.

Hyaena dubbah Meyer, 1793. Uebersicht der Entdeckungen in Neu-Holland und Africa, p. 94. Type from Atbara, Sudan. Valid subspecies: *H. h. dubbah*. Perhaps enters Palestine from the Sinai.

Hyaena syriaca Matschie, 1900. Sitzungsber. Ges. Naturf. Fr. Berl., 1900:54. Type from Antioch, Syria. Valid subspecies in the Holy Land: *H. h. syriaca.*

Diagnosis.—Hyenas are distinctive carnivores of medium size. The weight of adults can be 20–50 kg; head and body length is 85–130 cm; tail length is 25–40 cm, and the greatest length of the skull is 23–26.5 cm. They are similar to dogs in facial aspects. The body is, however, very different, with a strong, thick neck and strong forelegs relative to hind legs (Fig. 55). This gives the posture of weak hindquarters, which earlier observers described as dragging its hind quarters. The color is also distinctive in being gray-whitish with long black stripes (hence the name striped hyena). In summer months, the stripes are more visible because the hairs are shorter. The dorsal region has a mane of elongated hairs. This gives the animal a more formidable appearance, especially when the hairs are erected. The fur is coarse and shows distinct stripes along the flanks that break up into spots on the legs. Females have two pairs of nipples. Perhaps two subspecies occur in Palestine if the smaller *H. h. dubbah* of Egypt enters Sinai and the Negev. *Hyaena hyaena syriaca* has a more elaborate crest and tail hairs and the stripes on the flanks and legs number about 10 (Tohme and Tohme, 1983). Skinner and Ilani

Fig. 55. The striped hyena (*Hyaena hyaena*). Photo: D. M. Shafi.

(1981) compared the skeletal morphology of the brown hyena with the striped hyena.

Range of species.—From Nepal through India, southwestern Asia, southern USSR, eastern Mediterranean, and northern Africa. It also occurs in savanna regions of Africa south to Tanzania.

Local status.—The hyena appears to be found in all major habitats in the Holy Land including coastal regions, uplands, the Jordan Valley, and the deserts (Fig. 56). Tristram (1866*b*, 1867, 1884) obtained specimens on Mount Carmel, Nazareth, and Tabor, which today are heavily populated by humans and where few hyenas, if any, remain. Pocock (1934) refers all eastern Mediterranean forms to *H. syriaca* (= *H. h. syriaca*). Bodenheimer (1958) reported that hyenas were widespread in Palestine and included a photograph of one obtained at Elbiran (= El Bira). Older specimens were reported from Bab el Wad, Tel El Shorier, Beersheba, Mount Carmel, Capernaum, Wadi Nimr (near Wadi Kelt), Safje, Wadi Swenit, and Deir Dosi (near Jerusalem) (Harrison, 1968*a*). Hyenas were reported at Bethlehem in the 1940s (Anonymous, 1945), Jebel Uweinid, between Ma'an and Hasa, Azraq, and Shawmari in the 1960s (Atallah, 1967*a*, 1967*b*), and at Jawa in the early 1980s (Searight, 1987*b*). Specimens at JUNHM are from Wadi Araba (1984) and Qatrana (1983). Other records are from Jebel Musa

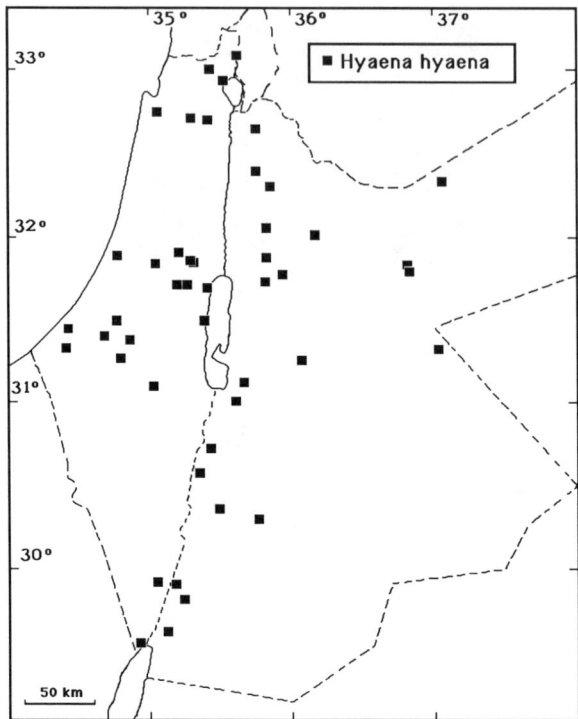

Fig. 56. Distribution of the striped hyena (*Hyaena hyaena*).

(Palmer, 1871), and Gaza (Howells, 1956). Specimens at the JUNHM were collected at Azraq in 1969, Qatrana in 1982, Wadi Finan in 1981, Jubeiha in 1982, Na'ur, Suweilih in 1988, Al Kastal in 1987, and West Madaba in 1987 (Amr and Disi, 1988). Specimens at JNHM were collected in 22 km S Irbid (1983), Madaba (1987), Ishtafayna (near Ajlun, 1980s), and Fo'ara (observed by Darwish Al Shafi'i in 1988). Specimens at TAUM are from many localities in Israel and the occupied territories including Safad, Eilat, Yotvata, Ain Gedi, Ain Fashkhah, Lahav, Mount Meron, Neot Hakikar, Nir'Oz, Rehovot, Hula, and Dimona.

Biology.—Striped hyenas eat carcasses of almost any medium-sized to large mammals, including dead hyenas. Their very powerful jaws enable them to crack large bones. They have a habit of bringing food to their den, which thus is usually surrounded by large piles of bones. They are also opportunistic feeders; their food includes not only carrion and small animals but also large amounts of some fruits and vegetables. They also will occasionally kill large prey if it is accessible, even domestic animals or young of large herbivores (Bird, 1946; Osborn and Helmy, 1980).

The striped hyena is a social animal that protects its territory and has a complicated system of communication. Territorial boundaries are marked by secretions from glands situated around the anus. Ilani (1975) and Bouskila (1985) made an interesting observation on the biology of hyenas near feeding stations. It appears that pregnant females dominate in certain feeding situations. The anal glands play a role in individual recognition.

Males and females have superficially similar external genitalia because of the extremely enlarged clitoris of the female. Copulation usually occurs in the early spring. Litter size is one to five (usually two to four), born after about a 90-day gestation (Rieger, 1979, 1981). Zoo animals were reported to live up to 24 years.

Genetics.—2N = 40 was reported by Hsu and Arrighi (1966) from a zoo animal.

Human interactions.—The name hyena is from *huaina,* a Greek name for this animal (derived from *hus,* meaning a hog because of the mane on the hyena). Hyenas may stray near villages and agricultural areas in search for food and water. In natural habitats, hyenas must have access to water. Bird (1946) shot hyenas in Palestine when they came for a staked-out goat carcass. Atallah (1967*b*) reported the bones of dogs, camels, donkeys, goats, sheep, and large birds at the entrance of hyena dens in the Azraq area. Hyenas are also known to dig up freshly buried corpses in graveyards in the region (Tristram, 1884). There is a well-established myth in the area that hyenas will spray their urine on isolated humans who will then become drugged and either will follow or be dragged by the hyena to its den to be devoured. Many other myths persist about these animals among the locals, including the idea that the hyena changes its sex.

FAMILY MUSTELIDAE   Badgers, otters, skunks, weasels

In mustelids, the facial region of the skull is shortened. The legs are usually short in relation to the length of the body. Mustelids are distributed worldwide with about 23 genera and 63 recent species.

Key to the Mustelidae of the Holy Land

1. Feet webbed. Adapted for swimming. Total number of
    teeth 36. . . . . *Lutra lutra*

    Feet not webbed. Terrestrial. Total number of teeth 32, 34,
    or 38. . . . . , , ... 2

2. Total number of teeth 34. Smaller animals with skull length
   less than 60 mm.   3

   Total number of teeth 32 or 38. Larger animals with skull length
   greater than 70 mm. 4

3. Pelage with patches of brown and yellow giving a marbled
   appearance. Tail length more than 150 mm. Pterygoid process
   connected with bulla. . . . . *Vormela peregusna*

   Pelage uniform brown. Tail length less than 100 mm. Pterygoid
   process not reaching bulla. . . . . *Mustela nivalis*

4. Total number of teeth 32. . . . . *Mellivora capensis*

   Total number of teeth 38. 5

5. Pelt predominantly gray on body with contrasting black and
   white pattern on head and legs. Skull length greater than
   100 mm. Tail short (usually less than one-third head and body
   length). . . . . *Meles meles*

   Pelt predominantly brown in color. Skull length less than 90 mm.
   Tail longer (usually at least one-half head and body length).
   . . . . *Martes foina*

Fig. 57. The common otter (*Lutra lutra*). Specimen from Acre. Photo: H. Mendelssohn.

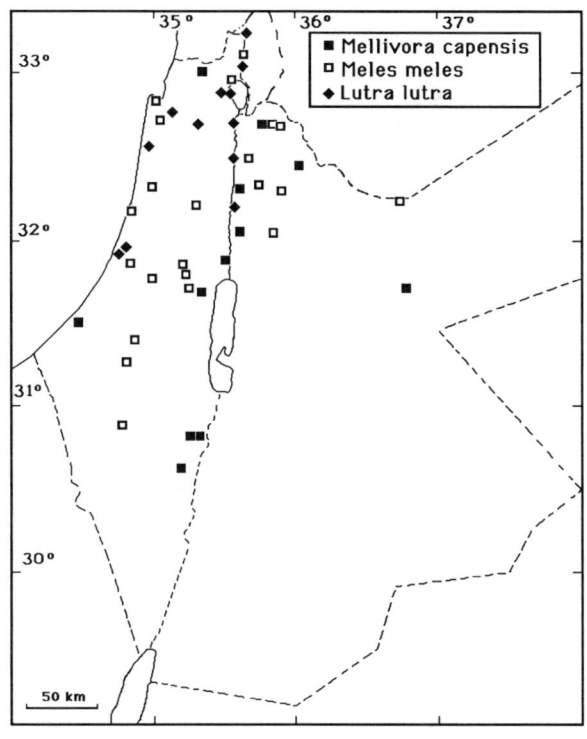

Fig. 58. Locality records of the honey badger (*Mellivora capensis*), the common badger (*Meles meles*), and the common otter (*Lutra lutra*).

Genus *LUTRA*   Otters

Lutra lutra   Common otter   V

Mustela lutra Linnaeus, 1758. Syst. Nat., 10th ed.,1:45. Type from Upsala, Sweden.
Lutra lutra seistanica Birula, 1912. Annu. Mus. Zool. Acad. Sci. St. Petersb., 17:274. Type from Gilmend (Helmand?) River, Seistan, Iran. Valid subspecies.

Diagnosis.—The otter is easily distinguished by its aquatic adaptations: the legs are short with webbed feet and the body is torpedo shape (Fig. 57). The tail is muscular, thick at the base, and tapers at the end, which helps achieve a body shape offering the least water resistance. The fur is short and dense with a heavy undercoat that is waterproof. The skull appears flattened dorsally (especially in lateral view). Otters weigh 5–12 kg, with a head and body length of 50–80 cm and a tail length of 29–60 cm. The greatest length of the skull is 105–125 mm. Males are larger and heavier than females.

Range of species.—Europe, the Palearctic regions of Asia, Morocco, Algeria, the eastern Mediterranean region, and Iran.

Local status.—The otter is found near lakes and rivers in the Holy Land (Fig. 58). Tristram (1866*b*, 1884) reported it from the shores of Lake of

Galilee (Lake Tiberias) "where it has abundant foods." Aharoni (1930) found it in streams and estuaries in the Jordan Valley and other areas (for example, Nazareth, Rishon le Zion, Nahr Rubin). Most other recent records in the Holy Land are from the Lake Hula Basin in the north. Localities reported include Tabgha, Hulata, Dafna, Ramat Yohanan, May'an Zevi, and Bet Zera (Harrison, 1968a).

Biology.—Otters have playful habits that may be directly or indirectly related to foraging for food. They are graceful animals and agile swimmers. Food consists mainly of fish obtained in rivers, small lakes, and streams. They are also known to take poultry, eggs, frogs, and small rodents (Harrison, 1968a). Otters are mainly nocturnal and hide in shelters dug in the banks of rivers or in small islands. The entrances to these shelters are frequently submerged. In Armenia, two to five young are born in May (Dahl, 1954) and in Pakistan young are born in the spring or early summer (Roberts, 1977). These otters have delayed implantation and thus the gestation period is long (more than 280 days).

Genetics.—2N = 38, AA = 60 was reported for a male specimen from Germany (Günther and Gebauer, 1982).

Human interactions.—*Lutra* is Latin for otter. Bodenheimer (1958) stated that the otter is increasing in numbers because it is thriving on fish ponds in Palestine. However, according to a study by B. Shalmon, S. M. MacDonald, and C. F. Mason, otters were exterminated in the coastal plains and remain only in the Jordan Valley and the Golan Heights (Shalmon et al., 1986; Ilani, 1988b).

Genus *MARTES*   Martens

Martes foina   Syrian marten, rock marten, stone marten   **R**

Mustela foina Erxleben, 1777. Syst. Regn. Anim., 1:458. Type from Germany.
Martes foina syriaca Nehring, 1902. Sitzungsber. Ges. Naturf. Fr. Berl., 1902:145. Type from Wadi Sir or Syr, Jordan (specimen is at the Zoological Museum in Berlin). Valid subspecies.

Diagnosis.—The rock marten (Fig. 59) is a medium-sized mustelid with a long tail. The head and body length is 34–50 cm, the tail length is 17–26 cm, and the skull length is 77–88 mm. Adult males weigh between 700 and 1600 g. The color is uniform chocolate brown except for a lighter (almost white) throat patch. The skull is elongated with a smooth and slender build.

Fig. 59. The rock or stone marten (*Martes foina*). Mounted animal at the JNHM. Photo: M. B. Qumsiyeh.

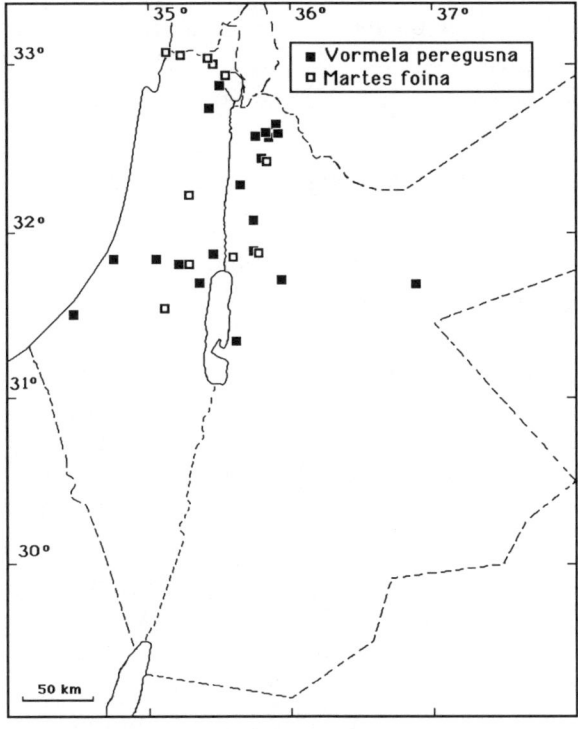

Fig. 60. Locality records of the marbled polecat (*Vormela peregusna*) and the rock or stone marten (*Martes foina*).

Range of species.—Southern and central Europe, Turkey east to the Himalayas and south to the Holy Land.

Local status.—The distribution of the rock marten in the Holy Land is not well studied (Fig. 60). Tristram (1876) obtained a specimen at Beirut and stated that it occurs in the "Taurid." Aharoni (1930) reported that it was common in Jerusalem, Hebron, Nablus, and villages in the Palestinian hillside. Bodenheimer (1958) stated that "about 1929 fresh skins were sold by local fellahin (villagers) in the streets of Jerusalem." He stated that it was rare at the time of his publication (1958) and was just reported from Eilon. The type specimen of *syriaca* came from Wadi Syr (Wadi es Sir), and another specimen was also reported from Wadi Kafrein (Nehring, 1902*a*). A specimen at the JUNHM with little data is said to come from north Jordan (Amr and Disi, 1988). Another specimen from Al Mazar is at the JNHM. The HUJ and TAUM have specimens from many localities west of the Rift Valley including Wadi Karkara, near Sasa, Rosh Pinna, and Mount Meiron. I observed a foraging specimen 1 hour after dark in the area of Wadi As Sir.

Biology.—Little is known of the biology of this species in the region. It is more common at higher altitudes and apparently feeds on small vertebrates (Lewis et al., 1968). In Armenia three to six young are born in May (Dahl, 1954).

Genetics.—2N = 38 was reported from Europe (Renzoni, 1970).

Human interactions.—The range in the Holy Land is probably expanding. The name *Martes* is the classical Latin name for this animal. The name *foina* is derived from an Italian dialect for polecat.

Genus *MELES*   Common badgers

Meles meles   Common badger   V

Ursus meles Linnaeus, 1758. Syst. Nat., 10th ed.,1:48. Type from Upsala, Sweden.
Meles canescens Blanford, 1875. Ann. Mag. Nat. Hist., ser. 4, 16:310. Type from Abadeh, between Shiraz and Isfahan, Iran.

Diagnosis.—Badgers are heavily built mustelids (Fig. 61). Males are significantly larger than females. Adult females weigh 5–10 kg and adult males weigh 7–14 kg. The head and body length is 55–85 cm, the tail length is 10–20 cm, and the skull measures 115–140 mm in total length. The feet are short, with heavy muscles and long claws adapted for burrowing. The tail is short and stubby, measuring between 10 and 15 cm. The fur is coarse and rough with little underfur. The color is distinctive with two longitudinal black stripes beginning at the muzzle and broadening posteriorly. A median white stripe provides a contrasting pattern on the head. The back is variable in color (from brown to buff gray), but the sides and feet are usually black. Harrison and Bates (1991:134) regard the Holy Land form as *M. m. canescens*, distinguished from *M. m. meles* by "average smaller size and by the shape of m1 which is characteristically more elongated and narrow."

Range of species.—Most wooded areas of the Palearctic region south to Palestine, Iraq, and Iran.

Local status.—The common badger is known from the northern regions of the Holy Land (Fig. 58). Tristram (1866*b*, 1884) stated that this species was common in all the hilly and wooded parts of Palestine but not in the Jordan Valley. Aharoni (1930) mentions its presence in mountain regions and suggested that the numbers fluctuate. More specific locality records are available from Mount Gerizim, near Nablus (Tristram, 1866*a*), Lake Hula (Anonymous, 1945), Nebi Samwili (Anonymous, 1946*b*), Jerusalem, Mount Carmel, Haifa (Harrison, 1968*a*, 1972), Suweileh (Clarke, 1977, unpubl. report), and Aqraba (Amr et al., 1987). Specimens at JUNHM were collected at Harta, and Irbid. Other specimens at JNHM are from Jerash, Mansura, and Al Munaysah. I obtained a specimen near Beit Sahur. Specimens at TAUM are available from many localities in Israel and the occupied territories including Ekron, Hartuv, Rosh Pinna, Herzelia, Beer-sheba, Lahav, Sede Boqer, and 6 km N Tulkarm.

Fig. 61. The common badger (*Meles meles*). Specimen at JNHM. Photo: M. B. Qumsiyeh.

Biology.—The following observations are of badgers in Lebanon taken from Lewis et al. (1968). Badgers lived near cultivated areas, building extensive burrow systems to depths of 1 m or more. Digging is done with the strong front legs, which are equipped with sizeable claws. The badger is primarily nocturnal and feeds on rodents, reptiles, and insects with some intake of plant material (fruits, buds, etc.). Two to four young are born in March or April and remain with their parents until the fall. Similar observations on litter number and breeding season were reported in Armenia (Dahl, 1954).

Genetics.—2N = 44 was reported for specimens from Europe and Japan (review in Zima and Král, 1984).

Human interactions.—The name *meles* is Latin for badger.

Genus *MELLIVORA*   Honey badgers

Mellivora capensis   Honey badger, ratel   E

Viverra capensis Schreber, 1776. Die Säugethiere, pl. 125 (1777), text pp. 450, 588. Type from Cape of Good Hope, South Africa.

Mellivora wilsoni Cheesman, 1920. J. Bombay Nat. Hist. Soc., 27:335. Type from Ram Hormuz, southwestern Iran. Valid subspecies for the Holy Land: *M. c. wilsoni.*

Diagnosis.—Honey badgers (Fig. 62) have a head and body length of 60–75 cm and a tail length of 11–18 cm. The greatest length of the skull measures 120–135 mm. Males are larger and heavier than females with adult weights of 4–10 kg for females and 6–12 kg for males. Like the common badger (*Meles meles*), the honey badger is also heavily built with short, strong legs. The coloration of the honey badger is very distinctive: the ventral coloration is black and is well separated from the dorsal coloration, which is white or light gray. The light areas extend from the dorsal regions of the head to the neck, the back, and the tail.

Range of species.—Semiarid regions from Nepal and India through southwestern Asia to Arabia. Most of the dry savannah and semiarid regions of Africa.

Local status.—The honey badger is probably more widespread than the available records indicate (Fig. 58). Aharoni (1930) mentions these animals as rare in the hills west of the Jordan River and "more common to the north" as one approaches the Anti-Lebanon mountains. The species was recorded from the hills between Bethlehem and the Dead Sea, the lower Jordan Valley, Gaza, Ain Hussub, and at Umm Falik (Bodenheimer, 1935, 1958).

Fig. 62. The honey badger (*Mellivora capensis*). Photo H. Mendelssohn.

A specimen at JUNHM was collected at Aqraba in 1983 (Amr and Disi, 1988), and another at JNHM is from Marsa'. Specimens at TAUM include those from Ain Hussub and Ain Yahav. The honey badger was also reported from Badanah in northern Saudi Arabia (31°N, 41°E) close to the Jordanian border and in an area with habitat similar to that of the deserts of eastern Jordan (Lewis and Atallah, 1966).

Biology.—Little is known of the biology of the species in the region. Animals were reported in sparsely vegetated areas feeding on lizards (Harrison, 1968a). Honey badgers were reported to be nocturnal but occasional individuals are seen during the day and they feed on rodents, birds, insects, and occasionally vegetable matter and snakes (Blanford, 1888–1891; Pocock, 1941). Harrison (1968a) quoted Doughty (1888) as stating that Bedouin told him that the primary enemy of the desert spiny lizard or *thab* (*Uromastix*), is the honey badger. Honey badgers appear to have good climbing ability in trees and on rocks (Blanford, 1888–1891). This species emits a very powerful smell to repulse any attackers. This smell and the black and white coloration are reminiscent of those of the skunks of North America, which are believed to have the color pattern as a warning to would-be aggressors. Probably because of this adaptation, honey badgers are occasionally seen not to fear humans but are on the contrary aggressive and daring. Two young are born after a gestation period of about 180 days

(Nowak and Paradiso, 1983). In Russia, the young are usually born in April and May (Ognev, 1935).

Human interactions.—It is believed by Bedouin in Arabia and by locals in India that honey badgers will dig up and eat human corpses (Cheesman, 1926; Pocock, 1941) but this has never been documented. *Mellivora* is derived from *mel,* L., genitive *mellis,* honey, and *voro,* L., devour (honey-eater). The name *capensis* refers to the Cape of South Africa. Bedouin call this animal *thurban* and villagers call it *al 'oqseh.*

Genus *MUSTELA*   Weasels

Sixteen species are known in this genus, of which one, *M. nivalis,* was reported in the Holy Land (perhaps locally extinct). The dental formula is i 3/3, c 1/1, p 3/3, m 1/2 = 34. The word *Mustela* is Latin for weasel.

Mustela nivalis   Least weasel   ?X

Mustela nivalis Linnaeus, 1756. Syst. Nat., 12th ed., 1:69. Type from Vesterboten, Sweden.

Mustela subpalmata Hemprich and Ehrenberg, 1833. Symb. Phys. Mamm., 3:folio k, p. 2. Type from "Houses of Cairo and Alexandria," Egypt. Possibly a valid subspecies: *M. n. subpalmata.*

Diagnosis.—The least weasel is the smallest mustelid in the Holy Land (head and body length about 250–300 mm) with a relatively short tail (less than one-third head and body length). The body is slender and cylindrical with short limbs. The color is brown above and whitish below.

Range of species.—The Palearctic region except Arabia, Ireland, Iceland, and the Arctic. Perhaps the form in North America (*rixosa*) is also conspecific (Corbet, 1978).

Local status.—The status of this animal in the Holy Land is not clear. Zoologists (Aharoni, 1930; Bodenheimer, 1958) of the first half of this century failed to confirm Tristram's listing of this species (as *M. boccamela*) as a member of the Palestinian fauna. *Mustela nivalis* is reported from Holocene fossils (11,000 to about 5000 years before present) from Jericho, Beersheba, and the Galilee (Tchernov, 1988). It probably became extirpated in the Holy Land due to increasing aridity. However, relict populations survived around the Nile Valley in northern Egypt (Osborn and Helmy, 1980), and two specimens are known from Lebanon (Harrison and Lewis, 1964). Thus, a population perhaps still survives in the Holy Land. Indeed, Harrison and Lewis (1964) reported undocumented skins in the collection of Salah Merrill, who made most of this collection while an American consul in Jerusalem.

Biology.—In Egypt, this species appeared to be more commensal than feral and was mostly obtained around human habitations and near cultivated areas (Setzer, 1958*a*). Harrison (1968*a*) stated that the weasel is a diurnal hunter feeding on small rodents, birds, reptiles, and amphibians. Stomach contents in Egypt included insects, small birds, and fish (Osborn and Helmy, 1980). A litter of five was noted born in December in Egypt (Flower, 1932). Flower also remarked that in Egypt, these animals frequented clubs, restaurants, homes, and other buildings. Such habitat choice was not seen in Egypt later by Osborn and Helmy (1980) or in other parts of the range. In Armenia, Dahl (1954) reported three to nine young born in the late spring and summer.

Genetics.—2N = 42 was reported for specimens from Europe (review in Zima and Král, 1984).

Human interactions.—The name *nivalis* is derived from *nix*, L., genitive *nivis*, snow. Hence, also, the common name snow weasel.

Genus *VORMELA*   Polecats

Vormela peregusna   Marbled polecat   R

Mustela peregusna Güldenstädt, 1770. Nova Comm. Acad. Sci. Imp. Petropoli., 14(1):441. Type from banks of the Don River in southern Ukrainian SSR.

Vormela peregusna syriaca Pocock, 1936. Proc. Zool. Soc. Lond., 1936:720. Type from near Lake Tiberias (Palestine). Valid subspecies.

Diagnosis.—The marbled polecat is a very small and colorful carnivore (Fig. 63). The weight in adults is less than 0.5 kg. The head and body length is 260–340 mm and the tail measures 160–200 mm. The skull length is 46–56 mm. These animals have variegated brown and yellow patches and stripes of irregular shape on the back. The head is dark brown with a white band across the forehead.

Range of species.—The northern Arabian Peninsula to southeastern Europe and east to western China.

Local status.—The marbled polecat is found in diverse habitats in the Holy Land (Fig. 60). Tristram (1884) reported *Genetta vulgaris* from Mount Carmel. Aharoni (1930:334) noted that "*Viverra genetta*" was very rare and that in 29 years of collecting he only had two specimens from Schedschera (Sejera, Galilee) and Wadi Fauar (= Ain el Fawwar). These records were

Fig. 63. Marbled polecat (*Vormela peregusna*), photo by D. M. Shafi.

suggested to be those of *Vormela peregusna* rather than the genet (Kock, 1983). This is partially based on the observation that Aharoni actually based his record from the locality of Wadi Fauar on the observations of Schmitz made in 1908. According to Schlawe (1981) and Kock (1983), this locality lies near the *lisan* of the Dead Sea, the mouth of the wadi referred to now as Wadi Jarra. However, there is at least one other Wadi Al Fawwar locality in the Holy Land and it is not entirely clear where the specimen came from. In addition to describing the type of *V. p. syriaca* from Tiberias, Pocock (1936) also mentioned specimens from the Jordan Valley and Gaza at BM. Records are also available from Gedera (Bodenheimer, 1935), and specimens at the BZM are from Mar Saba, Es Salt, Jerusalem, Bir Zeit, Jericho, and Bab el Wad (Harrison, 1968a). Specimens at HUJ are from Jerusalem, near Acre, Oranim, Rehovot, Tantura, and Wadi Bira. The species is common in northern Jordan; specimens at the JNHM are from Al Barha, al Mazar Al Shamali, Al Tayyibah, Harima, Kafr Ja'iz, Irbid, Salt, and N of Amman Airport; and those at the JUNHM are from Wadi As Sir and Umm Al Hiran (collected in 1984).

Biology.—Behavior of captive animals was described by Akhtar (1945). In Lebanon, the marbled polecat lived near cultivated areas and a captive specimen killed and ate *Acomys* (Lewis et al., 1968). Like many mustellids, the marbled polecat produces an offensive smell. The striking (and beautiful) coloration may act as a warning for other animals. Howells (1956) describes the attempted preying of a hyena on a polecat and how this warning and smell may ward off potential predators. Marbled polecats appear to be well adapted for digging (Roberts, 1977). In Baluchistan, they feed on Libyan jirds (*Meriones libycus*), hamsters, and house mice. In Armenia four to eight young are born in April and May after a gestation period of 8 weeks (Ognev, 1935; Dahl, 1954).

Genetics.—2N = 38, AA = 70 was reported from Derra District, Syria (Peshev and Al Hossein, 1989) in agreement with European specimens (Zima and Král, 1984).

Human interactions.—The name *peregusna* is Latinized from *pereguznya,* Ukranian for polecat.

FAMILY URSIDAE   Bears

Bears, like canids, have unspecialized incisors and long powerful canines. However, bears have lost the carnassial (shearing teeth) adaptation. Instead, all premolars and molars have tuberculated low crowns. This adaptation allows them to have an omnivorous diet. The dental formula is i 2–3/3, c 1/1, p 4/4, m 2/3 = 40-42. The Ursidae is a small family of eight species, of which one species formerly inhabited the Holy Land. *Ursus* is the classical Latin name for bear.

Genus *URSUS*   Bears

Ursus arctos   Syrian brown bear   X

Ursus arctos Linnaeus, 1758. Syst. Nat., 10th ed., 1:47. Type from Sweden.
Ursus syriacus Hemprich and Ehrenberg, 1828. Symb. Phys. Mamm., 1:sig. a, pl. 1. Type from near Bischerre, Mount Makmel, Lebanon. Valid subspecies: *U. a. syriacus.*
Ursus schmitzi Matschi, 1917. Sitzüngsber. Ges. Naturf. Fr. Berl., p. 33. Type from Mount Hermon, Palestine. Synonym.

Diagnosis.—Bears are easily distinguished from other carnivores of the region by their large size, stocky build, and short tail. See under family.

Range of species.—Formerly widespread in the Palearctic and Nearctic regions. The Syrian bear is now absent from most of its previous range.

Local status.—The bear is referred to frequently in the Scriptures "in the days when the Judean hills were still clothed with wood, and the primeval forests crowned the rugged heights of the Galilee" (Tristram, 1884). Tristram (1866*b*) encountered it near Gennesaret and stated that it was still seen on Mount Hermon. Tristram further reported (1884) that "the bear has become very rare in Palestine, though still not uncommon on Hermon and the wooded parts of Lebanon." However, he actually only saw it once in Galilee at Wadi Hammam, near Arbel. Carruthers (1909) described the habits of this bear on Mount Hermon in Palestine and the Anti-Lebanon Mountains. Schmitz (1912:174) cited a killing on Mount Hermon close to the Palestinian border, perhaps the last authentic record of the bear in this region (Aharoni, 1930; Bodenheimer, 1958).

Biology.—In Iran, bears fed on wild fruits and each female gave birth to two young (Lay, 1967). Little is known of the biology of the eastern Mediterranean subspecies, which became extinct at the turn of the century. Lewis et al. (1968:527) quoted the unpublished diary of Carruthers (September 1904) as follows: "they are found in the Mount Hermon district and in the Antilebanons north of the railway. This bear is not a light sandy color, but a tawny color, with a dark patch on the neck and rump. They live in great shelving precipices and travel great distances. This makes them very hard to hunt. They feed in the summer on hummis, a sort of pea [chick pea, *Cicer arietinum* L.], and if the natives do not guard their fields during the whole summer, they would be destroyed by the bears. Later on in September, when the hummis is all gathered, they feed on grapes. In winter they are partly hibernative since the region they inhabit is partly covered by snow." This is probably the most recent account of this bear in the area.

Genetics.—2N = 74, AA = 80 was reported for *U. a. syriacus* (Hungeford and Snyder, 1966).

Human interactions.—The bear was extirpated from the Holy Land (see discussion under local status). It probably is not possible to reintroduce this species because its habitat is no longer available. *Ursus* and *arctos* (*arktos*) are the words for bear in Latin and Greek, respectively.

ORDER PERISSODACTYLA   Odd-toed ungulates

These are hoofed animals with the middle (third) digit bearing the most weight. The other digits are reduced or rudimentary. The skull is large but without enlargement of the braincase. The cheekteeth are high crowned with fused cusps (termed lophs), which gives the appearance of a continuous row with the separation between teeth not apparent. The lophs form the grinding surface of the teeth that is used for crushing vegetation. These teeth are termed lophodont. If horns are present, they lack bony cores and are composed of dermal outgrowths of the dorsal regions of the skull.

FAMILY EQUIDAE   Horses, asses, zebras

Equids (*equus* is Latin meaning a horse) are characterized by the reduction of all toes except the third, which bears the hoof. The skull has an elongated rostrum, small orbits, and a large gap (diastema) between the canines and the cheekteeth. There is one genus with nine living species distributed in areas of Africa, Asia, and the southwestern Soviet Union. Possibly two subspecies of *Equus hemionus* occurred in the Holy Land in historic times.

Genus *EQUUS*   Horses, asses, zebras

Equus hemionus hemippus   Syrian wild ass, half ass   x

Equus hemionus Pallas, 1775. Nova Comm. Acad. Sci. Imp. Petropoli., 19:394, pl. 7. Type from Tarei-Nor, Transbaikailia, USSR.

Equus onager Zimmerman, 1780. Geogr. Gesch. Mensch. Vierf. Thiere, Band 2, p. 80. Nomen nudum.

Equus onager Boddaert, 1785. Elench. Anim., 1:160. Type from "mountains about Kasbin," more likely the desert south of Kasbin (Qasvin) Caucasus.

Equus hemippus I. Geoffrey 1855. C. R. Acad. Sci. Paris, 41:1214. Type from Syria.

Equus hemionus var. *syriacus* Milne-Edwards, 1869. Nouv. Arch. Mus. Hist. Nat. Paris, Bull., 5:40, pl. 4. Type from Damascus, Syria. Synonym.

The Syrian wild ass is a small subspecies (height 1 m at the shoulder) with a tawny olive color in the summer changing to pale sandy-yellow in the winter (Groves and Mazak, 1967). It was common in the Syrian Desert in historic times. The Sumerian civilization of Syria domesticated the wild ass and Sumerian art shows these animals as early as 2700 BC (Wooley, 1934). In the Holy Land, a herd was reported by Felix Fabri near Jericho and by another traveler near Ain Gedi in 1660 (Paz, 1982). Harrison (1972) stated that Burkhardt found them to be numerous in Shararat country in Jordan and northern Saudi Arabia in the first decade of the 19th century. Tristram (1866*b*) stated that "the Syrian wild ass, though most common in Mesopotamia, is still found in the Ledjah and the Hawran." Tristram (1884) stated that this species occurred in "north Arabia," Syria (present-day Syria and Lebanon), and Mesopotamia (Iraq) but rarely enters north of Palestine.

Carruthers (1909) reported that this species is rarely seen in the area southeast of Jebel Druz (perhaps near Azraq Oasis). Musil (1927) reported that they were found in the mid-1800s in the Azraq area at the Sirhan Depression with the last one killed near there at Al Ghamr Wells, 34 mi SE

Fig. 64. The onager (*Equus hemionus onager*), at Shawmari Wildlife Reserve. Photo: M. B. Qumsiyeh.

Azraq. Bodenheimer (1958) stated that the species had been extinct in the Syrian Desert for at least 100 years. Aharoni (1930) also commented that most accounts of the existence of wild horses and asses in this region in the past two centuries were dependent on information given by the Bedouin, which maybe unreliable. A live animal was collected in 1909 in the Syrian Desert and died at the Vienna Zoo in 1927. A well-preserved image of the hunt for the wild ass can be seen in the frescoes of Qasr Amrah (near Azraq), one of the several Umayyed palaces built in eastern Jordan during the 8th century AD.

The second species occuring in the region is perhaps a variety of the Syrian wild ass. The relationship of the Syrian wild ass to the **onager** (*Equus hemionus onager*) is controversial. The onager is a medium-sized ass with pale yellowish brown color with tinges of reddish or pink shades. Tristram (1884) was assured that this species enters areas of Hawran occasionally. He also stated that they are rare, but not absent in Arabia and that he had seen wild onagers in the Sahara (presumably of Egypt).

Onagers recently have been reintroduced in the Shawmari Wildlife Reserve (Fig. 64) and in the Yotvata Nature Reserve (Markowitz, 1967).

They were not as successful as the oryx and Arabian gazelle, which were reintroduced into the same area in Jordan, and their future is uncertain (Nelson, 1985). The scientific name for the species derives from *hemi-*, Gr., prefix meaning half, and *onos*, Gr., meaning ass. *Hemionos* is also the Greek word for mule, which is misleading because a mule is a hybrid between a male ass and a female horse. *Onager* comes from the Greek *onos*, an ass, and *agrios*, wild or savage, hence meaning "wild ass."

ORDER HYRACOIDEA   Hyraxes

FAMILY PROCAVIIDAE   Hyraxes

The hyraxes or conies are placed in separate order, Hyracoidea, with a single family, the Procaviidae. This taxonomic separation is due the uniqueness of this group. Hyraxes are small rabbit-sized animals that superficially resemble rodents. Yet, the closest living relatives of hyraxes are the ungulates (including elephants). The tail is exceedingly short. The legs are also short with unique digit morphology. The major weight axis passes along the third digit of the foot. The second and fourth digits are also strong but the fifth is small and the first is rudimentary (hidden under the skin). The soles on the feet have soft pads that allow good footing on rocks and cliff sides. The dental formula in adults is i 1/2, c 0/0, p 4/4, m 3/3 = 34. There are seven species in this group; they are found only in Africa and Arabia. The word *Procavia* derives from *pro*, Gr., before, hence *protos*, first, and *cavia*, new Latin derived from *cabiani*, a Brazilian word used in South America for the guinea pig.

Genus *PROCAVIA*   Hyraxes

Procavia capensis   Coney, hyrax   V

    Cavia capensis Pallas, 1766. Misc. Zool., p. 30, pls. 3, 4. Type from Cape of Good Hope, South Africa.

    Hyrax syriacus Schreber, 1784. Die Säugethiere, 4(1792):923, pl. 240B. Type from Mount Lebanon, Lebanon. Valid subspecies: *P. c. syriaca*.

    Hyrax sinaiticus Gray 1868. Ann. Mag. Nat. Hist., 1:45. Type probably from Mount Sinai, Egypt. Synonym.

    Procavia sinaitica ehrenbergi Brauer, 1917. Schriften Berl. Ges. Naturf. Fr. Berl., 1917:301. Type from El Tor, near Wadi Timar, Sinai, Egypt. Synonym.

    Procavia sinaitica schmitzi Brauer, 1917. Schriften Berl. Ges. Naturf. Fr. Berl., 1917:302. Type from Bteha Plains, north of Lake Galilee, Palestine. Synonym.

Fig. 65. The hyrax or coney (*Procavia capensis*). Male in distress or giving a warning call. Photo: G. Rubin.

Diagnosis.—The hyrax (Fig. 65) belongs to a distinct order of mammals and thus is easily separated from all other mammals of the region. It is the size of a domestic rabbit, with adults weighing 2–5 kg. The tail is vestigial and the feet have fleshy black pads suited for the mountainous rocky habitat of this species. The hind feet have three functioning toes and the forefeet have four. All toes have distinct black nails. The fur is short and rough and is dark reddish brown with a distinct yellowish spot dorsally in adults (characteristic of the subspecies). The neck, shoulders, and legs are usually lighter than the back. Throughout the dorsal fur are longer black hairs, resembling vibrissae or whiskers. The skull is very distinctive with a notably wide mandible bone and tusklike upper incisors that are triangular in cross section (Fig. 66).

Range of species.—Most of Africa except western and equatorial regions, eastern Mediterranean regions, and Arabia.

Local status.—This species is found in the rocky mountains of Wadi Araba, southeastern regions of the Syrian Desert, and around the Dead Sea, Mount Carmel, and upper Galilee (Fig. 67). Tristram (1866b:84) stated that "the Coney is not uncommon by the shores of the Dead Sea, in rocky gorges, rare in the rest of the country, but is occasionally found in the mountainous

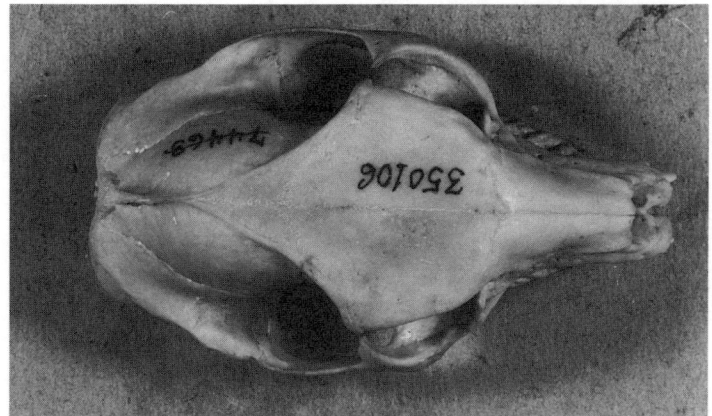

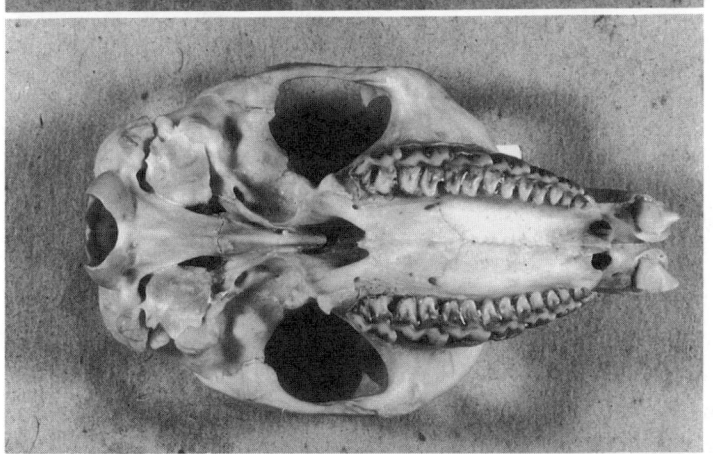

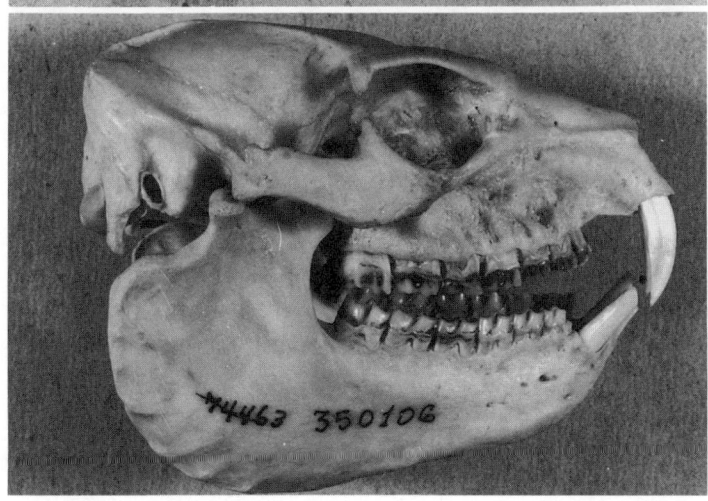

Fig. 66. Dorsal, ventral, and lateral views (from top to bottom) of the skull of the hyrax (*Procavia capensis*). The mandible is also included in the lateral view.

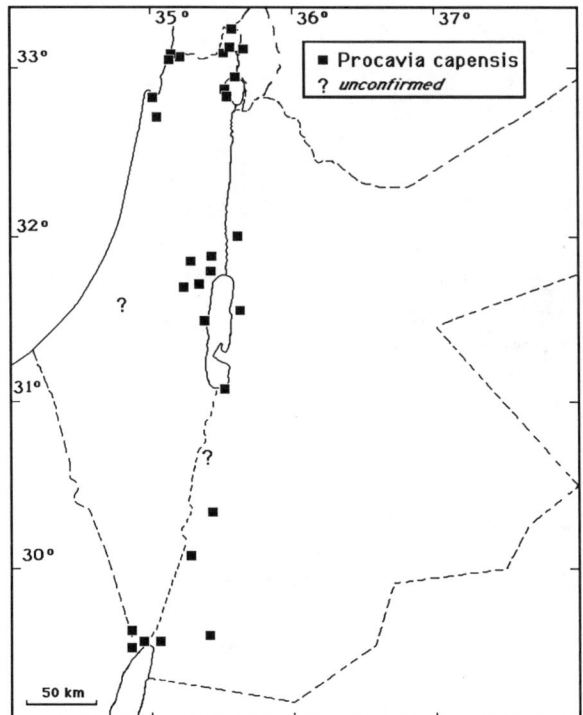

Fig. 67. Distribution of
the hyrax (*Procavia
capensis*).

ridges north of the plain of Acre." Aharoni (1930) recorded it as present in
the hills from Der Mar Saba to the hills of the Dead Sea, and mentioned
two color phases, yellowish brown and gray. I observed these animals on the
rocky outcrops near Deir Mar Saba, and it is my experience that the gray
individuals are immatures.

The species was also recorded from Beit Sahur, the mountains of Moab
(Hahn, 1934), Eilon, Wadi Araba, near Eilat (Mendelssohn, 1965), Wadi
Rum (Mountford, 1965), Petra, Wadi Ram, Haifa, Nahal Oren (Mount
Carmel), Tabgha (Tiberias), Wadi Kelt, mountains SE Dead Sea (Harrison,
1968a), Safje, Nabi Musa, and Ain Fara (near Jerusalem) (*fide* Harrison).
Recent records are from Ghor as Safi (Amr and Disi, 1988), and Deir Mar
Saba (M. B. Qumsiyeh, pers. data). Specimens at the HUJ are from
Mogharet Khureitun, Kefar Gileadi, Lahavot Habashan, Mount Carmel,
Dead Sea region, Al Magor, and Wadi Fallah. It was also recorded from
Ain Netafim, Ain Gedi/Nahal Arugot, Nahal Amud, Nahal Dishon (Wadi
Ouba), Nahal Qedesh, near Metullah, Nahal Bezet (Wadi Karkara), Nahal
Keziv, and Nahal Kelekh (questionable locality) (Anonymous, 1976), and
Wadi Taba (Sinai) near the border (Osborn and Helmy, 1980).

Biology.—The hyrax is well adapted to climbing, both on rocky cliffs and in trees. It has been reported to feed on many plants including *Acacia, Rumex vesicaris,* grasses, herbs, berries, and oak acorns (Mendelssohn, 1965; Osborn and Helmy, 1980). Hyraxes are colonial with a hierarchical social system. I observed a colony in the rocks overlooking the wadi of Der Mar Saba in the Judean hills in January, February, and July (1978–1983). The animals were diurnal and active throughout the year. I observed immature animals (usually three to five per family unit) in the late spring and early summer and a young animal was collected in January. The hyrax is timid and wary (hence the biblical name Shafan meaning "the hider") and generally responds to a high-pitched alarm signal (sharp bark) given by any member of their group. Escape is accomplished by retreating into deep natural crevices in the rocks and between boulders. Conies are quite vocal, with other sounds including growling in threat (Haltenorth and Diller, 1980; Fig. 65). Within their home territories, specific sites are used as lookout areas, sunbathing areas, grazing grounds, and drinking sites (Haltenorth and Diller, 1980).

Predators include eagles, leopards and other cats, wolves, jackals, and foxes (Gasperetti, 1978*a;* Haltenorth and Diller, 1980). Hyraxes appear to tolerate extensive heat and require little if any water. I have observed individuals feeding on plants and basking on rocks for extended hours in the midday summer sun.

In Egypt, captive hyraxes have produced litters of up to five young. No definite breeding season was noted but most litters were born in March (Flower, 1932). Haltenorth and Diller (1980) reported that the breeding season is from August to September in northern Africa, with the litters of one to six born after a gestation period of about 8 months. Young hyraxes leave the den at 3 months of age and achieve sexual maturity at 16–17 months. In the Holy Land, hyraxes were noted to have a gestation period of 205–245 days with usually three to four young born in the spring (Mendelssohn, 1965).

Genetics.—2N = 54, AA = 62 was reported for specimens from Africa (Hungeford and Snyder, 1969). The X is submetacentric and the Y is a small acrocentric.

Human interactions.—The name coney (cony) of the earlier Greek and Latin translations of the Bible comes from the Latin *cunniculus* for rabbit, and hyrax is from the Greek *hurax,* genitive *hurakos,* meaning shrew-mouse (see also Gasperetti, 1978*a*). Hardy (1947*a*) stated that a substance called "hyraceum" comes from the feces and urine of the hyrax. It is not very clear what this substance is used for; although Hardy claims that it was used as a "fine perfume." I have heard that it was used in folk medicine for certain neurological conditions.

The Bible speaks frequently of the hyrax or *shaphan,* a Phoenician word later used by the Hebrews. The word Spain is believed to be derived from a Phoenician name meaning the land of the hyrax (probably in reference to abundant rabbit populations on the Iberian coast; Kingdon, 1990). Some Western translations of the Bible use the name coney, whereas others refer to it as a rock "badger." The latter translation can obviously be confused with the carnivorous badgers. Some examples of statements about the hyrax follow. "The conies are but a feeble folk yet make they their houses in the rocks" also translated "the badgers are not mighty folk, yet they make their houses in the rocks" (Proverbs, 30:26). "Likewise, the rock badger, for though it chews cud, it does not divide the hoof, it is unclean to you" (Leviticus, 11:5). "The cliffs are a refuge for the rock badgers" (Psalms, 104:18).

ORDER ARTIODACTYLA   Even-toed Ungulates

Much variation in morphology occurs in this order. The main distinguishing feature is that the weight is born on both the third and fourth digits. These two digits are fused along most of their axes forming the so called "cannon" bone of the leg. In most species the other digits are rudimentary. In pigs and hippopotami, four digits are functional on each limb. The arrangement of the digits is what gives the artiodactyls their common name of "even-toed" ungulates.

True horns occur in the family Bovidae (sheep, goats, gazelles, cattle, etc.). The horn consists of an inner bony core and an outer sheath of fibrous proteins called keratin. Horns are evergrowing and are not shed like antlers. Antlers occur in members of the family Cervidae (deer) and usually are more pronounced in males. Antlers consist of bone that grows before the breeding season; they are first covered by a layer of skin with hair (velvet) that then dries and is shed once growth is complete. Antlers usually have a main beam and a variable number of side branches or points, which are called tines.

There are nine recent families with 185 known species of Artiodactyla. Three families occur in the Holy Land.

Key to the Families of Artiodactyla of the Holy Land

    1. Horns present. . . . . Bovidae

       Horns absent. 2

Fig. 68. The wild boar (*Sus scrofa*). Female with piglets. Photo: H. Mendelssohn.

2. Antlers present, especially in males. . . . . Cervidae

   Antlers always absent. . . . . Suidae

FAMILY SUIDAE   Pigs, boars

In Suidae, the teeth are low crowned with four cusps. Such teeth are termed bunodont and produce mostly a crushing action. The canines are long and continuously growing, with the upper ones forming tusks (which are especially prominent in males). Nine species are known in this family, of which only one occurs in the Holy Land.

Genus *SUS*   Wild boars

Sus scrofa   Wild boar, wild hog   I

Sus scrofa Linnaeus, 1758. Syst. Nat., 10th ed., 1:49. Type from Germany.
Sus libycus Grey, 1868. Proc. Zool. Soc. Lond., 31. Type from Xanthus, near Günek, Turkey. Valid subspecies: *S. s. libycus*.

Diagnosis.—The wild boar (Fig. 68) is easily recognized by all who know the domestic pig. These are large animals with a strong build and an

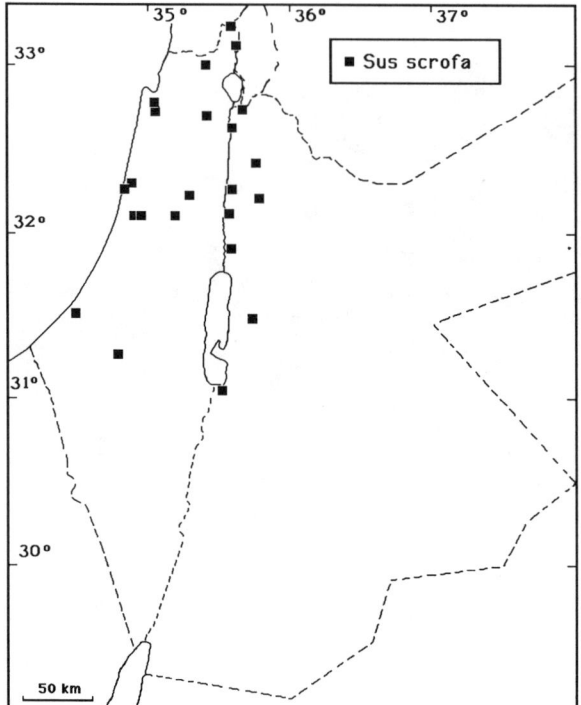

Fig. 69. Selected marginal and other locality records of the wild boar (*Sus scrofa*).

elongated forehead. The muscular snout has a broad, flattened anterior. The neck is short and thick. The legs are slender and short. The ears are usually erect, pointed, and large. The tail is relatively short. Males are larger than females. Weights for males range from 100 to 125 kg and for females from 50 to 90 kg.

Range of species.—Europe and southwestern USSR, south to northern Africa, Arabia, and east through China to Japan and Vietnam. Feral populations can be found worldwide.

Local status.—Wild hogs are common in the Holy Land, as in many other parts of the world. This adept animal managed to survive in many areas despite extensive hunting and is found in most areas having a permanent source of freshwater (Fig. 69). According to Tristram (1866*b*, 1884), the wild boar was abundant in every part of the country, even in desert habitats, but "swarms in all the thickets by the Jordan and the Dead Sea." Specific localities were given (Tristram, 1867) as near Mount Meron, Wadi Arnon, Mount Tabor, Jabbok, Mount Carmel, Kishon River, and Sharon Plain.

Wild boars are reported from Beersheba, the Ghor, and west to Gaza (Hart, 1891), Petah Tikvah (near Jaffa), Ras el Ain, Wadi Hanna (near Nablus), Salfit, Falik, Shooni (= North Shuna), Damia Bridge (Anonymous, 1946a; Harrison, 1968a). Bodenheimer (1958) claimed that this species was absent at his time in the coastal regions and only inhabited selected areas of the upper Galilee and the Wadi Araba. Harrison (1968a) reported on specimens from Hula and Kefar Gileadi. The wild boar was seen frequently around North Shuna, Mashari', and the Yarmouk River area (Amr and Disi, 1988). A specimen from the latter area is at the JNHM.

Biology.—Wild boars occur in many habitats from forest to desert but seem to prefer forested areas or thickets near abundant water sources. Wild boars are omnivorous but feed mainly on buds, fruits, roots, tubers, bark, insects and their larvae, and eggs of birds and reptiles. However, there is surprisingly little that hogs do not eat. Even a carcass of a domestic animal may be consumed.

In the marshes in Iraq, which are similar to those in the Jordan Valley, wild boars forage mainly at night, resting during the daytime in "nests" of plants (Thesiger, 1964). Wild boars are aggressive animals and will charge humans or other animals and are able to inflict serious injury. In Armenia, 7–10 young are reported in a litter (Dahl, 1954). A study in the Galilee (Canaani, 1976) indicated similar observations and added additional insights about social behavior of these animals. Births take place from March to June after a 4-month pregnancy. It appears that groups of females protect and suckle one another's piglets, forming a "kindergarden." It was noted that these groups do not defend their territories.

Genetics.—2N = 36-38, AA = 60 was reported for specimens from many areas in Europe as well as from feral populations in the USA and elsewhere (see Zima and Král, 1984).

Human interactions.—*Sus* is a Latin word, genitive *suis*, meaning a pig, and *scrofa*, L., means a breeding sow. Large populations of wild hogs were common in the Holy Land until the early part of this century, when extensive mechanized hunting ensued. There are many localities that used to derive their local (Arabic) names from wild boars. Many of these names were later changed by settlers. As an example, "Jeziret el Hanzir" (= Pig Island) on the Alexander River has disappeared from the maps of the area, which is now in Kibbutz Maabarot (Dar, 1975). The species is still common in the Jordan Valley and as far south as Ghor as Safi. In a survey conducted in 1982, 1193 signs/events (which may correspond to number of animals) were counted as animals fed on citrus and other crops in the Jordan Valley (Rahamat, 1982).

In Jordan, the wild boar is hunted and no restrictions by the Royal Society for the Conservation of Nature apply to hunting it. Although the pig was hunted extensively early in the century, it is not allowed as a food item by either Islam or Judaism. The *khanzir* is considered a "dirty" animal by many locals who despise them and kill many when possible.

<h2 style="text-align:center">FAMILY CERVIDAE   Deer</h2>

Deer are placed in a distinct family, the Cervidae (from *cervus,* Latin for a stag or a deer) because of several specializations. The elongated cusps on the teeth form crescent-shaped ridges (such teeth are termed selenodont). The teeth are also low crowned and excellently adapted for grinding plant material. Vacuities occur between the nasal and lacrimal bones in the skull. The stomach is four chambered and ruminating. Seventeen genera with 36 living species are known worldwide.

Key to the Reintroduced Cervidae of the Holy Land

1. Tail short, inconspicuous. Color in adults solid brown with little spotting or stripes, antlers almost parallel. . *Capreolus capreolus*

   Tail longer. Color brown with conspicuous spots fusing in some areas (for example, along the sides) to form stripes, antlers diverge laterally. . . . . *Dama mesopotamica*

Note: A third extirpated species, *Cervus elaphus,* has not been formally reintroduced into the Holy Land.

<h2 style="text-align:center">Genus CAPREOLUS   Roe deer</h2>

Capreolus capreolus   Roebuck, roe deer   X

Cervus capreolus Linnaeus, 1758. Syst. Nat., 10th ed., 1:68. Type from Sweden.

Capreolus coxi Cheesman and Hinton, 1923. Ann. Mag. Nat. Hist., 12:608. Type from Zakho, Kurdistan, northern Iraq. Valid subspecies: *C. c. coxi.*

Diagnosis.—The tail is very short and inconspicuous. Young have spotted coats, as in most cervids. Adult animals are a solid color. Antlers are small and terete.

Range of species.—Most of Europe east through southern Russia south to Palestine (formerly) and Iran. Also found in central Asia (China and south to Korea) to Japan.

Local status.—Tristram (1866*b*) stated that it is only found in the "bare hilly country of North-eastern Galilee." Tristram (1876) saw this species in Lebanon and stated that a specimen from Mount Carmel had been collected and was sent to Cambridge, England. He also stated that the species was present further south at Sheikh Iskander. Carruthers (1909) reported that roe deer occurred at that time in the forested regions in the northern part of the Jordan Valley and in the hills of northern Palestine (between Tyre and Lake Galilee). Two specimens at HUJ were collected by Aharoni (no date given) from Mount Carmel. Bodenheimer (1958) suggested that the roe deer has been absent since at least early in this century. The status of the roe deer in Iraq and Syria is uncertain (Harrison, 1968*a*). In Iran, the roe deer is still common in the forested regions in the north (Lay, 1967).

Genetics.—2N = 70, AA = 68 was reported for specimens from Europe (see Zima and Král, 1984). The X and Y chromosomes are submetacentric.

Human interactions.—This is a forest-adapted species and deforestation is the most important reason for its decline (Harrison, 1968*a*). Roe deer were exterminated in the Holy Land and in Lebanon. Remnant populations perhaps still exist in Iraq, but much conservation effort is needed to save them. This deer feeds on a variety of trees, shrubs, and grasses. The rich Mediterranean forests that spread across the northern parts of the Holy Land provided an ideal place for this species for millennia. Continuous decline of these forests in the past two hundred years was facilitated by human activity and the slow spread of arid conditions due to climatic changes. Wadi Yahmur near Sheikh Iskander derives its name from the Arabic and Hebrew name of the roe deer *(yahmur)*.

Genus *CERVUS*   Red deer

The dental formula in this genus is i 0/3, c 1/1, p 3/3, m 3/3 = 34. The word *Cervus* is derived from the Latin *cerva* meaning deer.

Cervus elaphus   Red deer, stag   X

Cervus elaphus Linnaeus, 1758. Syst. Nat., 10th ed., 1:67. Type from southern Sweden.

Diagnosis.—This is a large deer (Fig. 70) weighing more than 200 kg. The height at the shoulder is more than 1 m. The antlers of adult males (those more than 3 years old) are long and bear up to six tines (points). Molt occurs in the spring and results in a reddish brown pelage. In winter, the color changes to brownish, with dark brown on the head, neck, and legs and light grayish brown on the back and sides.

Fig. 70. The red deer
(*Cervus elaphus*).
Photo: M. B.
Qumsiyeh.

Range of species.—Much of the Northern Hemisphere until recently, when it was exterminated from much of its range.

Local status and human interactions.—This animal was the first of three species of deer to disappear in the Holy Land, according to Bodenheimer (1958). Remains of the red deer were excavated at Tel Hesbon in layers from the 12th to the 15th century AD (Boessneck and Von den Driesch, 1977). It is not known when the last of the red deer vanished from the forests of the Holy Land. In Iran, the red deer was common in the Caspian Forest (Lay, 1967). The specific name *elaphus* comes from the Greek *elaphos* and means stag or deer.

Genetics.—2N = 68, AA = 68 was reported for specimens from Europe (see Zima and Král, 1984).

Genus *DAMA*   Fallow deer

Dama mesopotamica   Mesopotamian fallow deer   **X**

Cervus (Dama) mesopotamicus Brook, 1875. Proc. Zool. Soc. Lond., 246. Type from Luristan Province, Iran.

Diagnosis.—The Mesopotamian fallow deer is distinguished from the roe deer (*Capreolus capreolus*), the other species of deer that has been reintroduced in the region, by the longer tail and the distinct spots and stripes of the summer pelage. Antlers in males have distinct trez tines and are flattened distally. Harrison (1968a) provided a detailed description of external and cranial morphology of this species. Ferguson et al. (1985) suggested that *Dama mesopotamica* is distinct from *Dama dama*. This conclusion was based on finding fossils of both in the same layers (late Bronze Age) in the Holy Land.

Range of species.—Eastern Mediterranean region to W Iran.

Local status.—Mesopotamian fallow deer were common in late Paleolethic caves in Palestine (Bate, 1937) and through the 7th to 6th century BC in Jordan (Boessneck and Von den Driesch, 1977). They were seen by Hasselquist (1757) in 1750 on Mount Tabor. Tristram (1866b) stated that *Dama vulgaris* (actually *D. mesopotamica*) was rare; a few still could be found on Mount Tabor and the woods between that mountain and the gorge of the Litany river, and in 1876 he noted a few animals still occurred north of Palestine. Tristram (1884) also stated that he saw fallow deer at 16 km W of the Sea of Galilee. In 1923, Bodenheimer (1958) saw fallow deer antlers in a shop that were said to come from Jerash, Jordan. It is clear then, that this species became extirpated in the past 100 years in the eastern Mediterranean region and currently survives only in western Iran. Reintroductions could reestablish the fallow deer in the Holy Land (Haltenorth, 1959).

Biology.—One to two young are born in the spring following a gestation period of 233–235 days. Males shed their antlers in February following the winter breeding season (Haltenorth, 1959).

Human interactions.—*Dama* is Latin for fallow deer and *mesopotamica* refers to Mesopotamia (modern-day Iraq), where these animals were collected. The common name (fallow deer) comes from the Old English *falu* meaning brownish yellow. Fallow deer probably were exterminated in the Holy Land before the turn of the century. Mesopotamian fallow deer were reintroduced in the Hai-Bar Reserve in 1978 and plans exist for reintroductions in other reserves in the Holy Land. European fallow deer (*D. dama*) have been raised by humans in a semiferal state (never actually domesticated) at least since the times of the ancient Greeks and they are common in parks in Scotland, England, and Germany (Fletcher, 1984).

FAMILY BOVIDAE   Cattle, goats, sheep, antelopes

Bovids have the characters of the Artiodactyla as discussed earlier. Bovids have high-crowned (hypsodont) cheekteeth with crescent-shaped ridges (termed a selenodont condition). The dental formula is i 0/3, c 0/1, p 2–3/3, m 3/3 = 30–32. A marked space (diastema) separates the incisiform (front) teeth from the cheekteeth. Horns are usually present, especially in males. They maybe reduced or absent in females. The horns consist of a keratin sheath covering a bony core. There are 45 genera with about 124 recognized species.

Aharoni (1930) mentioned the Egyptian auri (*Ammotragus lervia*), as living in Wadi Araba, probably based on information from Bedouin who used the name *el-Kebsch*. As pointed out by Bodenheimer (1958), the occurrence of the auri there is doubtful because at that time the species was almost extinct in Egypt and had not been reported in the Sinai.

Genera *ADDAX* and *ALCELAPHUS*

Addax nasomaculatus   Addax   X

Cerophorus (Gazella) or *Antilope nasomaculata* de Blainville, 1816. Bull. Sci. Soc. Philom., Paris, 1816:75. Type probably from Senegambia, western Africa (Ellerman and Morrison-Scott, 1951).

Alcelaphus buselaphus   Bubal hartebeest, bubale   X

Antilope buselaphus Pallas, 1766. Misc. Zool., p. 7. Type probably from Morocco.

The Pleistocene presence in the Holy Land of the addax and the bubale is documented (Bate, 1937). The exact times that these species became extirpated in the eastern Mediterranean region is unknown and remaining small populations may have existed during the Roman periods, with the bubale perhaps surviving to the 19th century (Tristram, 1876). Bodenheimer (1958) suggested that literature records, including those of Tristram (1866*b*), were perhaps misidentifications or relied on Bedouin, who may use the Arabic common name (*baqar al wahsh*) for more than one species. Aharoni (1930) also mentioned that these species were not seen in Palestine since the turn of the century. Bubales were introduced into the Hula Nature Reserve. The scientific names for these two species (*Addax nasomaculatus* and *Alcelaphus buselaphus*) are derived as follows: *addax*, L., a wild animal with crooked horns; *nasus,* L., the nose; *macula,* L., a spot, a mark; *atus,* L., suffix meaning provided with (the addax has brown patches on the nose); *alke,* Gr., the elk; *elaphus,* Gr., a deer; *bous,* Gr., a bullock or cow.

Genus *BOS*   Cattle, oxen

Bos primigenius   Aurochs, wild ox   X

Aurochs (*Bos primigenius* Boianus 1827) were reported from ancient strata (about 7000 BC) from Jericho (Zeuner, 1963) and biblical records of the *Re'em* may refer to the aurochs. There is no record of their domestication in the Jericho Tell strata. However, the species was domesticated in India where it gave rise to the domestic humped cattle or zebu (*Bos indicus*). Tristram (1876) reported that these animals are depicted in earlier dynasties (Nineveh) but not in the latter dynasties of the Assyrians at Kuyonjik. Thus, the aurochs probably became extirpated in the eastern Mediterranean region sometime during Assyrian rule.

The domestic forms of the ox have a 2N = 60, AA = 58 (reviewed in Zima and Král, 1984). *Bos* is Latin, genitive *bovis,* meaning an ox, and *primigenius* is Latin meaning original or primitive (in reference to its being the ancestor of domestic cattle). The word *aurochs* is from the Old High German.

Genus *CAPRA*

Capra ibex   Ibex   V

Capra ibex Linnaeus, 1758. Syst. Nat., 10th ed., 1:68. Type from Valais, Switzerland.

Capra nubiana F. Cuvier, 1825. *In* Geoffroy and Cuvier, Hist. Nat. Mamm., Bouc. Sauvage de la Haute Egypte, 50:2 and pl. 397. Type from Upper Egypt. Valid subspecies: *C. i. nubiana.*

Capra sinaitica Hemprich and Ehrenberg, 1828. Symb. Phys. Zool., 1:18. Type from Sinai.

Capra arabica Rüppell, 1835. Neue Wirbelth. Abyssinien, Säugethiere, 1835:17. Type from Sinai.

Aegoceros beden Wagner, 1835. Schreber Säugethiere, 5:1303. Type from Hejaz, Saudi Arabia.

Diagnosis.—Ibex (see Fig. 71) have normal "goat" characteristics, including a beard in the males. The shoulder height is around 85 cm. Males are significantly larger than females. The weight of adult females rarely exceeds 35 kg (not pregnant), whereas males can weigh up to 58 kg. The color of the upper parts is brownish but with characteristic contrasting black and white leg markings. The tail is short (shorter than the ear). The male's horns

Fig. 71. Ibex (*Capra ibex*). Photo by D. M. Shafi.

Fig. 72. Skull of the ibex (*Capra ibex*). Photo: M. B. Qumsiyeh.

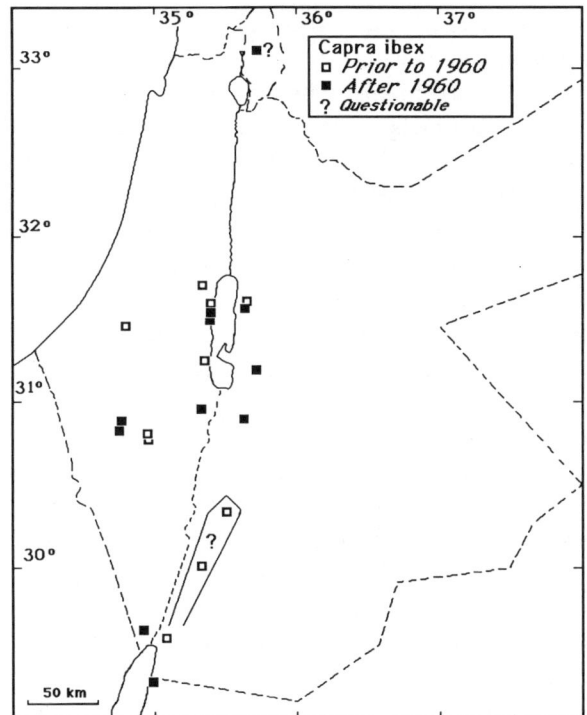

Fig. 73. Locality
records of the ibex
(*Capra ibex*).

are long and scimitar-shaped. The anterior surfaces of the horns in adults
have 20–24 ridges (Fig. 72). The tail is short (about 6–10 cm) with a terminal
tuft of hair.

Range of species.—Mountain regions from southern Europe through
eastern Europe, Turkestan, to Mongolia. Southern areas of the range include
southwestern Asia and India, Saudi Arabia, the eastern Mediterranean
region, Egypt, and the Sudan.

Local status.—The ibex is in dire need of protection. We are not able to
state with certainty how many populations have been decimated. The few
records from the past and recent records indicate extensive losses in the
mountainous regions of the Holy Land (Fig. 73). Tristram (1866*b*, 1876,
1884) reported this species as common from the Moab (east of the Dead
Sea) and the Judean hills near the Dead Sea. He also acquired a few
specimens from these localities and from Ain Gedi (Engedi). Tristram
(1866*a*) specified other localities where he saw ibex: near Mar Saba, Ain
Terabeh, Wadi Sudeir, and Jebel Hatrura (near Masada). Hart (1885)
reported seeing ibex on Mount Hor, and Palmer (1871) saw them at Wadi

Hanjurat al Gattar (35 mi SW Dead Sea). Carruthers (1909) noted these animals at Wadi Zarqa Ma'in, and Hart (1891) reported seeing them at Wadi Araba. Allen (1915) quoted Phillips, who supplied him with the specimens for his report on the mammals of Sinai and Palestine, as follows: "The Sinai Ibex persists over all the rugged parts of the Sinai peninsula, near Aqaba and up at least as far as the northeast end of the Dead Sea. Although undoubtedly greatly reduced in numbers since Tristram's time (mid 1800's), it manages to persist . . . . I hunted three days and saw only four smallish animals, but signs were fairly numerous. . . . The leopard hunts these Ibexes and presumably kills a good many, as various sportsmen have testified." Aharoni (1930) also mentioned that these animals were present in the mountains surrounding the Dead Sea and those in Beersheba. The species was reported from Wadi Kelt (Anonymous, 1945). Bodenheimer (1935, 1958) gave anecdotal remarks on the presence of ibex near Jerusalem, Ain Gedi, and the Negev. Harrison (1968a) gave measurements for a specimen at the British Museum collected from Ain Gedi in 1864. Specimens at HUJ are from Nahal Ze'elim (Wadi Siyal), Sede Boqer, Avdat (Abde), and Ain Gedi. Ilani (1987) includes a record from the Golan Heights.

Specimens available at the JUNHM were collected from near Mazar in 1981 and Mazra'ah in 1983, as well as a live specimen kept at Shawmari Wildlife Reserve that was originally from Tafileh (Amr and Disi, 1988). These authors also reported seeing two mummified specimens from Wadi Araba and Karak. A specimen from the latter locality (collected in the late 1980s) is at the JNHM. There are verbal records from Haql, Tabuk, and other localities in northwestern Saudi Arabia near the Jordan–Saudi border (Gasperetti, 1978b).

Biology.—The ibex is restricted to rocky habitats in wadis and mountains and rarely ventures into plains or other habitats. The ibex's color is an excellent camouflage on the mountains and they occasionally escape detection by remaining motionless. Ibex were found to feed on *Acacia raddiana, Lindenbergia sinaica, Lycium shawii, Capparis spinosa, Ficus pseudosycomorus, Phragmites, Imperata, Juncus,* and *Ahagia* (Osborn and Helmy, 1980). Ibex require water for survival and surface water is always available in the areas in which they occur. For example, ibex move down from the mountains to Ain Gedi twice daily for water. This behavior is exploited frequently by human hunters, who are the main enemy of this magnificent mountain goat. The ibex needs urgent conservation efforts in the area. Leopards were reported to eat ibex in Sinai (Buxton et al., 1895), and this is also doubtless true in the mountains surrounding the Dead Sea.

Ibex are social animals; each group is led by a dominant male. Vagrant young males are also seen and occasionally they succeed in driving an older male from its harem (M. B. Qumsiyeh, pers. obs. near Ain Gedi). Males display to females by extending the tongue and tipping their horns back. Tristram (1866*b*) reported that ibex give birth to one young in March or April. The young are well camouflaged when among rocks. In Egypt, the gestation period is 150–163 days, with one young born from February to April (Flower, 1932).

Genetics.—2N = 60, AA = 58 was reported for specimens reared in zoos and whose origin is not known (reviewed in Zima and Král, 1984).

Human interactions.—*Capra* is Latin for the she-goat, *ibex* is Latin for this species. The ibex or mountain goat is referred to in the Bible (Samuel, 1:24): "when Saul returned from pursuing the Philistines, he was told, saying 'Behold, David is in the wilderness of Engedi' then Saul . . . went to seek David and his men in front of the Rocks of the Wild Goats."

Genus *GAZELLA*   Gazelles

Gazelles are relatively small slender antelopes with short tails. The horns have transverse rings, at least in males. The color is variable but the dorsal color is always distinct from the lighter ventral color and the two are usually separated by a dark stripe. The head has characteristic white and black stripes extending from the nose to the eyes. Gazelles were common in most areas of the Holy Land and were a staple food item for the Stone-Age cultures of the eastern Mediterranean region. Sophisticated ancient trapping and hunting methods included construction of elaborate stone corrals that were used to capture whole herds (Rees, 1929; Harding, 1954; Mendelssohn, 1974; Legge and Rowley, 1987). Slaughters of hundreds of gazelles caught by such elaborate corrals were witnessed and vividly described by Aharoni in 1915 (quoted in Mendelssohn, 1974), by Burckhardt in 1831 (quoted in Rees, 1929), and by other travelers. However, gazelles remained the only large game that was common in every part of the region south of Lebanon (Tristram, 1876). Authors from Tristram to Aharoni (1930) commented on the abundance of gazelles throughout the Holy Land. With the introduction of armies and conflicts associated with the First and Second World Wars, intensive hunting left very few viable populations (Mendelssohn, 1974). This is an observation that applied to all large mammals, as any individual with knowledge of the wildlife at the time can attest to. Unfortunately, it was not until the middle of this century that the systematics of the genus was advanced to a point where species could be separated confidently. Thus, it is difficult to know (except perhaps by habitat at the locality) which species was being discussed by the earlier authors. The first serious taxonomic treatment of gazelles was that of Brooke (1873). This was followed by additional studies that were more restricted in scope (Blaine, 1913; Dollman, 1927; Groves and Harrison, 1967; Groves, 1969; Mendelssohn, 1974; Saleh, 1987). I am indebted to C. Groves for sending me reprints, which I used extensively in writing this section on *Gazella*. The words gazelle and *Gazella* derive from the Arabic *ghazal*, the name for these animals.

Key to the Species of *Gazella* of the Holy Land

It is very difficult to construct a key that works for every specimen because of the variability of the individuals in each species. Generally, examination of multiple characters is necessary (Harrison and Bates, 1991). The following is only to be used as a guideline. More details are given in the diagnosis section for each species.

1. Heavily built gazelles with very pale sandy color. Side stripes
   and other markings poorly defined. Horns in male well developed
   and lyrate, widely divergent at the tips but with their bases less
   than 15 mm apart (Fig. 75). Both sexes with visible swelling
   around the larynx, especially noticeable in breeding males.
   · *G. subgutturosa*

   More slender gazelles. Color and horns not as above. No visible
   swelling around the larynx in males. · · · · · · · · · · · · · · · · 2

2. Horns of adult males long and straight or slightly lyrate (Fig. 74).
   Horns of adult females slender and long. Posterior nasal margin
   triangular. Premaxillary bones abut the nasal bones.
   · *G. dorcas*

   Horns of adult males short, nonlyrate or semilyrate (Fig. 75).
   Horns of adult females slender and short. Posterior nasal margin
   round. Premaxilla do not touch the nasal bones. · · · · *G. gazella*

Fig. 74. Dorcas gazelle (*Gazella dorcas*). Photo by D. M. Shafi.

Gazella dorcas   Dorcas gazelle, afri   **V**

Capra dorcas Linnaeus, 1758. Syst. Nat., 10th ed., 1:69. Type from "Lower Egypt." Nominate subspecies in Egypt.

Gazella dorcas saudiya Carruthers and Schwartz, 1935. Proc. Zool. Soc. Lond., 1935:155. Type from Dhalm, 240 km NE Mecca, Saudi Arabia. Valid subspecies in Arabia (perhaps a valid species). May enter eastern Jordan.

Gazella dorcas isabella Gray, 1846. Ann. Mag. Nat. Hist., 18:214. Type from "Abyssinia." Valid subspecies, extralimital.

Diagnosis (Fig. 76).—This is a small (weight 10–19 kg, shoulder height less than 70 cm), slender gazelle. Males are larger than females. The color is brownish red with distinct facial markings. The flank stripe is poorly delineated. The male horns are either parellel or slightly spreading and are S-shaped. The horns are well developed in both sexes and have 15–25 rings. The tips of the horns are not hooked or are slightly hooked forward and in both sexes they are curved toward the midline. The ears are long and when laid forward almost reach the nostrils. See under *G. gazella* for differences between the two species. *Gazella dorcas* has longer horns, and in males, straighter horns. The posterior margin of the nasals is narrow in this species, almost V-shaped. In *G. gazella*, the posterior margin is broad, never triangular, and is wider than the anterior part. In dorcas gazelles, the fenestrae are lacking in the infraorbital fossa.

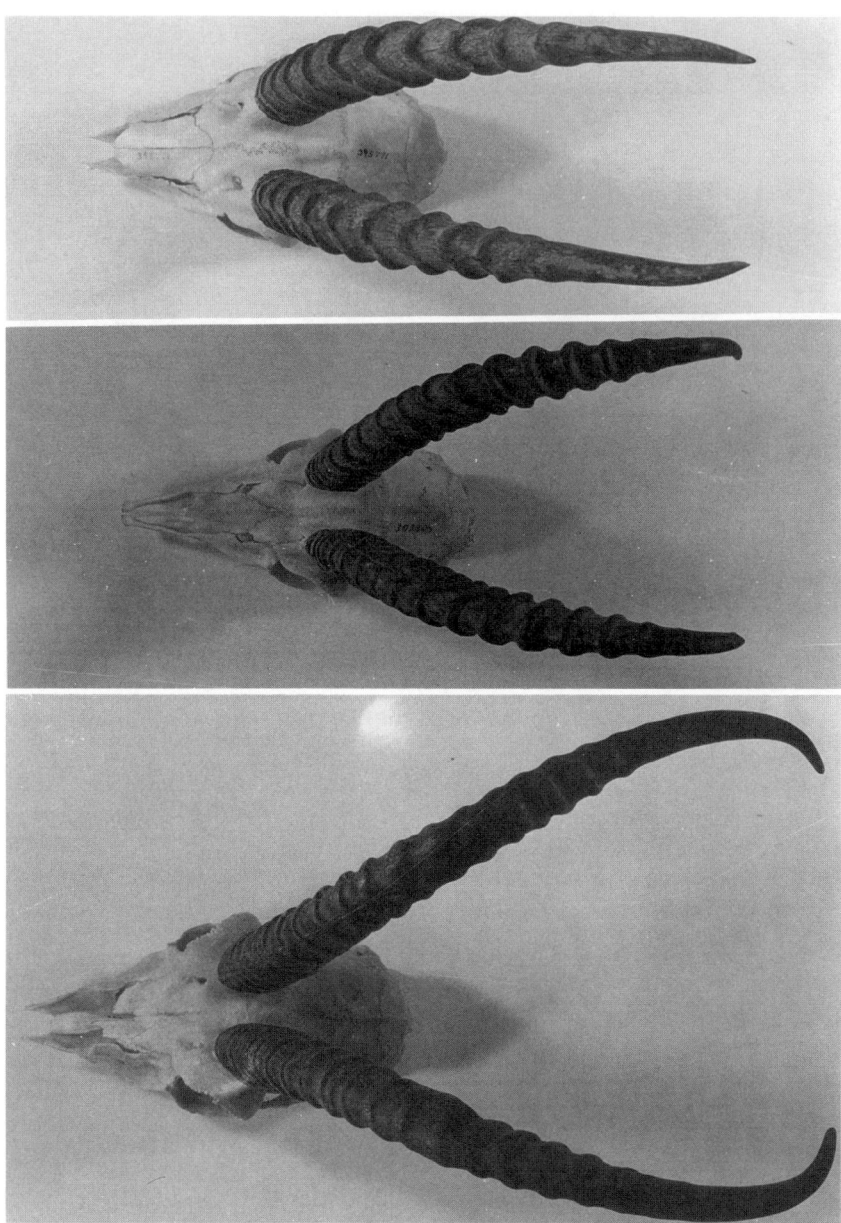

Fig. 75. Skulls of three species of gazelles. (Upper) Mountain gazelle (*Gazella gazella*), (Middle) dorcas gazelle (*Gazella dorcas*), (Lower) goitered gazelle (*Gazella subgutturosa*).

Fig. 76 (facing page). Two gazelles from the Holy Land. (Upper) The mountain gazelle (*Gazella gazella*), male in front and female. Photo: H. Mendelssohn of animal at Tel Aviv University. (Lower) The dorcas gazelle (*Gazella dorcas*). Photo: D. M. Shafi of animal at Shawmari Wildlife Reserve.

Gazella d. dorcas is found in Egypt. Ferguson (1981*b*) stated that *G. dorcas dorcas* and *G. d. isabella* interbreed in areas in Sinai, the populations around the Dead Sea are hybrid animals that arrived from Sinai, and *G. d. saudiya* is not found in Israel (pre-1967 borders). However, the differences between these subspecies are minimal and clinal in nature. Many of the subspecies identifications are based on few specimens without genetic studies or proper statistical analysis of morphological differences. Speculations about migrations, isolation, and colonization are best avoided until further data are collected. Groves (1983) postulated that all *G. dorcas* from the Sinai belong to *G. d. isabella*. The limits of *G. d. isabella* and *G. d. saudiya* are yet to be delineated. The two subspecies may intergrade in eastern Jordan.

The following characters of the two subspecies are tentative and follow Groves (1985):

G. d. isabella: Has a deeper, redder color. It has a poorly marked flank stripe but strong facial markings (however, lighter bands are not white); many specimens with black nose spot, black pygal band. Horns of males simple, curve out and then inward near the tips. According to Brooke (1873) *G. d. isabella* also has shorter hair, indistinct facial stripes, faint dark lateral bands, short massive horns, and overall smaller size (except longer ears).

G. d. saudiya: Smaller, has shorter legs, is lighter in color with faint or absent flank and pygal stripes, ears are pale buff, has no nose spot, and the light face stripes are buffy white.

Range of species.—Eastern Mediterranean region, Arabia, and in northern Africa south to northern Ethiopia and Chad.

Local status (Fig. 77).—These gazelles were hunted extensively during the two World Wars and are decimated from much of the area that they used to occupy (see discussion under genus). Bodenheimer (1958) reported this gazelle from Ain Radian. The Schmitz collection has a specimen from Amman (Anonymous, 1946*b*). This species was seen in Wadi Araba by Amr and Disi (1988), who also reported two dead gazelles from Wadi Fidan. The dorcas gazelle is common in arid regions of the country. Specimens are available at the TAUM from many localities in the Negev and Wadi Araba.

Biology.—The previous abundance of the dorcas gazelle in varied habitats is well documented. This gazelle used to occur in most arid and semiarid areas of the Holy Land. Range overlap is found between the dorcas gazelle and the mountain gazelle in intermediate habitats (those receiving about 100–150 mm rainfall). Hunting and human activity has reduced the distribution of the dorcas gazelle to areas of desert and subdesert where they are not disturbed. Observations suggest that dorcas gazelles must drink water

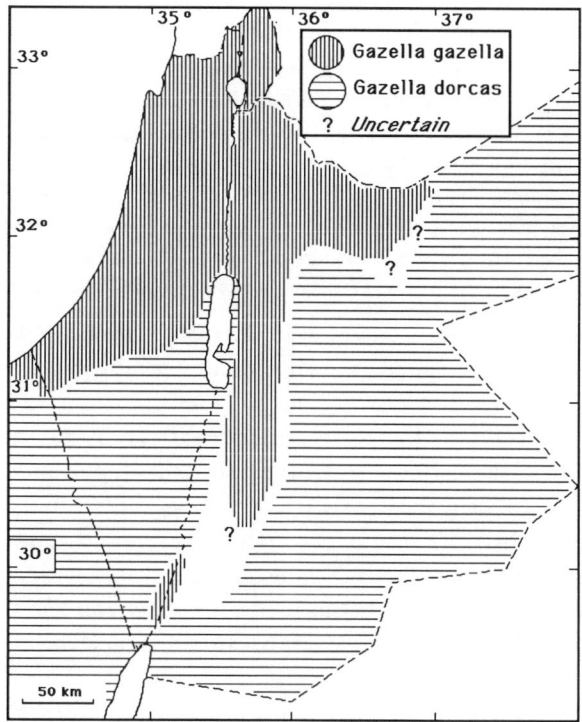

Fig. 77. Distributions of the mountain gazelle (*Gazella gazella*), and the dorcas gazelle (*Gazella dorcas*).

and that they will travel great distances to get to water sources (Ghobrial and Cloudsley-Thompson, 1966; Ghobrial, 1976). Dorcas gazelles feed on *Acacia ehrenbergiana, A. raddiana, Nitraria retusa, Phoenix dactylifera, Cuscuta* sp., *Psoralea plicata, Astragulus vogelli*, and other plants (Osborn and Helmy, 1980). Normally one young is born after a gestation period of 5.5 months. Natural predators include jackals, wild cats, leopards, and cheetahs. However, the most efficient slaughter is by humans, which has reduced the numbers of dorcas gazelles to remnants. This reduction in available prey in turn has resulted in the decrease or absence of some predators, such as the cheetah. For more on human impact, see the discussion under the genus and the chapter on conservation.

Genetics.—A diploid number of 31 (males) and 32 (females) was reported by Wurster (1972). This is due to an autosome sex chromosome translocation.

Human interactions.—See the discussion under the genus. The word *dorcas* is from *dorkas,* Gr., meaning a gazelle.

Gazella gazella   Mountain gazelle, common Arabian gazelle   V

Antilope gazella Pallas, 1766. Misc. Zool., p. 7. Type from Syria.

Antilope cora Hamilton Smith, 1827. *In* Griffith, The animal kingdom, Vol. 5, Geo
B. Whittaker, London. "Eastern Arabia, Persian Gulf." Valid subspecies, extralimital.

Antilope arabica Lichtenstein, 1827. Darstellung neuer oder wenig bekannter Säugethiere, C. G. Luderitz, Berlin, pl. 6. Type fixed by Neumann (ref. below) to Farasan
Island, eastern Red Sea coast, Saudi Arabia. Possibly valid species: *G. arabica* (Groves, 1983).

Gazella arabica rueppelli Neumann, 1906. Sitzungsber. Ges. Naturf. Fr. Berl.,
1906:244. Type from Sinai. According to Groves (1983) this is a synonym of *G. g. arabica*.

Diagnosis.—The mountain gazelle (Fig. 76) is a rather large gazelle with
a body weight of 18–30 kg. The height at the shoulders is usually greater
than 60 cm. The color is pale reddish brown, especially on the haunches.
Distinct facial markings are observed. The horns are short with 10–15 rings.
Compared to *G. dorcas*, the following characters are critical for distinguishing
G. gazella (see Fig. 76): the ears are shorter (when laid forward reach only
halfway to the nostrils); the horns are shorter, strongly curved, and diverge
gradually for two-thirds of their lengths and then more markedly with a
slight turn forward; the anterior side of the horn core in *G. gazella* has a
single groove whereas in *G. dorcas*, the posterior side has two grooves (Ducos,
1968); there is a distinct light facial streak running from the base of the horns
and over the eyes almost to the nose and a similarly distinct black facial streak;
there are lateral and rump bands that are distinct and brown; the infraorbital
fossa is usually lacking fenestra; the posterior margin of the nasals is round
and the nasals do not make contact with the premaxilla; the breadth across
the bases of the horns measures 62–69 mm; and the bases of hairs are smoky
to slate in color (lighter in *G. dorcas*) (Morrison-Scott, 1939).

Gazella g. gazella in Syria averages slightly larger and darker than *G. g.
cora* of Arabia. Populations in southern areas of the Holy Land are intermediate in both characters. The distribution of the Palestine mountain gazelle
(*G. g. gazella*) relative to the Arabian mountain gazelle (*G. g. cora*) is yet to
be resolved (Groves, 1983). These and other systematic problems in mountain gazelles (for example, status of *arabica*) may be resolved by using genetic
(chromosomal and/or electrophoretic) techniques. The following characters
can be used to distinguish the various subspecies that may inhabit the Holy
Land (Groves, 1985):

G. g. gazella: Large (male weighs 24–28 kg), dark brown-gray color, light
flank stripe, well marked pygal stripe, midface deep rufous with dark brown
forehead, rather straight horns. A mountainous form.

G. g. cora: Smaller (height 61 cm), markings as in *G. g. gazella* but face markings different. Midface and forehead dark chestnut with broader and blacker dark face stripe.

G. g. ssp.: The relict population in southern Wadi Araba is either distinct or related to *G. g. cora*. It is a small gazelle (males weigh around 16 kg) with elongated limbs, sandy-gray in color with large black nose spot.

Range of species.—The Arabian Peninsula including the eastern Mediterranean region and Sinai.

Local status.—The mountain gazelle occupied most of the Holy Land during the Holocene Epoch. Remains of mountain gazelles (apparently used extensively as food) were excavated at the Neolithic sites of Mount Carmel (Bate, 1937), Gilgal in the Jordan Valley (Noy et al., 1980), Jericho (Clutton-Brock, 1971, 1979, as *G. dorcas*; Tchernov et al., 1987*a*), and Sinai (Tchernov et al., 1987*a*). The expansion of the Sahara combined with hunting caused the range of *G. gazella* to retreat gradually until the distribution was limited to areas having a temperate Mediterranean climate; it was replaced by the dorcas gazelle in many areas. Relict populations were left, such as a small pocket of *G. gazella* in the southern Wadi Araba surrounded by *G. dorcas* (Mendelssohn, 1974; Tchernov et al., 1987*a*; Fig. 77).

Within the residual range of the mountain gazelle, populations were decimated by hunting during the two World Wars. Similarly, the mountain gazelle was once common in Lebanon but had been exterminated by the end of World War II (Lewis et al., 1968). For the effect of hunting and other human activities on this species, see the discussion under the genus.

The Arabian mountain gazelle was reported from the hills of Jerusalem, the Galilee, Mount Carmel, and "south of the Dead Sea" (Bodenheimer, 1958). Specimens were reported from Midieh (Dollman and Burlace, 1935), Hizmeh, mountains between the Dead Sea and Jerusalem, Katana, and Jerusalem (Harrison, 1968*a*). Recently, this gazelle was reintroduced into the Shawmari Wildlife Reserve in Jordan (Nelson, 1985). A specimen was killed in the Salt Mountains in the summer of 1986 (Amr and Disi, 1988) and is now at JUNHM. The largest population of mountain gazelles in the Holy Land appears to be at Ramot Yissakhar, where there is little or no control of gazelle populations by predators or hunters (Baharav, 1976).

Biology.—The mountain gazelle is well adapted to mountainous habitats, as its English name implies. However, the habitats used by this species in the Holy Land are variable, from the cold mountain regions of the Galilee and Irbid, to the plains of the central regions, to the hot and humid habitats of the Jordan Valley. Mountain gazelles forage on a wide variety of plants and are much less specialized than other ungulates. The following description of the habits of these gazelles is taken primarily from Mendelssohn (1974) and Baharav (1981, 1983). A preferred food is the grass *Cynodon dactylon*. The mountain gazelle and other species of gazelles have similar habits. For example, gazelles deposit their feces in regular places. Males defend a territory during the reproductive season. Mountain gazelles give birth (usually in the spring) to one young following a gestation period of 6 months. Fawns spend the first month in hiding, usually in darker places such as between rocks, logs, and shrubs.

Genetics.—A diploid number of 31 (males) and 32 (females) was reported by Wurster (1972). This is due to an autosome sex chromosome translocation.

Human interactions.—See under genus.

Gazella subgutturosa   Goitered gazelle, Arabian sand gazelle, rhim   E

Antilope subgutturosa Güldenstädt, 1780. Acta Acad. Sci. Petropoli., 1778, 1:251. Type from northwestern Persia. According to Groves (1969) from Tiflis (now Tbilisi) in the Caucasus.

Gazella marica Thomas, 1897. Ann. Mag. Nat. Hist., 19:162. Type from Ibri, Najd Desert, Saudi Arabia. Valid subspecies: *G. s. marica.*

Diagnosis.—This is a large, heavily built gazelle; adult males weigh from 20 to 43 kg and females from 18 to 33 kg (Heptner et al., 1988). The male has a prominent swelling on the throat (goiterlike enlargement of the larynx). The females of this species have rudimentary horns or are hornless. Horns in the male are long, annulated (with 20–27 rings), and lyrate, with their points slightly projecting inwards. The facial region in adult males is distinctly white. There is a faint dark facial streak but otherwise body markings are faint. The hair is rough and coarse in the winter but soft and smooth in the summer. The lateral and pygal bands are very faint. Ears and tail are slightly smaller relative to those of *G. gazella*. The tail is black. *Gazella. s. marica* is a very pale subspecies that is present in the Arabian Peninsula. The skull of *G. subgutturosa* is heavy and is distinguishable from the other two species of *Gazella* in the Holy Land by its larger size, broader palate, and greater orbital width (Groves and Harrison, 1967).

Range of species.—Arabian Peninsula west through Russian Turkestan, Iraq, Iran, Afghanistan, Pakistan, to Tibet and Mongolia.

Local status.—See the discussion under the genus for the effect of hunting. A specimen at the BZM is from El Katrana (Harrison, 1968a). A specimen at the Philadelphia Zoological Gardens came from 160 km SE Hibar

(Groves and Harrison, 1967), a picture of this beautiful specimen can be found in Harrison (1968*a*) and Harrison and Bates (1991). A skull was reported from Safawi (H 5 station) in 1950, and another from Qa'a Dhuweila in September 1983 (Amr and Disi, 1988). A specimen is available at the BZM from Al Busayta in northern Saudi Arabia (Green, 1986).

Biology.—This gazelle is an inhabitant of deserts. The following information is mainly from Vesey-Fitzgerald (1952), Roberts (1977), and Heptner et al. (1988). Goitered gazelles feed on desert shrubs including *Phanterium eppaposum* and *Calligorum* ssp. and may consume 6 kg of food per day (about 30% of their weight). They are able to subsist for long periods of time without water. A gestation period of 5–6 months is followed by the birth of one or two calves (occasionally three or four) in March, April, or early May. Male goitered gazelles are territorial during the reproductive season. At other times of the year, herds of 50–100 animals are the usual. Discussion of human impact on gazelles can be found under the genus and in the chapter on conservation. This species is likely near extinction in Jordan (East, 1992).

Genetics.—The chromosome number in goitered gazelles ranges from 30–33. This is because of one fusion event and one sex chromosome translocation with an autosome. In animals obtained from a zoo, the diploid number for males was 31 and for females was 30 (Wurster, 1972; Hsu and Benirschke, 1977). This is due to the presence of an XY1Y2/XX sex-determining mechanism. Sokolov et al. (1975) reported a 2N = 30 for both sexes in Mongolia. A polymorphism was reported whereby males had a 2N = 31–33 and females 30–32 in populations from Jordan (Kingswood and Kumamoto, 1988) and Qatar and Saudi Arabia (Vassart et al., 1993).

Human interactions.—See discussion under the genus. The word *subgutterosa* derives from *sub-*, L., prefix meaning underneath, *guttur,* L., the throat, *osus,* L., suffix meaning full of. This is an allusion to the enlarged neck, which is especially noticeable on the males of this species in the breeding season.

Genus *ORYX*

Oryx leucoryx   Oryx, white antelope (reintroduced)   **X**

Antilope leucoryx Pallas, 1777. Spicil. Zool., 12:17. Type from "Arabia."

Oryx beatrix Gray, 1857. Proc. Zool. Soc. Lond., 1857:157. Described from "Shores of Persian Gulf, or of the Red Sea." Synonym.

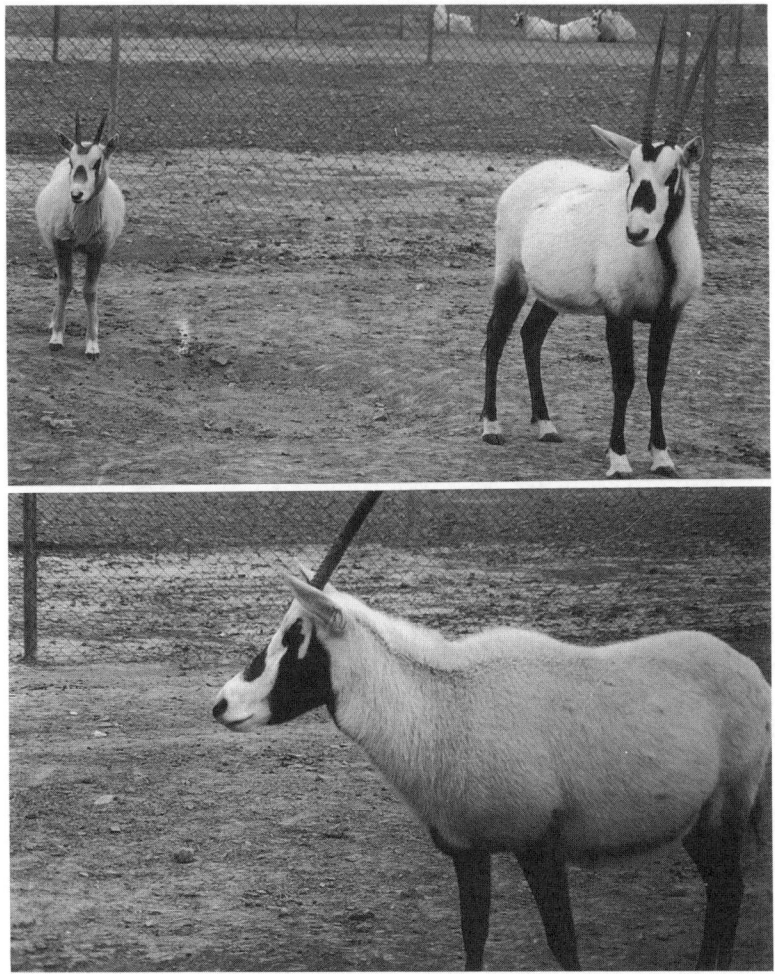

Fig. 78. The oryx or white antelope (*Oryx leucoryx*), at Shawmari Wildlife Reserve.

Diagnosis.—The oryx (Fig. 78) has very long horns that are slightly curved backwards (appearing almost straight). The horns are rounded in cross section and have rings (annulations) close to the base. The color of the back is distinctive, being white or slightly gray. There are characteristic black markings on the face. The legs are black and this coloration extends to varying degrees to the thighs, neck, and underside. The horns arise posterior to the level of the eyes.

Range of species.—Previously widespread in the Arabian Peninsula as far north as Syria.

Local status.—In the 19th century, this beautiful antelope was still common in northern Arabia and in Belka and Hawran in Jordan (Tristram, 1866*b*, 1876) but it was rare or absent in Palestine. Talbot (1960) stated that it was already becoming rare in Sinai and the southern deserts of Palestine in 1800. Schmitz collected this species in Jordan around the turn of the century (Anonymous, 1946*b*) and I have seen the specimen collected by Schmitz at Amman in 1910. According to Mountford (1965), a hunter shot three animals at Qatrana in the 1920s. In southern Jordan, the species may have persisted into the 1930s as a British Army Unit kept one there (Dollman and Burlace, 1935), but by the 1940s the oryx probably was exterminated throughout Jordan (Talbot, 1960). Populations persisting early in this century were reported near Jebel El Tubaiq (Carruthers, 1935) and in Al Busayta and Wadi Sirhan (Raswan, 1935) in northern Saudi Arabia near the border with Jordan.

Biology.—The oryx inhabited most of the Arabian and Syrian deserts until very recently. Its habitat included hammada deserts as well as wadis, sandy deserts, and plateaus. Wild oryx ate succulent plants such as *Aristidia* and *Cynomonium* and buds of tamarisk and other shrubs (Carruthers, 1935). Many other juicy desert plants are eaten, including the desert melon, *Citrullus colocynthis* (Stewart, 1963). The biology of the oryx in the wild was studied by Stewart (1963), and in reintroduced populations by Lloyd (1965) and Hatough and El-Eisawi (1988). Primary plants consumed in Jordan include *Artiplex halimus*, *Artemesia hebaalba*, *Achillea fragrantissima*, *Salsola*, *Ciotrullus colocynthis*, *Fargonia*, *Zygophyllum*, and others.

Genetics.—G- and C-banded karyotypes of the oryx from Saudi Arabia show a diploid number of 58 with a large acrocentric X and a small acrocentric Y (Cribiu et al., 1989). Individuals occasionally are found with diploid numbers of 56 and 57 (for example, from Jordan; Vassart et al., 1992). Recent protein studies showed higher variability than anticipated; this genetic diversity indicates a better chance of survival for the oryx populations in Arabia (Vassart et al., 1991, 1992; Greth et al., 1992).

Human interactions.—The name *Oryx* derives from *orux*, Gr., a gazelle or antelope, and *leucoryx* derives from *leokos*, Gr., white, and *oryx* (hence white antelope). Carruthers (1935) stated that the legend of the unicorn may have been based on this animal. Sometime between the First and Second World Wars, populations of the oryx were decimated in the Arabian and Syrian deserts. This was accomplished by intensive hunting using modern weapons and vehicles, especially near the newly discovered oil fields. In the early 1960s, several international organizations began to cooperate in saving the oryx. These organizations included the International Union for the Conservation of Nature (IUCN), the World Wildlife Fund, and the Shikar-Safari Club. A breeding population was established at the Phoenix Zoo in Arizona, USA, with animals collected from a trip to Saudi Arabia in 1962 and donated animals from holdings in Kuwait, Saudi Arabia, and the London Zoo. This "world herd" began to multiply and was the nucleus to be used to "repopulate the desert." The oryx was successfully reintroduced to Jordan in 1978, where a herd of about 70 now lives in the Shawmari Wildlife Reserve (Anonymous, 1978; Fitter, 1984; Lamb, 1984; Nelson, 1985; Hatough and El-Eisawi, 1988).

ORDER RODENTIA   Rodents

The Rodentia has the largest number of species of all orders of mammals (at least 50 families, 418 genera, and 1750 species). Rodents are also the most numerous mammals in most areas of their distribution. Despite their abundance and explosive radiation that has resulted in great variation (from tiny mice to the porcupines), rodents share many morphologic similarities. The structure of the teeth is characteristic: each jaw has a single pair of ever-growing incisors with enamel only on the anterior and lateral surfaces. This condition allows continuous wear to maintain a sharp, chiseled edge. The canines are absent, and a distinct gap (diastema) occurs between the incisors and the chewing teeth (the molars and, in some species, premolars). There are many other features that are distinctly characteristic of rodents but it is sufficient here to mention that rodents are clearly a unique group of mammals that has succeeded in exploiting almost every terrestrial habitat. Nine families are found in the Holy Land.

Key to the Families of Rodentia of the Holy Land

1. Large animals with average head and body length of 600 mm. Body covered dorsally with long, round, hollow quills. Skull distinctive (Fig. 115): frontal and nasal regions inflated giving the skull a rounded profile dorsally; crowns of cheekteeth flat, complexly folded; angular process of mandible arising lateral to an imaginary line along the outside border of incisive alveolar length. . . . . Hystricidae

Smaller animals with head and body length less than 400 mm. Pelage not as above. Skull not as above: frontal and nasal regions not inflated; crowns of cheekteeth not flat nor complexly folded;

angular process of mandible arising medial to an imaginary line
along the outside border of incisive alveolar length. 2

2. External ear absent. Tail not visible externally. Eyes small,
   covered with hairy skin. Supraoccipital large and sloping
   forward so that occipital crest reaches the level of zygomatic
   process of squamosal. Spalacidae

   External ear and tail present. Eyes normal. Supraoccipital small
   (except in genus *Nesokia*) but with occipital crest never reaching
   level of zygomatic process of squamosal. 3

3. "Squirrel" morphology (Fig. 79): tail always fully haired and
   bushy; all digits with long claws; long hind legs with characteristic
   posture. Postorbital process of the frontal present and sharply
   pointed. . . . . Sciuridae

   Mouse- or ratlike morphology. Tail and posture variable; digits
   with or without long claws. Postorbital process of the frontal
   absent or small and blunt. 4

4. Hind limbs elongate for jumping movement (tibia and fibula
   fused). Functional hind toes three. Tail long (about 150% of
   head and body length). Infraorbital foramen greatly enlarged,
   with upper root of zygomatic process posterior to lower.
   A large foramen present in the angular region of the mandible.
   . . . . Dipodidae

   Hind limbs not greatly elongated. Functional hind toes five.
   Tail length averages less than 150% of head and body. Infraorbital
   foramen not greatly enlarged. Upper root of zygomatic process
   anterior to lower. No large foramen present in the angular region
   of the mandible. 5

5. Occlusal pattern of cheekteeth prismatic (enamel forming sharp
   reentrant folds and angles). Teeth high crowned (Fig. 99).
   . . . . Arvicolidae

   Occlusal pattern of cheekteeth not prismatic. Teeth not high
   crowned (Figs. 82 and 83). 6

6. Tail bushy nearly to base. Black facial markings present.
   Zygomatic plate of maxilla completely below infraorbital
   foramen. . . . . Gliridae

   Tail not bushy, except in genus *Sekeetamys*. Black facial
   markings absent. Zygomatic plate of maxilla not completely
   below infraorbital foramen, frequently tilted upward. 7

7. Cheek pouch present. Tail very short (less than one-half
   length of head and body). . . . . Cricetidae

   Cheek pouch absent. Tail long (more than one-half length
   of head and body). 8

8. Auditory bulla greatly enlarged (length more than one-fourth
   skull length) (Fig. 82C). . . . . Gerbillidae

   Auditory bulla not greatly enlarged (length less than one-fourth
   skull length) (Fig. 82A and 82B). . . . . Muridae

FAMILY SCIURIDAE   Squirrels and marmots

The squirrel family includes nine genera with almost 300 species. One species, the Syrian squirrel (*Sciurus anomalus*), occurs in the Holy Land. The souslik (*Spermophilus citellus*) was reported in the uplands of Moab and Gilead (Tristram, 1877, 1884) but this is perhaps an erroneous record (Atallah, 1977). However, the high mountains in southern Jordan are little explored and a relict population of sousliks from the last glaciation may still persist there.

Genus *SCIURUS*   Squirrels

Sciurus anomalus   Persian or Syrian squirrel   E

Sciurus anomalus Schreber, 1758. Die Säugethiere, 4: 781. Type from Sabeka, 25 km SW Kutais, Georgia, Caucasus.

Sciurus syriacus Ehrenberg, 1828. Symb. Phys. Mamm., 1, pl. 8. Type from Lebanon Mountains. Valid subspecies *S. a. syriacus* in southern Turkey, Syria, Lebanon, and the Holy Land.

Sciurus russatus Wagner, 1842. *In* Schreber, Die Säugethiere, Suppl. 3:155. Synonym.

Sciurus historicus Gray, 1867. Annser. 3, 20:273. Type from Syria. Synonym of *S. a. syriacus*.

Fig. 79. The Persian or Syrian squirrel (*Sciurus anomalus*). Photo: A. Shoob.

Diagnosis.—See under family. This mammal is easily identified by the presence of a long tail with bushy hairs extending laterally, creating a flattened appearance (Fig. 79). The skull morphology is distinctive from other rodents, as in the key (infraorbital foramen very small). An adult male from Dibbine Forest had a total length of 353 mm, tail length of 140 mm, and greatest length of skull of 50.9 mm.

Range of species.—Iran, Caucasus, Iraq, Turkey, and the eastern Mediterranean region.

Local status (Fig. 80).—Squirrels were common in the forests of the Holy Land in the last century. Tristram (1884:15) reported this species to be "extremely abundant in woods south of Hermon and throughout the Lebanon" but "never south of Banias." The species was reported from Upper Galilee and the Golan (Bodenheimer, 1935), Mount Hermon (Harrison, 1972), Wadi Assal (Nahal Siyon) (Paz, 1982), and Dibbine Forest in northern Jordan (Atallah, 1977). Specimens at HUJ collected by Aharoni are from Mount Hermon and Kufrein. The JUNHM and JNHM have specimens from Kufranja and Dibbine Forests (Amr and Disi, 1988). The species is also known from the wooded regions of Lebanon (Lewis et al., 1967).

Biology.—According to Lewis et al. (1967) this species is abundant in certain areas of Lebanon. It appears to be active throughout the day and feeds on a diet that includes acorns and pine and cedar seeds. Nests are

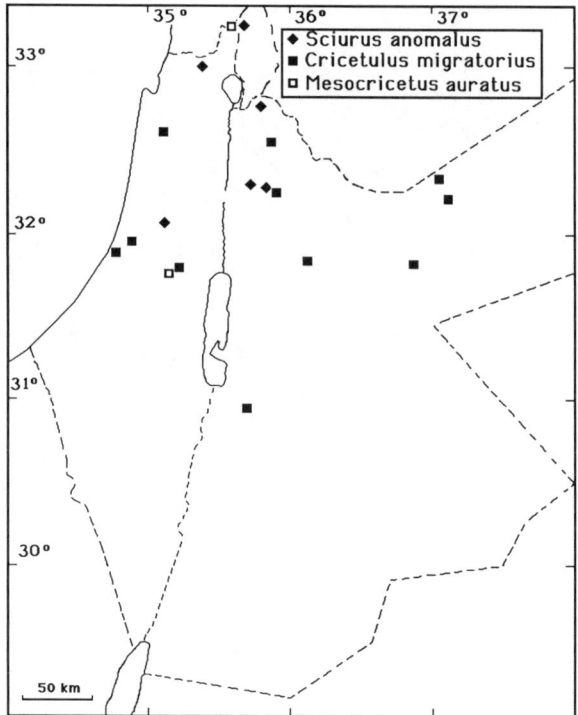

Fig. 80. Locality records of the Persian or Syrian squirrel (*Sciurus anomalus*), the golden or Syrian hamster (*Mesocricetus auratus*), and the gray hamster (*Cricetulus migratorius*).

constructed in trees or crevices and young are born April and May. In Russia, three to seven young are born in April (Ognev, 1940).

In the Holy Land, as elsewhere, the Syrian squirrel spends most of its time in trees (arboreal). Thus, its distribution is limited by the extent of forested regions. Because these forested regions have declined considerably in the past 200 years, the range of the species has similarly declined. In fact, the some populations are now isolated from other populations by uninhabited terrain. According to locals in northern Jordan, these animals are hunted frequently by children with slingshots for sport and food.

Genetics.—2N = 40, AA = 76 was reported for a specimen from Iran (Nadler and Hoffmann, 1970). The X chromosome is submetacentric and the Y is a small submetacentric.

Human interactions.—The squirrel or *sinjab* is familiar to many residents of northern Jordan, who find it to be edible. Habitat destruction and excessive hunting pressure are placing this species in grave danger. The name *Sciurus* is from the Greek *skiouros*, a squirrel, which is derived from *skia* (shade or shadow) and *oura* (tail).

FAMILY CRICETIDAE   Hamsters

The Cricetidae (derived from *criceto*, Italian for hamster) is here used in the restrictive sense to include the hamsters and their relatives. The older literature treats the Gerbillidae, Arvicolidae, and other groups as subfamilies of the Cricetidae. Hamsters are small rodents with short tails (less than 45% of head and body length). They have stout bodies, a short rostrum, and usually cheek pouches. The dental formula is i 1/1, c 0/0, p 0/0, m 3/3 = 16. The family occurs in Europe and northern Asia and contains 22 species in five genera. The Holy Land represents the most southern extension for two species.

Key to the Cricetidae of the Holy Land

1. Small animal but with relatively longer tail. Color almost uniform. Eight nipples in females. . . . . . . *Cricetulus migratorius*

    Larger animal but with relatively short tail (less than hind foot length). Color patchy brown, black, and white. Sixteen nipples in females. . . . . *Mesocricetus auratus*

Note: A third species of hamster, *Calomyscus bailwardi*, may occur in northern Jordan, as it was reported by Peshev (1989) from Hammer and Thafas in Derra District, southern Syria. It is distinguished from the other two hamsters by the absence of cheek pouches and by its elongated tail. Peshev described the form from Syria as a new subspecies, *C. b. tsolvi*, that is significantly smaller than *C. b. bailwardi*.

Genus *CRICETULUS*

Cricetulus migratorius   Gray hamster   I

Mus migratorius Pallas, 1773. Reise Pro. Russ. Reichs., 2:376, 703. Type from mountains of Lower Ural, western Siberia.

Mus phaeus Pallas, 1778. Zoog. Ross.–Asiat., Glires, i:163. The name used by Tristram.

Hypodaeus cinerascens Wagner, 1848. Archiv. Naturgesch., 1:184. Type from Syria. Perhaps a valid subspecies.

Cricetus isabellinus de Filippi, 1865. Viaggio in Persia, p. 344. Type from Iran.

Diagnosis.—Hamsters are easily distinguished from other rodents in the region by the overall stout body with short tail (Fig. 81) and the presence of cheek pouches. Also, the teeth have parallel rows of cusps unlike any other family of rodents in the region. The gray hamster is small with the greatest skull length of 24–31 mm (a specimen from Wadi Rajil at JNHM has

Fig. 81. The gray hamster (*Cricetulus migratorius*). Photo: A. Shoob.

unusually large skull, 31.5 mm in length). The gray hamster is a handsome rodent with a gray back and a conspicuous dark stripe along the back. *Mesocricetus auratus*, the other hamster in the region, differs from this species by its pronounced facial markings, relatively shorter tail, and by the presence of a supraorbital ridge on the skull.

Range of species.—According to Corbet (1978:90) this species occurs from the "steppes from southern European Russia . . . through Kazakhstan to southern Mongolia and Sinkiang. North almost to Moscow; south to Palestine, most of Iran and central Afghanistan."

Local status.—This is a northern species that reaches its southernmost distribution in the Holy Land (Fig. 80). Tristram (1884) reported this species (as *C. phoeus* Pall.) from cultivated areas in the north. Reported as rare by Aharoni (1932), who listed Biliramun (Rehobot) and Jerusalem. It is however reported as common by Bodenheimer (1935, 1958), who did not give specific localities or specimens. Harrison (1972) reported it from Lod (BM) and Ramat Hashofet (HZM) and Atallah (1977) reported it from Jerusalem (specimens at the BZM). Specimens are available at HUJ from Rehovot, Dan, Huldah, and Sasa; at JUNHM from Ghazaleh and Al Muwaqqar; and at JNHM from Irbid, H5, and Wadi Rajil. The species was also reported at Jawa (Searight, 1987*b*) and from Wadi Zarqa, S. Jerash (Bates and Harrison, 1989). In Lebanon, the species was not difficult to trap in many parts of the country near cultivated fields (Lewis et al., 1967; Atallah, 1977).

Biology.—In Lebanon, specimens were collected during winter and spring along grain fields or on rocky grounds. These nocturnal animals were associated with *Meriones tristrami* at low altitudes and with *Microtus guentheri* at higher altitudes in mountain slopes (Lewis et al., 1967; Atallah, 1977). The diet consists of grass and wheat seeds in Lebanon (Atallah, 1977). Lay (1967) made some additional observations on the biology of these animals in Iran. He reported that these animals readily attacked and ate jerboas and frogs in captivity.

Hamsters breed year-round, giving birth to 2–11 young per litter both in Armenia (Dahl, 1954) and in Iran (Lay, 1967). Birth occurs in March and April in Lebanon (Atallah, 1977). Four to seven embryos were noted in Lebanon. Similarly, females collected in Afghanistan contained six embryos in August (Hassinger, 1973). However, in Turkey, animals were collected with four to six embryos in May (Felten et al., 1971).

Genetics.—2N = 22, AA = 40 was reported for a domesticated lab animal (Lavappa, 1977).

Human interactions.—*Cricetulus* is derived from *criceto,* Italian for hamster, and *ulus,* L., diminutive suffix (little hamster). *Migratorius* is Latin for migrating.

Genus *MESOCRICETUS*

Mesocricetus auratus   Golden or Syrian hamster   **K**

Cricetus auratus Waterhouse, 1839. Proc. Zool. Soc. Lond., 1839:57 (Mag. Nat. Hist, p. 276). Type from Aleppo, Syria.

Cricetus (Mesocricetus) brandti Nehring, 1898. Zool. Anz., 559:331. Type from Georgia SSR.

Diagnosis.—The golden hamster has a head and body length of 155–165 mm and a tail length of 10–15 mm. See under *Cricetulus migratorius* for differences between these two species. This is the common pet hamster of Europe and North America. The available pets are of the nominate subspecies and are characterized by a golden yellow color, distinct facial markings, and short tail (about the length of the hind foot). The color of the local subspecies (*M. a. brandti*) is dominated by dark gray and is thus much darker than that of the nominate subspecies.

Range of species.—Rumania and Bulgaria, southwestern republics of the USSR, Iran, Turkey, Syria, Lebanon, and the Holy Land.

Local status (Fig. 80).— Tristram (1884) reported seeing this species in northern Palestine. Aharoni (1930) reported that this species is known from Metullah and later (1932) listed three specimens collected by Siehe at Mersina (southern Lebanon). A specimen at HUJ is from Qiryat Saide.

Biology.—In the Near East, Syrian hamsters were found near grain fields in varied habitats from steppes to the hill country (Aharoni, 1932; Neuhäuser, 1936). They are much rarer than the gray hamster. The Syrian hamster is nocturnal and omnivorous (feeding on plants as well as insects and other small animals). The Syrian hamster is used extensively in laboratory experiments throughout the world.

Genetics.—2N = 44 was reported for domesticated animals (Hsu and Arrighi, 1971; Popescu and Di Paolo, 1980).

Human interactions.—*Mesocricetus* is derived from *mesos,* Greek for middle, and *criceto,* Italian for hamster (a medium-sized hamster).

FAMILY GERBILLIDAE   Gerbils, jirds, sand rats

Gerbillids (derived from the Arabic word *jarbil)* are the most conspicuous rodents in the desert and subdesert regions of northern Africa and southwestern Asia. The family includes some 16 genera and 83 species distributed throughout Africa and southwestern Asia and to Mongolia. Gerbillids are distinguished from other rodents mainly by skull characteristics. The most obvious difference is the presence of enlarged ear bones (forming a structure called the auditory or tympanic bulla) (Fig. 82). Also, the infraorbital canal is rounded dorsally but narrows to a slit ventrally. The dental formula in most species is i 1/1, c 0/0, p 0/0, m 3/3 = 16. Because of the significance of the ear bones in the identification of species of this group, a brief description of the morphology of the ear bones follows.

Morphology of the ear bones in Gerbillidae.—The ear bones of gerbillids are very specialized and diversified as a primary adaptation of hearing in desert environments (Lay, 1972). This diversification allows for the use of these structures in distinguishing between gerbil species that are otherwise very difficult to separate. Because of this, a brief discussion of the structure of the gerbil ear is provided here. The reader is referred to the excellent work of Douglas Lay for more complete discussion (Lay, 1972; 1983).

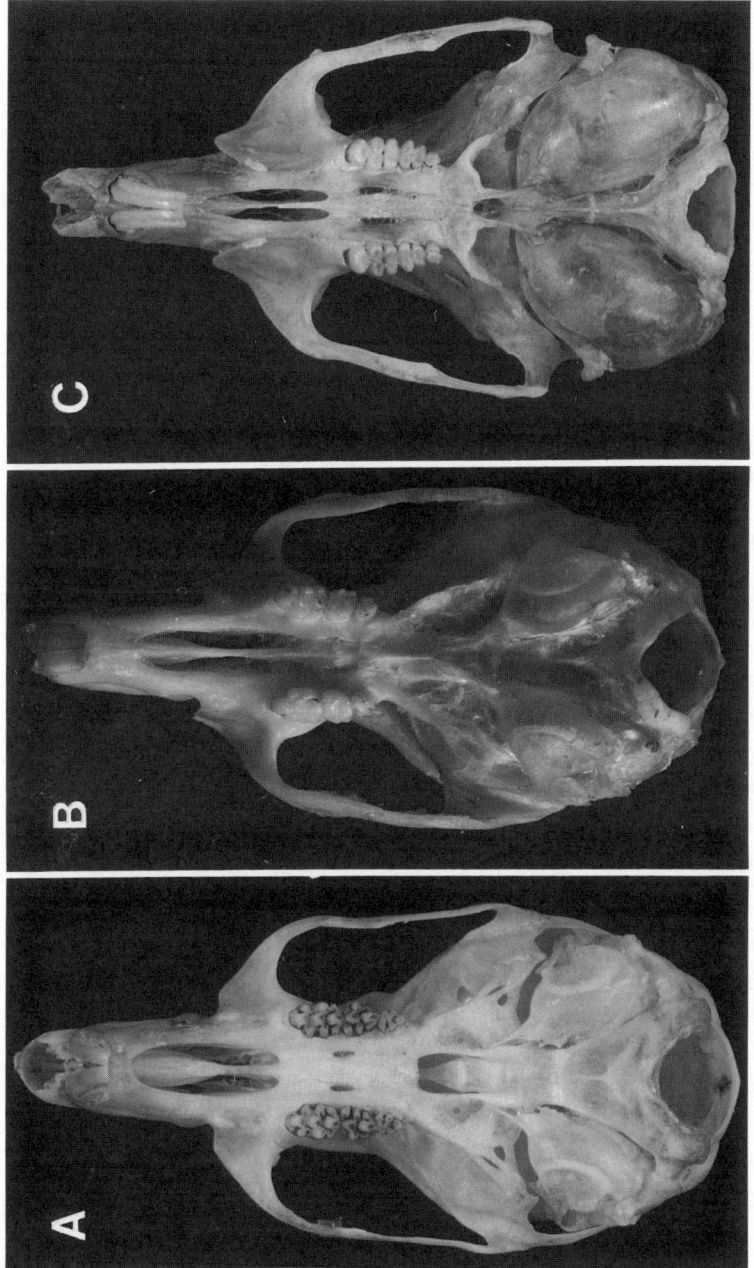

Fig. 82. Ventral views of skulls of some muroid rodents showing different morphologies. (A) *Apodemus*, (B) *Acomys*, (C) *Gerbillus*. Note the difference in teeth, palatal emargination, and enlargement of the tympanic bulla (in *Gerbillus*).

A bony capsule called the tympanic bulla houses the middle and inner ear bones in gerbils, as in many rodents. However, the tympanic bulla in gerbils is significantly expanded. This expansion affects the associated structures, as shown in Figure 89.

The degree of expansion of the tympanic bulla is variable among species of gerbils and jirds. The expansion has two effects on other structures in the skull. First, as the tympanic bulla expands, the occipital condyles become partially hidden from lateral view and as expansion continues, completely hidden. Thus, in species with small tympanic bullae (for example, *Meriones tristrami*), the occipital condyles are the most posterior bones in the skull, whereas in species with larger bullae (for example, *M. crassus*), the tympanic bullae protrude significantly behind the condyles. Second, the suprameatal triangle (the area between the supraoccipital bone and the hamular process of the tympanic bone) can be closed posteriorly in species with small bullae and can be wide open in species with larger bullae.

Another useful character in gerbil ears is the presence or absence of the membranous accessory tympanum. The primitive and widespread condition in gerbils is to have a bony shelf of the tympanic bone visible in the ear canal of the skull. The accessory tympanum is so well developed in many gerbils that it replaces a major portion of the tympanic bone. Skulls prepared from species with an accessory tympanum allow a clear view of the middle ear bones from a lateral view through the auditory meatus (see Fig. 89).

Key to the Genera of Gerbillidae of the Holy Land

    1. Upper incisors not grooved; tail short, less than 80% of head and body. . . . . *Psammomys*

       Upper incisors each with one longitudinal groove; tail longer. . . 2

    2. Tail longer than head and body length and distinctly bushy for more than 80% of the length. . . . . *Sekeetamys*

       Tail not as above. 3

    3. Cheekteeth becoming moderately to very hypsodont in adult animals showing no remnants of the original tubercles (Fig. 83C); larger forms with skull length more than 30 mm. . . . . . *Meriones*

       Cheekteeth not or only slightly hypsodont (Fig. 83B); smaller animals. . . . . *Gerbillus*

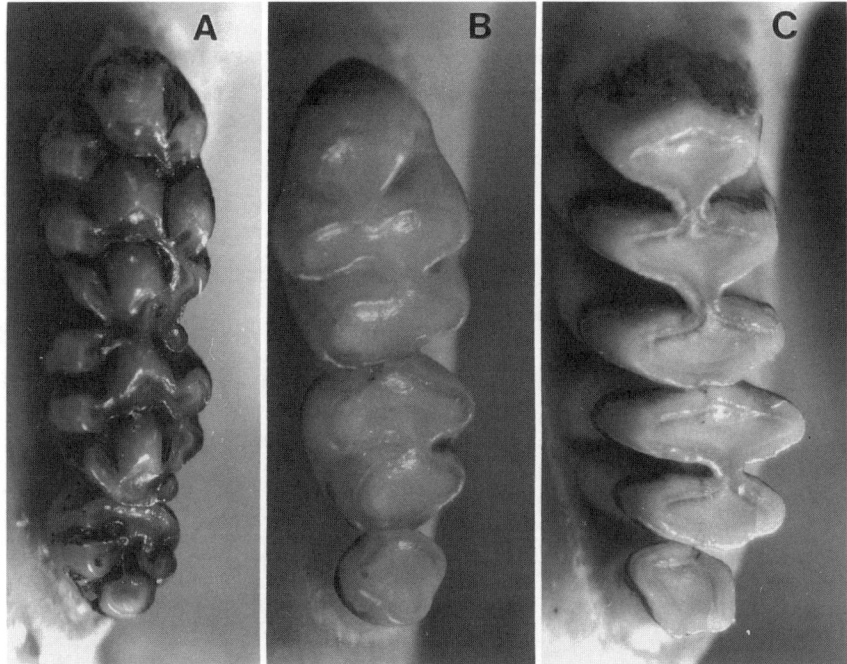

Fig. 83. Right upper cheekteeth of (A) *Apodemus*, (B) *Gerbillus*, and (C) *Meriones*.

Genus *GERBILLUS*   Gerbils

The genus *Gerbillus* (from *jarbil*, Arabic, and *illus*, L., diminutive suffix, meaning little gerbil) is a large and diversified group of small gerbils. Gerbils are distinguished from jirds (*Meriones, Sekeetamys, Psammomys*) by their tooth morphology (Fig. 83) and more slender body (and tail) and relatively longer hind legs. In some species of *Gerbillus* (for example, *G. gerbillus*), the hind legs allow an almost saltatorial movement. Many authors divide this group into two genera, *Dipodillus*, with hairy soles, and *Gerbillus*, with naked soles. The situation, however, is more complex from an evolutionary perspective and this simplification is unwarranted (Lay, 1983).

Key to the Species of *Gerbillus* of the Holy Land

1. Soles of hind feet naked (subgenus *Hendecapleura*). The posterior margin of the second tubercle on the first upper cheektooth is

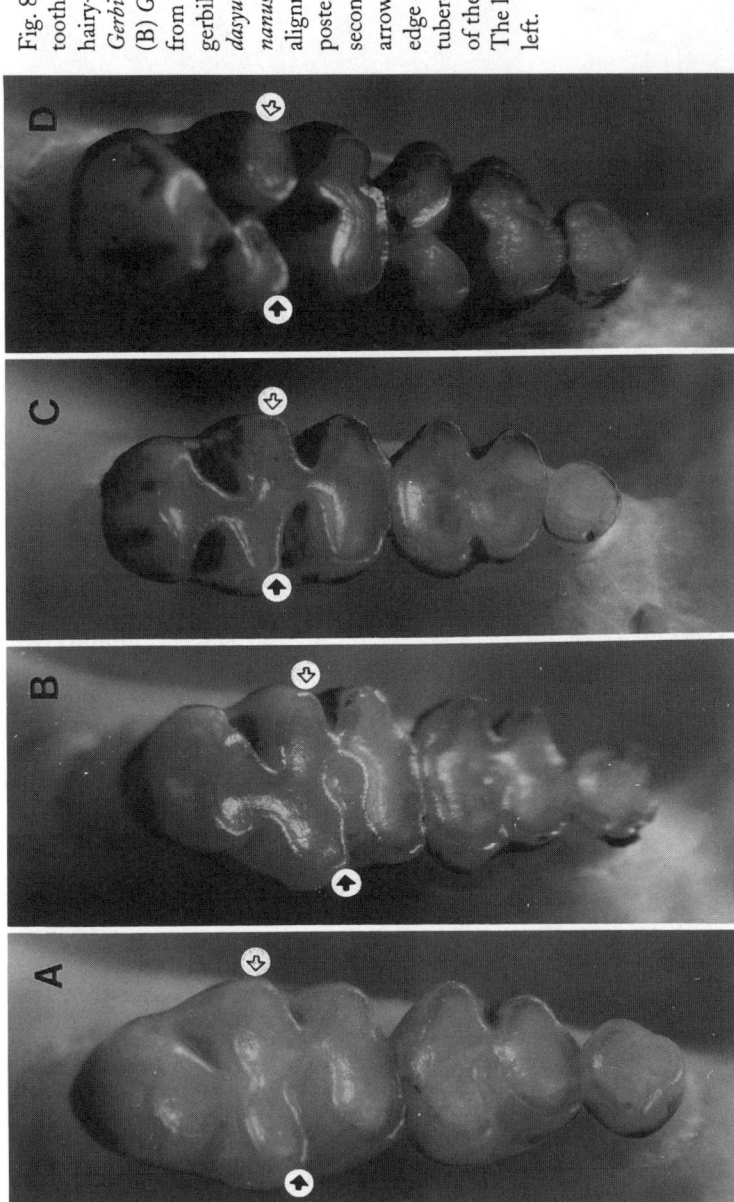

Fig. 84. Left upper toothrow from two hairy-footed gerbils, (A) *Gerbillus cheesmani* and (B) *G. gerbillus*, and from two bare-footed gerbils, (C) *Gerbillus dasyurus* and (D) *G. nanus*. Note the alignment of the posterior margin of the second tubercle (black arrow) with the lateral edge of the third tubercle (outline arrow) of the first cheektooth. The labial side is to the left.

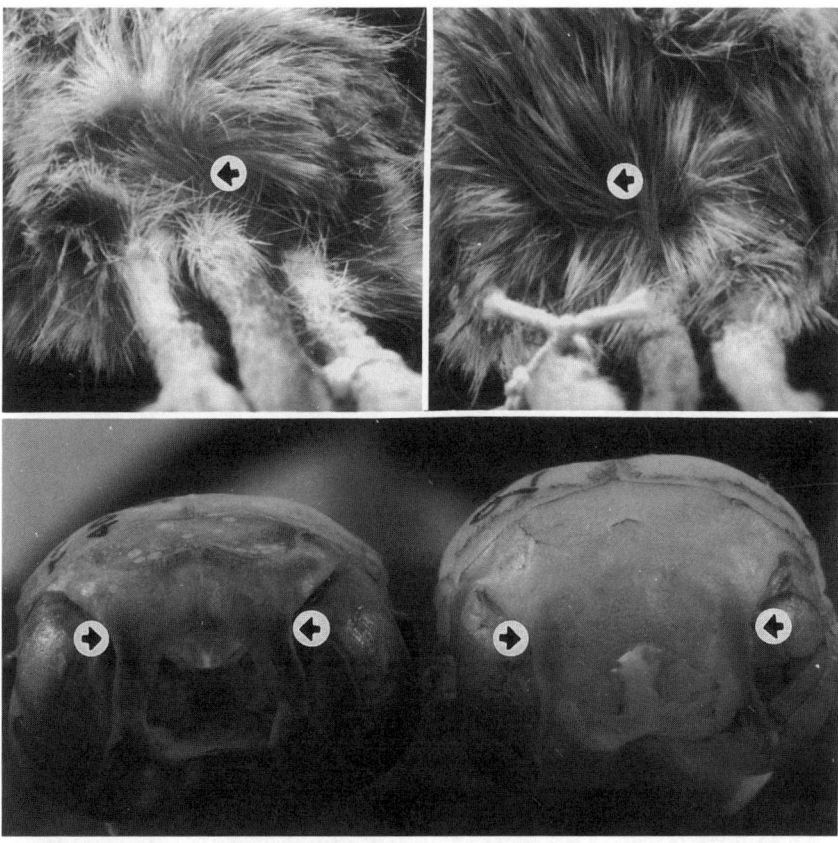

Fig. 85. Two characters to distinguish the Baluchistan gerbil (*Gerbillus nanus*, left) and the rough-tailed dipodil (*Gerbillus dasyurus*, right). In *G. nanus*, the bases of hairs on the rump are whitish and the bulla are expanded, causing shortening of the distance between the margins of the occipital region (arrowheads).

nearly in alignment with the outer edge of the third tubercle (Fig. 84C and 84D). 2

Soles of hind feet hairy (subgenus *Gerbillus*). The posterior margin of the second tubercle on the first upper cheektooth is displaced posteriorly (Fig. 84A and 84B). 4

2. Very small, hind feet less than 20 mm long, greatest length of skull less than 23 mm. . . . . *G. henleyi*

Larger than above. 3

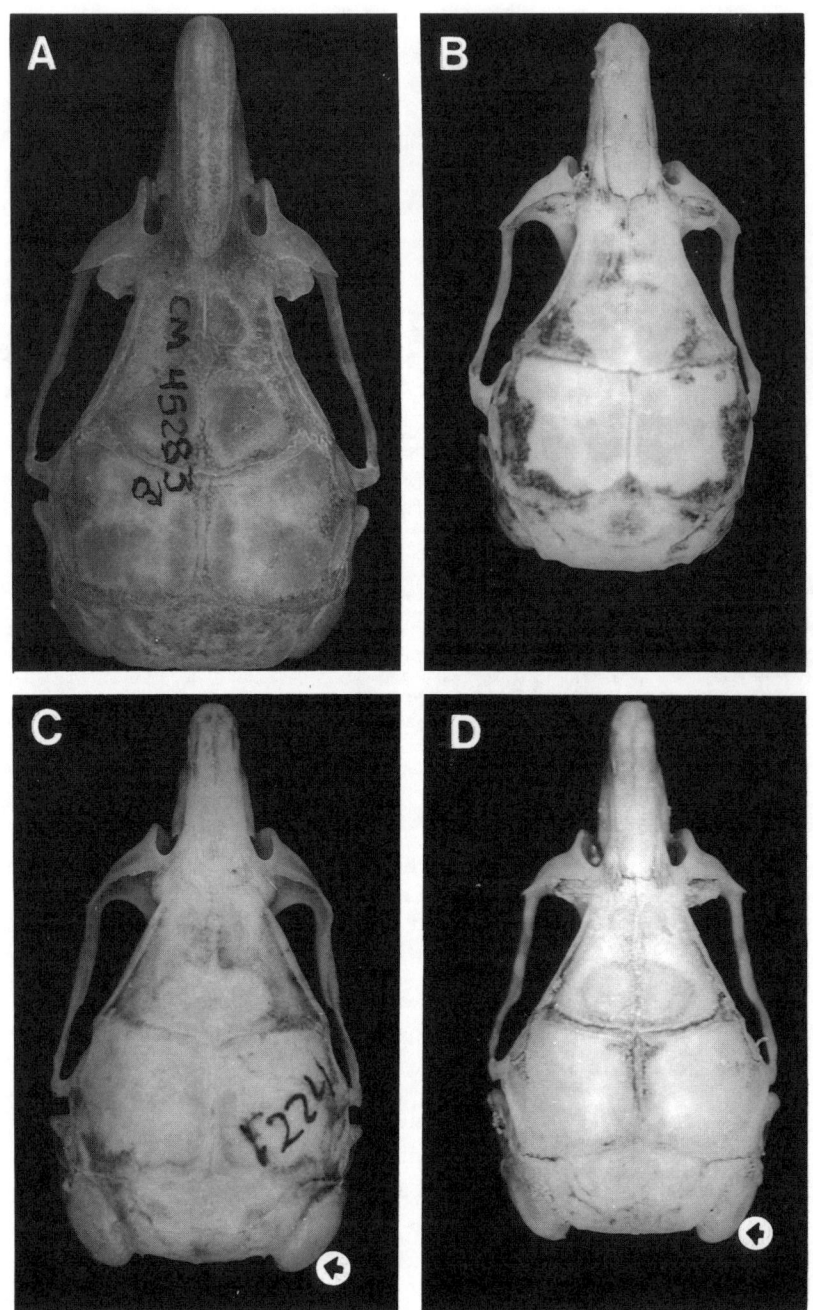

Fig. 86. Dorsal views of skulls of two bare-footed gerbils, (A) *Gerbillus pyramidum* and (B) *Gerbillus dasyurus*, and two hairy-footed gerbils, (C) *G. cheesmani*, and (D) *G. nanus*. Note the expansion of the tympanic bulla beyond the occiput in B and D (arrows).

3. Bases of hairs at rump (above tail) grayish (Fig. 85); auditory
bulla small, not extending beyond occiput (Fig. 86); auditory
meatus with anterodorsal rim round; accessory tympanum
absent (a curtain of bone within the meatus blocking its
upper half and concealing the ossicles). . . . .   *G. dasyurus*

Bases of hairs at rump (above tail) white (Fig. 85); auditory
bulla large, extending beyond occiput (Fig. 86); auditory meatus
with anterodorsal rim inflated and reflexed; accessory tympanum
present (no curtain within the meatus). . . . .   *G. nanus*

4. Bulla large with mastoid parts projecting behind occipital
condyles (Fig. 86D). . . . .   *G. cheesmani*

Bulla smaller and not projecting behind occipital condyles
(Fig. 86C).   5

5. Larger, head and body generally greater than 100 mm, hind
foot length 30–36 mm. . . . .   *G. pyramidum*

Smaller than above.   6

6. Mid-dorsal pelage clear yellowish brown, sides of feet and digits
with prominent fringes of hair, tail distinctly tufted. .   *G. gerbillus*

Mid-dorsal pelage grayish, sides of feet and digits lacking fringes
of hair, tail lacking terminal tuft. . . .   *G. andersoni* (incl. *allenbyi*)

Gerbillus andersoni   Anderson's gerbil   V

Gerbillus andersoni de Winton, 1902. Ann. Mag. Nat. Hist. ser. 7, 9:45. Type from
El Mandara, east of Alexandria, Egypt.

Gerbillus allenbyi Thomas, 1918. Ann. Mag. Nat. Hist., ser. 9, 2:146. Type from
Rehobot, near Jaffa, Palestine. Valid subspecies.

Gerbillus eatoni Thomas, 1902. Proc. Zool. Soc. Lond., 2(pt. 1):6. Type from El
Cusher, Libya.

Gerbillus bonhotei Thomas, 1919. Ann. Mag. Nat. Hist. ser. 9, 3:560. Type from
Khabra Abu Guzour, SE Al Arish, Sinai. Valid subspecies: *G. a. bonhotei*.

Diagnosis.—This is a medium-sized gerbil with brownish orange dorsal
color and a white underside. The head and body length in adults is 70–110
mm, the tail length is 90–135 mm, and the greatest skull length is 26–30
mm. The ear is conspicuous with dark pigmentation. *Gerbillus andersoni* and
G. gerbillus are similar in body proportions and in having completely haired

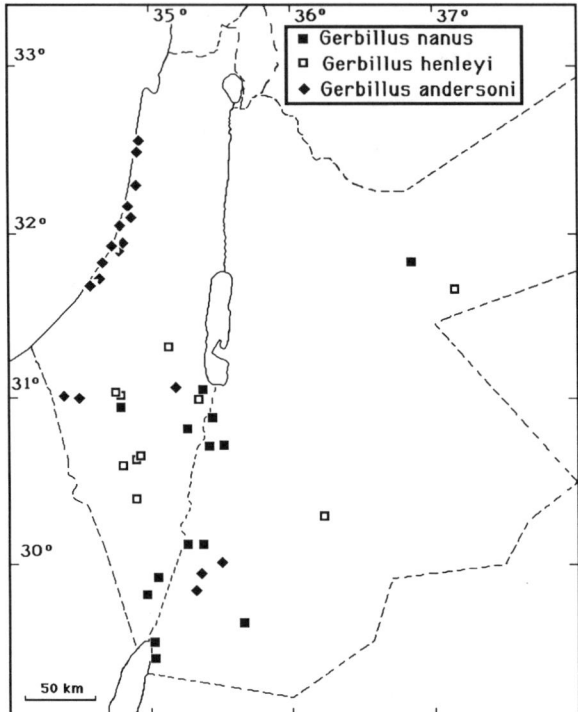

Fig. 87. Distributions
of the Baluchistan
gerbil (*Gerbillus nanus*),
the pygmy gerbil
(*Gerbillus henleyi*), and
Anderson's gerbil
(*Gerbillus andersoni*).

soles of the footpads. Based on this similarity, earlier authors placed *G. andersoni* as a subspecies of *G. gerbillus* (Setzer, 1958*b*). *Gerbillus gerbillus* is distinguished from *G. andersoni* by lighter coloration, larger ears (more than 50% of hind foot length), more rounded (rather than flat) posterior margin of the nasals, and a larger ratio of the length of the palatal/incisive foramina (Harrison, 1972; Osborn and Helmy, 1980). Also, the cusps on the first lower molar are distinct in *G. gerbillus*, whereas they fuse anteriorly to form a trifoliate structure in *G. pyramidum* and *G. andersoni* (Wassif et al., 1969). Osborn and Helmy (1980) provided an excellent comparison between the species of hairy-footed gerbils of Egypt, which can be used for the Holy Land gerbils. Chromosomal (see below) and morphological data do not substantiate the specific status of *G. allenbyi* (Petter, 1956; Harrison, 1972; Atallah, 1978). In fact, F. Petter originally believed *allenbyi* to be distinct (1956) but later (1971) changed his views.

Range of species.—Coastal sand dunes of Palestine, Jordan, Egypt, Libya, and Tunisia.

Local status (Fig. 87).—This species was reported (as *G. gerbillus allenbyi* or *G. allenbyi*) from Rehobot, Jaffa (Aharoni, 1932), Herzenja (misspelling for Herzelia), Ashdad (Ashdod), Hadassim, Holon (Petter, 1956), Ma'agan Mikhael, Qeisariya, Hadassim, SW Herzelia, Benei Beraq, Rishon le Zion, Palmahim, Nitsanim, near Ashqelon (Zahavi and Wahrman, 1957), Tel Aviv (Atallah, 1978), Agur, Holot Shunra, and Mishor Rotem (Abramsky, 1984). Records of *G. gerbillus* from Guierah (Al Quweira), 22 km N Al Quweira (Allen, 1915) and of *G. g. bonhotei* from 5 km S Ras en Negeb (Atallah and Harrison, 1967) are referable to *G. andersoni bonhotei* (Harrison, 1972).

Biology.—Along the coastal zone, this species occupies sandy areas but little is known of its biology in the region (Zahavi and Wahrman, 1957). Abramsky (1984) studied the population biology in northern Israel and reported small home ranges averaging 32–34 m². In Jordan, the species was an inhabitant of sandy habitats in wadis or mountain sides (Atallah and Harrison, 1967, as *G. gerbillus*). Food items in Egypt included *Thymelea hirsuta*, *Plantago*, and *Cutardia* (Wassif and Soliman, 1979). In Egypt, three to seven young are born during a breeding season extending from January to May (Wassif and Hoogstraal, 1959; Osborn and Helmy, 1980).

Genetics.—*Gerbillus andersoni* has a 2N = 40, AA = 76 with an XY sex determining system; whereas *G. gerbillus* has a diploid number of 42 in females and 43 in males and thus an XY1Y2 sex-determining system (Wassif et al., 1969). Specimens from the Holy Land identified as *G. allenbyi* were reported to have an indistinguishable karyotype from *G. andersoni* (Wassif et al., 1969; Lay et al., 1975). *Gerbillus allenbyi* was maintained as a distinct species from *G. andersoni* based primarily on geographic grounds and on the observation of a slightly larger Y chromosome in *G. allenbyi* (Wahrman et al., 1988). Because chromosomes in gerbillids can incorporate extra heterochromatic material with little phylogenetic significance (Qumsiyeh, 1988) and because of the allopatric distribution of *allenbyi* and *andersoni*, I consider them to be conspecific, in agreement with Petter (1971) and Cockrum et al. (1976).

Gerbillus cheesmani   Cheesman's gerbil   K

Gerbillus cheesmani Thomas, 1919. J. Bombay Nat. Hist. Soc., 26:748. Type from Mesopotamia (perhaps from near Basra, Iraq).

Gerbillus arduus Cheesman and Hinton, 1924. Ann. Mag. Nat. Hist., 14:551. Type from Jebel Dharabin, Jafura, central Arabia. Perhaps a valid subspecies for northern Arabian desert forms.

Diagnosis.—This species is similar to *G. andersoni* and *G. gerbillus* in size and form and in possessing hairy soles on the feet. It differs from both species

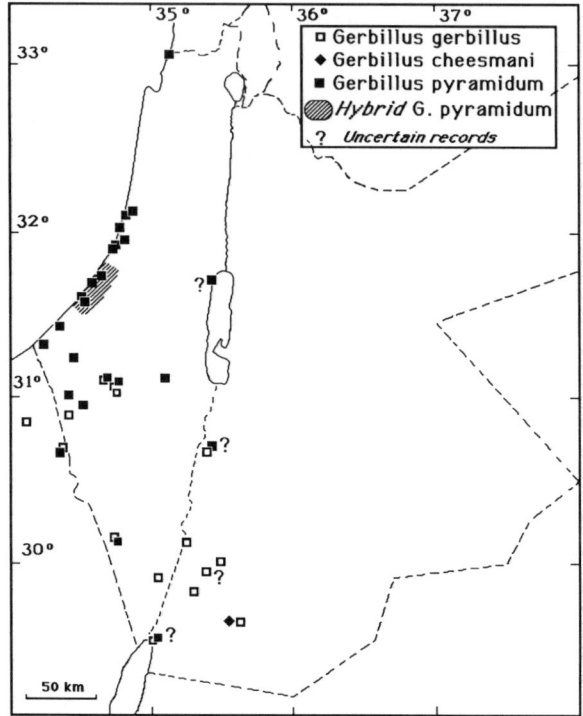

Fig. 88. Marginal and significant locality records of the lesser Egyptian gerbil (*Gerbillus gerbillus*), Cheesman's gerbil (*Gerbillus cheesmani*), and the greater Egyptian gerbil (*Gerbillus pyramidum*). A chromosomal hybrid zone (hatched lines) of *Gerbillus pyramidum* is found along the coastal region.

and from *G. pyramidum*, however, by the large size of the tympanic bulla, which is inflated beyond the level of the supraoccipital. The tail in *G. cheesmani* is relatively long, usually exceeding 120% of head and body length. *Gerbillus c. arduus* is supposed to be distinct from the nominate subspecies from Iraq by its paler sandy buff color and longer tail (nearly 150% of head and body length) (Harrison, 1972).

Range of species.—The Arabian Peninsula, Iraq, and Iran.

Local status (Fig. 88).—I collected four specimens of Cheesman's gerbil from near Disi in Wadi Rum recently. These are the first records from the Holy Land. The species is known from 30 km W Badanah and 5 km W Turaif, both in northern Saudi Arabia, close to the Jordanian border (Lewis et al., 1965). I examined skins of specimens from 22 km N Al Quweira reported as *G. gerbillus* by Atallah and Harrison (1967) and find them to have tails of 128–140% of head and body length, which would be agreeable with their being *G. cheesmani* rather than *G. andersoni* or *G. gerbillus*. Because the skulls are not available (probably deposited at the University of Teheran), I cannot confirm this identification.

Biology.—Cheesman's gerbils are found on sandy soils and mud flats in the eastern deserts. Plants associated with these habitats in Saudi Arabia are *Haloxylon salicornicum, Calligonum, Ephera alata,* and *Artemesia* (Vesey-Fitzgerald, 1953; Lewis et al., 1965; Harrison, 1972). The limited reproductive data available suggest an extended reproductive season in the Arabian Peninsula (Harrison, 1972). Up to eight young are born.

Genetics.—2N = 38, AA = 71–75 was reported from specimens from Kuwait (Badr and Asker, 1980), AA = 70 was reported from specimens from Iran (Lay and Nadler, 1975). Specimens from Jordan also have AA = 70 (unpublished data).

Gerbillus dasyurus   Wagner's gerbil, rough-tailed dipodil   O

Meriones dasyurus Wagner, 1842. Archiv. Naturgesch., 8th year, 1:20. Type from west coast of Arabia or Sinai, no exact locality.

Gerbillus dasyuroides Nehring, 1901. Schriften Berl. Ges. Naturf. Fr. Berl., 1901:173. Type from the mountains of Moab, Jordan. Perhaps a valid subspecies distinguished by yellowish rather than white above the eyes.

Gerbillus (Dipodillus) dasyurus palmyrae von Lehman, 1966. Zool. Beitr., 12(2):288. Type from Palmyra, Syria. Synonym.

Gerbillus (Dipodillus) dasyurus leosollicitus von Lehman, 1966. Zool. Beitr., 12(2):288. Type from Deir-el-Hajjar, 25 km SE Damascus, Syria. Synonym.

Diagnosis.—This species is similar in size and morphology (for example, both have naked soles) to *G. nanus.* The head and body measures 70–110 mm, the tail length is 85–145 mm, and the greatest skull length is 26.5–29 mm. The facial markings are indistinct and the bases of hairs at the rump are grayish. However, *G. dasyurus* is more robust and shows distinctly less inflated tympanic bulla than *G. nanus.* Also, the bulla is less specialized in the absence of development of the accessory tympanum. Thus, the middle ear bones are partially hidden by a bony shelf.

Harrison (1972) recognized two subspecies in the Holy Land, *G. d. dasyurus* in the mountainous regions and *G. d. leosollicitus* in the arid south and east. The latter is reportedly darker brown on the dorsal surface. However, the dorsal coloration is variable depending on the habitat. In this species, the underside is white. The bases of the dorsal hairs are gray, the middle portion is variable, and the tips are black. The tail is bicolored.

Range of species.—The Arabian Peninsula and marginally into the Eastern Desert of Egypt.

Local status.—Wagner's gerbil is present throughout the country and in all habitats. It is reported from as far north as Eilon and Naharia, as far south as 'Aqaba and many localities in Sinai, and as far west as Jawa, Azraq, and Shawmari. It is common in habitats ranging from the Mediterranean mountainous regions to the Ghor, Wadi Araba, and the arid Negev and Syrian deserts. Specific locality records for these areas are available in the literature (Allen, 1915; Aharoni, 1932; Ellerman, 1948; Wassif, 1956; Zahavi and Wahrman, 1957; von Lehmann, 1966; Atallah, 1966, 1967a, 1967b, 1977; Harrison, 1972; Qumsiyeh et al., 1986; Abu-Dieyeh; 1988). This species is also recorded from localities in Sinai near the Palestinian border: Jebel Dhalfa, El Quseimah, and Suweira (Osborn and Helmy, 1980).

Biology.—This is a very common gerbil that is present in diverse habitats in the region from the mountainous Mediterranean climates to the deserts. It does, however, seem to prefer rocky habitats in hill country and is not found in loose, sandy soils (these are usually inhabited by the competitor and closely related species *G. nanus*). As in most desert rodents, this gerbil plugs its burrow entrance in desert areas. In mountain regions, such as those around Bethlehem, Wagner's gerbils do not plug the burrow entrance. This species is reported to feed on succulent plants and some insects in Iraq (Buxton, 1923), and on seeds of annual herbs, especially *Medicago* in Saudi Arabia (Vesey-Fitzgerald, 1953).

In Egypt, captive-raised *G. dasyurus* could breed during an extended breeding season from October to June with multiple litters and a varied number of young per litter, ranging from two to six (Flower, 1932). In the Holy Land, two to six young are reported to be born in the spring or early summer (Harrison, 1972; Atallah, 1977; Searight, 1987b). In Saudi Arabia, Al-Khalili and Delany (1986) found a gestation period of 24–26 days, with a mean litter size of 5.1 (range 4–6 in the wild), and the main breeding period from February to July. They also discussed postnatal development in this species.

Genetics.—2N = 60, AA = 66,68 was reported from Azraq and AA = 70 from Ghor Nimrin (Qumsiyeh et al., 1986).

Gerbillus gerbillus   Lesser Egyptian gerbil

Dipus gerbillus Olivier, 1801. Bull. Sci. Soc. Philom. Paris, 2:121. Type from Giza Province, probably near the Pyramids, Egypt.

Gerbillus aegyptius Desmarest, 1804. Nouv. Dict. Hist. Nat. Paris Tab. Méth., 24:22. Type from near Alexandria, Egypt. Synonym.

Gerbillus gerbillus asyutensis Setzer, 1960. J. Egypt. Public Health Assoc., 35:3. Type from Wadi el Asyuti, 13 mi SE Asyut, Eastern Desert Governorate, Egypt. Valid subspecies (incl. *?longicaudatus*).

Diagnosis.—See comparison with *G. andersoni* under that account. The head and body measures 70–105 mm, the tail length is 95–125 mm, and the greatest skull length is 26–29 mm. According to Thomas (1919*b*), the tail in *G. gerbillus* is more tufted and the proximal part of the hind foot more slender than the *G. cheesmani* or *G. andersoni*. The ear length is less than one-half the length of the hind foot in this species. The posterior margins of the nasal bones are tapered and the incisive foramina are short, less than the length of the cheekteeth (Harrison, 1972). This species has a light, golden coloration that matches the sand areas it frequents. The feet are greater than 30 mm in length and the soles are covered with hair. The subspecies *G. g. asyutensis* differs from the nominate subspecies only in color, with the dorsal hairs having brownish tips, which are absent in *G. g. gerbillus* (Setzer, 1960).

Range of species.—Mainly in the Sahara Desert of northern Africa and with marginal penetration to the Holy Land.

Local status (Fig. 88).—Recent authors have confused this species with *G. andersoni*. See the discussion under that species. The following records are only tentative pending a reevaluation of these specimens. This species was reported from 'Aqaba, Wadi Araba, Guierah (Allen, 1915), Revivim, Ein Radian (Petter, 1956), the Sekher Dunes (S of Beersheba), 5 km S Halutsa, 1 km S Beer Mashabim (Bir Asluj), 3 km NW Nitsana (Auja), 1 km E Yotvata (Ain Ghidyan) (Zahavi and Wahrman, 1957), 22 km N Al Quweira (77 km N 'Aqaba on the 'Ma'an–'Aqaba highway), and 5 km S Ras En Negeb (Atallah and Harrison, 1967; Atallah, 1978). This species was also recorded from localities in northern Sinai near the border: Abu Aweigila and El Quseimah (Wassif, 1954*b;* Osborn and Helmy, 1980). Records of *G. gerbillus* from Guierah (Al Quweira), 22 km N Al Quweira, and 5 km S Ras En Negeb by Atallah and Harrison (1967) are perhaps referable to *G. andersoni bonhotei* (Harrison, 1972). Abu-Dieyeh (1988) reported collecting specimens from Wadi Araba at the following localities: 40 km S Khanzira, Fidan–Al Greigrah road, and 5 km E Al Reishah. Many other localities in the Negev Desert are represented by specimens at TAUM, but they do not add much distributional information than those listed above.

Biology.—Little is known of the biology of this species in the region. These rodents live in sand dunes and alluvial soil in northern Sinai (Osborn and Helmy, 1980) and are limited to elevations below 1100 m according to Haim and Tchernov (1974). In Wadi Araba, the species was found in salt flats (sebkha) with little plant cover (Zahavi and Wahrman, 1957) and in sand dunes (Abu-Dieyeh, 1988). The latter author reported on the structure of the burrow system of these rodents in Wadi Araba. These studies and others conducted in Egypt (Yunker and Guirgis, 1969) exemplify how rodents can create a distinct microenvironment (small niche) in their burrows having at least 10% higher humidity than that outside. There is no evidence for insect-eating behavior as suggested by Bodenheimer (1935). Instead, the food of these animals is composed mainly of desert plant seeds, leaves, buds, or fruits (Osborn and Helmy, 1980). This species is well adapted to live with a limited amount of moisture (Burns, 1956). In Mauritania a mean litter size of 4.4 is reported, with the breeding season from February to December (Klein et al., 1975).

Genetics.—The diploid number is 43 in males and 42 in females due to the presence of an XY1Y2 sex-determining mechanism with AA = 76, 78 (Wahrman and Zahavi, 1955; Lay et al., 1975).

Gerbillus henleyi   Henley's gerbil, pygmy gerbil, pygmy dipodil   K

Dipodillus henleyi de Winton, 1903. Novit. Zool., 10:284. Type from Zaghig, Wadi el Natron, Egypt.

Dipodillus mariae Bonhote, 1909. Proc. Zool. Soc. Lond., 1909(pt. 2m):792. Type from Jebel Mokattam, near Cairo, Egypt. Valid subspecies.

Diagnosis.—This is the smallest gerbil in the region (the total length measures 130–170 mm and the greatest skull length is 20–22.5 mm) and has a delicate build. The soles are naked. The dorsal color is buffy brown with contrasting white areas on the rump, above the eyes, and behind the ears. The tympanic bulla is inflated and is very large relative to the small and delicate skull. From the nominate subspecies, *G. h. mariae* is distinguished by having more prominent brownish hair on the dorsal region and the observation that the dark-tipped hairs of the hind limbs do not reach the heels (Setzer, 1958*b;* Osborn and Helmy, 1980). The diploid number for this species is 52, which is similar to that of *G. nanus*; however, *G. henleyi* has an extra submetacentric chromosome (Wahrman et al., 1988).

Range of species.—Western regions of the Arabian Peninsula, Egypt, and the northern regions of Libya and Algeria.

Local status.—The distribution of the pygmy gerbil is similar to that of the Baluchistan gerbil in the deserts of the Holy Land (Fig. 87). Records of the pygmy gerbil are available from 10 km E Dimona (Harrison, 1963*c*), Mashabei Sade, Har Arod, Makhtesh Ramon (Wadi Ruman), Mohila, Mezad Mukheila (= Mohila) (Harrison, 1972), Faidhat ed Dahikiya , and El Jafr (Atallah, 1967*a*, 1967*b*; Atallah and Harrison, 1967). Specimens at

HUJ are from Ein Tamar, and at TAUM from Tel Arad, Ein Bir, and Bir Asluj.

Biology.—In Sinai, this species was collected in stoney, gravelly wadis with vegetation of *Anabasis* spp. and *Lygos raetam* (Haim and Tchernov, 1974). In Egypt, females with four newborns were collected in June and August (Osborn and Helmy, 1980). One female examined by Atallah had six embryos and another had four and was lactating in June. Atallah (1967*a*) also commented on the small, 1–2 cm burrow entrances typical for this animal.

Genetics.—The diploid number is 52 based on specimens from Egypt (Wassif et al., 1969) and Morocco (Lay et al., 1975).

Gerbillus nanus   Baluchistan gerbil   K

Gerbillus nanus Blanford, 1875. Ann. Mag. Nat. Hist. ser. 4, 16:312. Type from Gedrosia, W of Gwadar, western Pakistan.

Dipodillus (Hendecapleura) garamantis Lataste, 1881. Le naturaliste, Paris, 3:506. Algeria. Possibly a valid species.

Dipodillus arabium Thomas, 1918. Ann. Mag. Nat. Hist. ser. 9, 2:61. Type from Tebuk, northwestern Saudi Arabia. Valid subspecies.

Dipodillus quadrimaculatus Lataste, 1882. Le naturaliste, Paris, 4:27. Type presumed from Nubia (Sudan). Questionable synonym of *G. nanus* but is used rarely as a synonym of *arabium*. This is the name used by earlier authors for the *G. dasyurus–nanus* group (in many instances without distinction) (Allen, 1915; Bodenheimer, 1935). The type specimen needs to be examined to determine the status of this form.

Diagnosis.—This species is similar to *G. dasyurus* in size and in the naked nature of the soles of the hind feet. The head and body measures 72–110 mm, the tail length is 85–150 mm, and the greatest skull length is 26.5–30 mm. The facial markings are distinct and the bases of hairs at the rump are whitish. *Gerbillus nanus* differs from *G. dasyurus* in having a more inflated bulla, the auditory meatus with anterodorsal rim inflated and reflexed, and no curtain within the meatus (Fig. 85). The accessory tympanum is well developed and thus the middle ear bones are visible in cleaned skulls. Petter (1957) considered *G. arabium* to be a valid species distinguished from *G. nanus* by an even further inflation of the mastoid portion of the bulla. Also, the color in *arabium* is much lighter. However, several studies have demonstrated that the two are conspecific (Ellerman and Morrison-Scott, 1951; Harrison, 1972; Atallah, 1978).

Range of species.—From Pakistan through Iran and the Arabian Peninsula and northern Africa.

Local status.—Baluchistan gerbils are found in arid regions of the Holy Land including the Negev, the Araba Valley, and the eastern deserts (Fig. 87). Records are available from 'Aqaba (now at the MCZ, originally reported as *Dipodillus quadrimaculatus*; Allen, 1915), 60 mi S Beersheba (Ellerman, 1948), Wadi Araba (Petter, 1957), 14 km S 'Aqaba (Kock and Nader, 1983), Sedom, Eilat, Hatseva (Ain Hussub), Nahal Hamda, Yotvata (Ain Ghidyan), and Nahal Timna (Wadi Meneya) (Zahavi and Wahrman, 1957; Wahrman et al., 1988). Specimens were also collected from 14 km S 'Aqaba (Kock and Nader, 1983), 'Aqaba, Wadi Al Khunayzir (tributary to Wadi Araba), and Ain Al Atmash (Qumsiyeh et al., 1986). The species is common in Wadi Araba; Abu-Dieyeh (1988) collected specimens at the following localities: 30 and 40 km S Wadi Al Khunayzir, Fidan–Al Greigrah road, 5 km E Al Reishah, 10 km E Gharandal, and Gharandal.

Biology.—The Baluchistan gerbil is common in desert regions of the Holy Land especially in wadis, plateaus, hammadas, and other habitats with loose soil. It has not been reported in rocky or mountainous terrains. The latter habitats are occupied by the closely similar Wagner's gerbil, *G. dasyurus*. The Baluchistan gerbil was collected from salt flats beneath *Nitraria retus* in Wadi Araba (Zahavi and Wahrman, 1957). They feed on grasses, seeds, and buds. They derive their moisture from their food and do not drink even if offered water. In Saudi Arabia and Kuwait, the species is reported from "silty wadis draining escarpment (limestone) country where varied perennial vegetation occurs" (Vesey-Fitzgerald, 1953). In Iran, Lay (1967) reported that the species was common and was collected in varied habitats from rocky hills to flood plains. However, specimens from the "rocky hills" need to be reexamined because of possible confusion with *G. dasyurus*. A recent study (Hatough-Bouran, 1990) on the ecology of "*Gerbillus dasyurus*" from the Shawmari Reserve in Jordan was actually a study of misidentified *G. nanus*. The two species come close together near the foothills, where they remain separated ecologically. For example, in the lava rocks south of Azraq Shishan, only *G. dasyurus* was collected, whereas at the soil areas at the periphery of the scattered lava rocks, I collected several *G. nanus*.

Both in Shawmari (Hatough-Bouran, 1990) and in Wadi Araba (Abu-Dieyeh, 1988), burrow systems are complex and average about 1.5 m in length. They are constructed both in hard soils and in looser soils. As in other species of *Gerbillus*, *G. nanus* is nocturnal with peaks of activity in the early and late hours of the night.

Lay (1967) reported that a female collected in Iran on 28 November had four embryos. In Oman, a female was obtained with five early embryos on

16 December and another had two embryos in 26 March (Harrison, 1968*b*). The gestation period is about 20 days. Baluchistan gerbils are preyed upon by owls, desert cats (for example, the sand cat and the caracal), and foxes.

Genetics.—2N = 52, AA = 60 was reported for specimens from the Holy Land (Wahrman and Zahavi, 1955; Qumsiyeh et al., 1986).

Gerbillus pyramidum   Greater Egyptian gerbil   K

Gerbillus pyramidum É. Geoffroy St.-Hilaire, 1803. Cat. Mamm. Mus. Nat. Hist. Nat. Paris, 1803:202. (See also I. Geoffroy St.-Hilaire, 1825. Dict. Class. Hist. Nat. Paris, 7:321.) Type from Giza Pyramid Area, Giza Province, Egypt.

Gerbillus floweri Thomas, 1919. Ann. Mag. Nat. Hist. ser. 9, 3:559. Type from Wadi Hareidin S Al Arish, northern Sinai. Valid subspecies in the Holy Land: *G. p. floweri.*

Gerbillus pygargus Cuvier, 1838. Trans. Zool. Soc. Lond., 2:142. Type from Upper Egypt. Synonym. Used by Tristram (1884).

Diagnosis.—This is the largest-hairy footed *Gerbillus* in the region, with a head and body length of 95–125 mm, a tail length of 105–160 mm, and the greatest skull length of 29–34 mm. In this regard it is very similar to *G. cheesmani* from southern Jordan. Indeed, based on available specimens it is now questionable whether *G. cheesmani* and *G. pyramidum* are distinct species. Typical *G. pyramidum* of Egypt and Sinai have smaller bullae than typical *G. cheesmani* of Iraq and Saudi Arabia. However, both species have an otherwise similar morphology and a chromosome number of 38 seems to be common to populations of both "species." Additionally, specimens of *G. cheesmani* from southern Jordan have large or intermediate bullae but are similar in their large size to those of *G. pyramidum.*

The upper color varies from orange-brown to tawny gray. The skull is relatively robust with heavy ridges. The subspecies *G. p. floweri* is distinguished from the nominate subspecies by its paler color, the absence of black hairs on the tail, the less distinct dorsal stripe, and some skull characters (Osborn and Helmy, 1980).

According to the original description of *G. p. floweri* (Thomas, 1919*c*), the palatal foramina are large, extending to the level of the front of the first upper molar; the skull is broader; and the tympanic bulla is larger than typical *G. pyramidum*. Osborn and Helmy (1980) provide the following characters for *G. p. floweri* (compared with *G. pyramidum*): dorsum pale, dorsal stripe indistinct or lacking, tail without a conspicuous black tuft (in *pyramidum* extends to greater than one-third of tail length), and nasals usually narrowly truncate posteriorly. Also, they stated that the the border between the two upper chambers of the bulla (anterior and posterior superior lateral chambers) is more posteriorly situated in *G. p. floweri*.

Range of species.—The range depends on the forms included in synonymy. If discussing *G. pyramidum* sensu lato (species in a wider sense to include other forms), then the range encompasses most of the Sahara Desert with marginal penetration into the Holy Land.

Local status (Fig. 88).—Originally reported by Tristram (1866*b*, 1884) as *G. pygargus* from the "wilderness" by Beersheba. The species was recorded from Wadi El Abiad, Ras el Feschiha (questionable locality, no specimens available, Aharoni, 1932), Negev, Holon, Wadi Rubin, Revivim (Petter, 1956), Tel Aviv, Revivim, and near Rishon le Zion (Harrison, 1972). This species was studied chromosomally by Jacob Wahrman and his colleagues (Wahrman and Zahavi, 1955; Zahavi and Wahrman, 1957; Wahrman and Gourevitz, 1973; Wahrman et al., 1988). The northern race with diploid number 50–52 was found at Holon, Rishon le Zion–Palmahim, and Yavna; hybrids were found at Nizzanim, Ashqelon, Ziqim, Yad Mordekhay, Tel Tin Fanis, Tel Al 'Ajjul; and the southern race at was found Irqeish and Sumeiri (both in Deir El Balah), Khan Yunis, NW Revivim, Beer Mashabim, Ain Turabi, Agur, Quseimah, and Sheikh Zuweid (Sinai). Other localities reported by Wahrman and his colleagues include: River Yarqon, Gevulot, Sekher Dunes (in Wadi Mishash), Halutsah, and Nahal Lavan (Wadi el Abiad). An isolated population of *G. pyramidum* with a diploid number similar to that of the Negev specimens was found at Nahal Bezet (Wadi Karkara) in the western upper Galilee and was suggested to be due to recent introduction by man (Wahrman et al., 1988). Abu-Dieyeh (1988) reported collecting specimens from Wadi Araba at the following localities: 10 km S Rahma, Fidan–Al Greigrah road, Al Reishah, and 'Aqaba. The latter records may actually be *Gerbillus cheesmani* (specimens not seen). The species was reported from localities in northern Sinai such as Al Arish and El Quseimah (Wassif, 1954*b*; Osborn and Helmy, 1980).

Biology.—Restricted to sandy areas in the Holy Land (Zahavi and Wahrman, 1957). In Sinai, this species was collected from palm groves and near cultivated areas (Hoogstraal, 1963). In northern Sinai, camel dung and seeds of *Citrullus colocynthis* were found in burrows (Wassif, 1954*b*). Burrow systems covering areas of 2–3 m^2 with two to three entrances are reported in Algeria (Petter, 1961). Petter (1961) stated that the gestation period is 25 days. In Sudan, the mean litter size is four with a breeding season from June to February (Happold, 1968).

Genetics.—The differences in diploid numbers reported for species in the *Gerbillus pyramidum* complex are striking. The diploid number is 38 in Egypt (Wassif et al., 1969; Lay et al., 1975) and 40 from Algeria (Matthey, 1953) and Tunisia (Jordan et al., 1974). On the other hand, two "cytotypes" exist in Palestine and Sinai; the northern form with 2N = 50–52 occurs in the coastal plains south of Jaffa, and the southern form with 2N = 64–66 is found in the Sinai, Negev Desert, and the coastal plains near Gaza (Zahavi and Wahrman, 1957; Wahrman and Gourevitz, 1973). The differences are thought to be due to Robertsonian rearrangements because the autosomal arm number is unchanged: AA = 76. Hybridization is known to occur between the southern form and the northern one in a 150-km^2 area in the coastal plains. The two forms are very close genetically (Nevo, 1982) and it is doubtful whether they should be considered taxonomically distinct. On the other hand, morphological differences between *G. p. pyramidum* (2N = 38) and *G. p. floweri* (2N > 50) are significant and genetic data are needed to check the distance between these forms.

Genus *MERIONES*   Jirds

This genus is rather widespread in the southern Palearctic with its center of distribution in southwestern Asia. Members of *Meriones* are distinguished from *Psammomys* by their grooved incisors, from *Sekeetamys* by their shorter, less bushy tails, and from *Gerbillus* by their more robust body and hypsodont teeth without traces of a cusp pattern (Fig. 83C).

"The genus *Meriones* is one which offers great difficulties to the systematic worker, as the different forms resemble each other very closely and have at various times been burdened with many names and but imperfect descriptions" (Bonhote, 1912:227). This statement is still valid today, as much more work is needed to resolve the taxonomy of this group. The last revision of the genus *Meriones* was published 50 years ago (Chaworth-Musters and Ellerman, 1947). The keys provided in that work are very inadequate and many problems arise when one is attempting to use the key for the jirds from Jordan. For example, *M. shawi* (including *tristrami*) is keyed as different from. *M. libycus* and *M. crassus* in having small bullae (less than one-third of the occipito-condylar length). Later, *M. s. shawi* was differentiated from other subspecies by having bullae "larger, averaging 32 percent of the occipito-nasal length." It is impossible to use a difference of less than 1% in these measurements because the coefficient of variation clearly exceeds that by several fold.

The domesticated Mongolian jird (*Meriones unguiculatus*) is a common pet animal in Europe and North America. It is frequently referred to as the Mongolian "gerbil." The common name "jird" was introduced by Shaw in 1738 and derives from the Berber name for the animal, *gherda*. Most recent authors use jird for members of the genera *Sekeetamys, Meriones,* and *Psammomys* (the subfamily Rhombomyinae). Gerbil is the common name used for other genera of gerbillids. *Meriones* and *Idomeneus* were companions in arms in the Trojan War of Greek mythology; the latter name was used and then abandoned for a subgenus of *Meriones.* A record of *Meriones vinogradovi* Heptner, 1931 from Palestine (Dobroruska, 1959) was probably erroneous (Harrison, 1972) although this species is known from Syria.

Key to the Species of *Meriones* of the Holy Land

1. Bulla small, length in horizontal plane less than or equal to diastema. Posterior margin of bulla not extending beyond level of paraoccipital (Fig. 89D). . . . . *M. tristrami*

   Bulla larger than above (Fig. 89A, 89B, 89C, and 89E). 2

2. Accessory tympanum present. Suprameatal triangle closed or almost closed posteriorly. Claws pigmented. . . . . . . *M. libycus*

   Accessory tympanum absent. Suprameatal triangle wide open posteriorly. Claws pale. 3

3. Bulla inflated with exoccipital and basioccipital constricted (Figs. 89E and 90). Ears not pigmented. Feet white. . . *M. crassus*

   Bulla not very inflated with exoccipital and basioccipital not constricted. Ears pigmented. Feet partly colored. . . *M. sacramenti*

Meriones crassus   Silky jird, sand jird, gentle jird   O

Meriones crassus Sundevall, 1843. Kongel. Svenska Vetensk. Akad. Handl., (Stockholm), p. 233, pl. II, fig. 4. Type locality in the Sinai Desert, restricted to Ain Musa by Allen (1915).
   Meriones pelerinus Thomas, 1919. Ann. Mag. Nat. Hist., ser. 9, 3:266. Type from Tebuk, on Hejaz Railway, northern Saudi Arabia. Synonym
   Meriones ismahelis Cheesman and Hinton, 1924. Ann. Mag. Nat. Hist., ser 9, 14:553. Type from Hufuf, Saudi Arabia. Synonym.
   Meriones longifrons Lataste, 1884. Proc. Zool. Soc. Lond., 1884:88. Type from Jedda, Saudi Arabia.

Diagnosis.—This is a medium-sized jird (head and body length 105–170 mm, tail length 100–150 mm, skull length 35–42.5 mm). The tail length is approximately the same as the head and body length. The hind feet are covered with white hair except a bare patch near the heel. The species usually

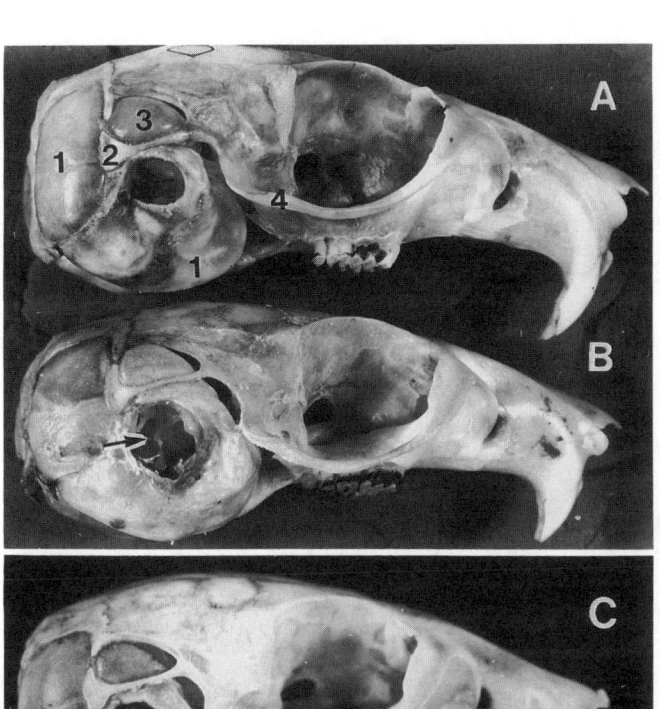
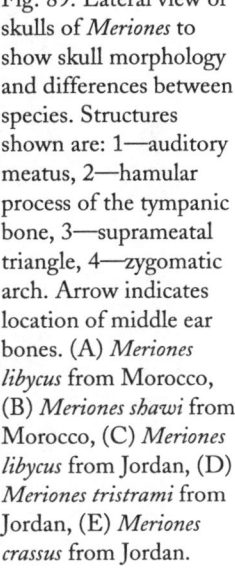
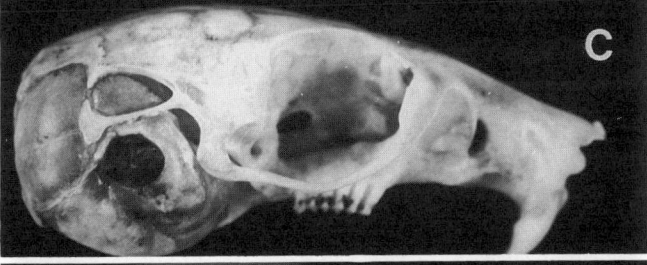
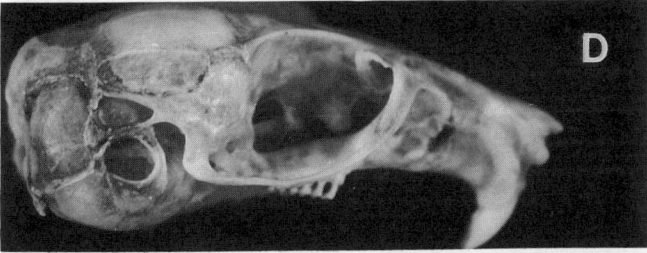
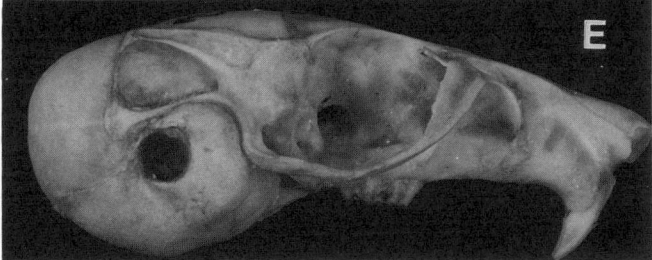

Fig. 89. Lateral view of skulls of *Meriones* to show skull morphology and differences between species. Structures shown are: 1—auditory meatus, 2—hamular process of the tympanic bone, 3—suprameatal triangle, 4—zygomatic arch. Arrow indicates location of middle ear bones. (A) *Meriones libycus* from Morocco, (B) *Meriones shawi* from Morocco, (C) *Meriones libycus* from Jordan, (D) *Meriones tristrami* from Jordan, (E) *Meriones crassus* from Jordan.

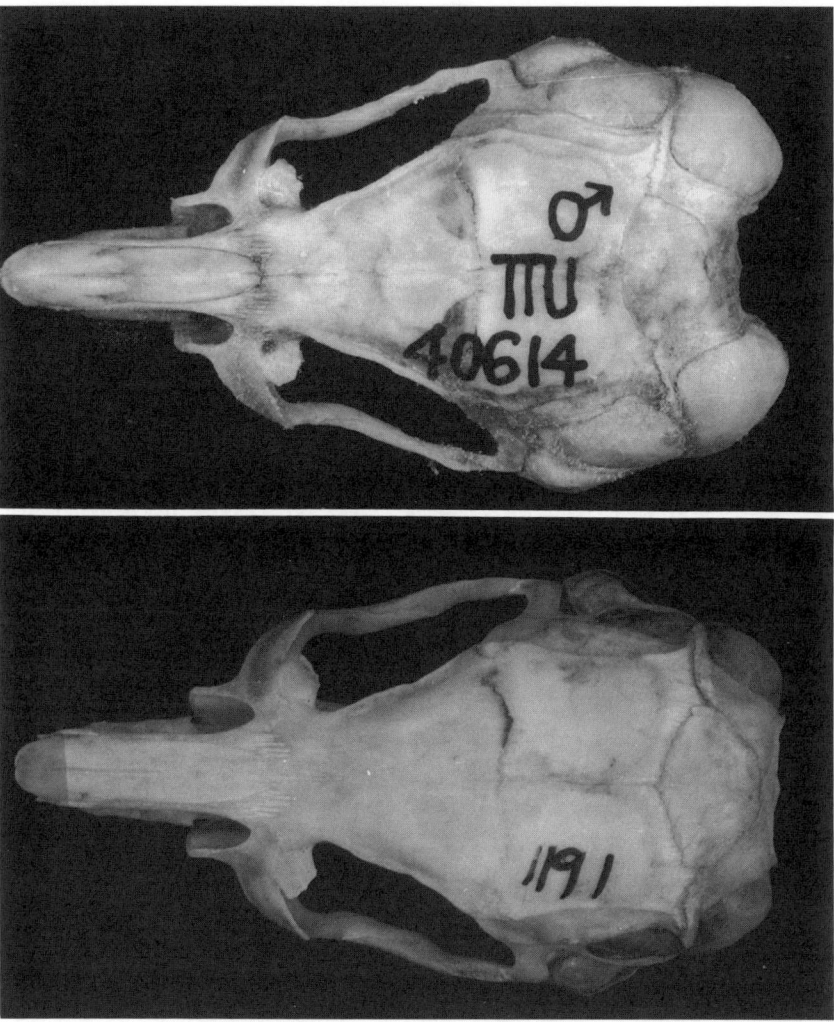

Fig. 90. Dorsal view of the skull of (Upper) *Meriones crassus*, showing extreme enlargement of the tympanic bulla as compared with (Lower) a more typical jird, *Meriones libycus*.

has pale claws. The skull is robust with strong ridges in adults. The bullae
are large, usually exceeding one-third the occipito-nasal length, with their
posterior margins extending well beyond the occipital condyles (Fig. 90).
The suprameatal triangle is not closed.

Range of species.—Southwestern part of the Palearctic region from Afghanistan through most of the Arabian Peninsula as far north as Turkey,

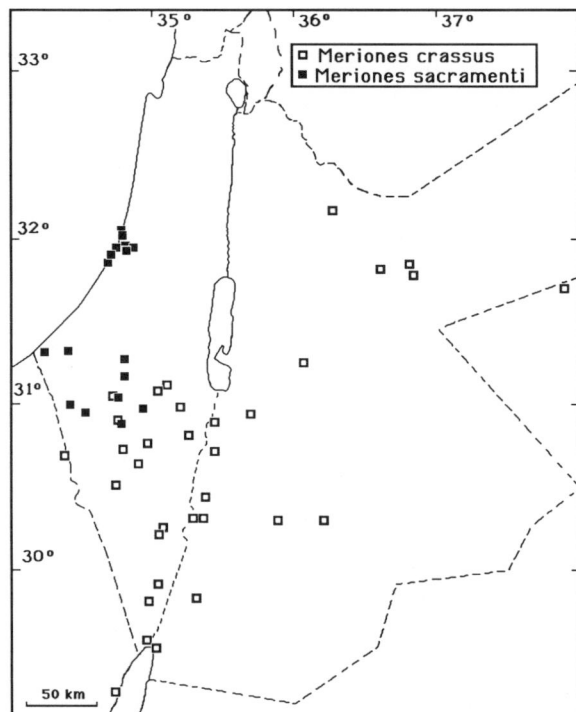

Fig. 91. Locality records of the silky jird (*Meriones crassus*), and Buxton's jird *(Meriones sacramenti)*.

and most of the arid and semiarid regions of northern Africa. This area represents the so-called Saharo-Sindian desert belt (Atallah, 1978).

Local status.—The gentle jird is common in the arid southern and eastern regions of the Holy Land (Fig. 91). It was reported from "Palestine" by numerous authors (Allen, 1915; Aharoni, 1930, 1932; Bodenheimer, 1935, 1958; Petter, 1957). Specific localities known include Revivim, Tureibe Plain, Makhtesh Hatsera, Sede Boqer, Nahal Nafha (Wadi Nafkha), Hatseva (Ain Husub), Makhtesh Ramon (Wadi Ramon), Beeret Oded (Jdeid), Nahal Hamda (Wadi Abu Hamda), Nahal Hiyon (Wadi Haiyani), Sede Avraham, Yotvata (Ghadyan), Nahal Timna (Wadi Meneya), Eilat (Zahavi and Wahrman, 1957), Wadi Ramon, Mount Hor (Maderah), near Eilat, 10 km E Dimona (Harrison, 1972), 45 km W El Jafr, El Jafr (Atallah, 1967*a*), Azraq Shishan, 2 km NW Azraq Shishan, 2 km S Shawmari Agricultural Research Station (Atallah, 1967*b*), 22 km N Al Quweira, and 3 km ENE Qasr Amrah (Atallah, 1978). I collected this species from 5 km W Azraq Shishan and from Wadi Khunayzir (Qumsiyeh et al., 1986). Abu-Dieyeh (1988) reported collecting specimens from the following localities in Wadi

Araba: (1) Bir Madhkur, along As Safi–'Aqaba main road, (2) 10 km N Fidan, (3) Al Rishah, (4) Fidan–Al Greigra road, and (5) 5 km E Al Rishah. Other specimens are available from 'Aqaba, Azraq Shishan, Ghazaleh, and Qatrana (Amr and Disi, 1988) and Rahma (JNHM). This species was also reported from Um Shomer, Suweira (specimens at the MCZ) (Allen, 1915), El Quseimah and other localities in Sinai near the border (Osborn and Helmy, 1980), and from 20 km W Turaif in northern Saudi Arabia near the Jordanian border (Lewis et al., 1965; Harrison, 1972).

Biology.—Gentle jirds prefer dry habitats in sandy or hammada deserts and are absent from mountainous regions with a Mediterranean climate. They are gregarious, but not highly colonial, and nocturnal animals. Atallah (1978) reported these jirds in both sandy areas and on bare hammada where the soil they dig contrasts with the black rocks on the surface. In suitable soils, they build elaborate tunnels up to several meters long with two or three food storage areas (Petter, 1961; Abu-Dieyeh, 1988). Although burrows are not covered during the daytime, their temperature and humidity appear to be more favorable than those outside (Briscoe, 1956). Reported food items include *Acacia* (Harrison, 1955), *Cassia acutifolia* (Hoogstraal et al., 1957), donkey melons (*Citrullus colocynthis*) (Lewis et al., 1965), *Peganum hormula* (Lay, 1967), *Medicago* (Vesey-Fitzgerald, 1953), *Zilla spinosa* and *Anabasis articulata* (Haim and Tchernov, 1974), and *Mesombryanthemum forskalii*, *Zygophyllum simplex*, *Plantago cylindrica*, *Neurada procumbens*, insects, and camel and donkey dung (Abu-Dieyeh, 1988). Among insects, the desert locust (*Schistocera*), and crickets (*Orthacantharis*), were reported to be eaten (Heim de Balsac, 1936; Vesey-Fitzgerald, 1953).

Meriones crassus has a large home range and a well-developed homing ability, returning to its burrow from distances up to 10 km (Petter, 1961, 1968). It has been suggested that this homing ability is acoustical because these animals have huge tympanic bullae and could home even when deprived of sight. The highest period of activity appears to be the first hour after twilight (Lewis et al., 1965). *Meriones crassus* seems to be more adapted to hot, dry conditions than other species of gerbils inhabiting the same area (for example, see Mermod, 1969). In laboratory tests, animals kept on a dry diet for a month lost only 5% of their weight and their urine showed a very high electrolyte concentration (Schmidt-Nielson, 1964).

Lataste (1884, 1887) provided the first detailed observations on reproductive behavior and other aspects of the biology of this species. The young are born blind, pink, and with long vibrissae. They are weaned at 1 month and can reproduce in captivity before 2 months of age. Gestation lasts 21 days but can be as high as 31 days (Petter, 1961). In Egypt one to five young (with an average of 3.3) are born during an extended breeding period from November to June (Osborn and Helmy, 1980).

Genetics.—The chromosome number and morphology are distinctive for this species, 2N = 60, AA = 70; this is the same for specimens from Egypt, Iran, and the Holy Land (Zahavi and Wahrman, 1957; Nadler and Lay, 1968; Qumsiyeh et al., 1986).

Meriones libycus   Libyan jird   O

Gerbillus libycus Lichtenstein, 1823. Verz. Doublet. Zool. Mus. Univ. Berlin, no. 9, p. 5. Type from Libyan Desert and fixed to near Alexandria, Egypt by Chaworth-Musters and Ellerman (1947).

Meriones melanurus Rüppell, 1842. Abhandl. Senckenb. Mus., 3, 2:95. Type from Alexandria, Egypt. Extralimital.

Merionus syrius Thomas, 1919. Ann. Mag. Nat. Hist., ser. 9, 3:268. Type from Karyatein, Syria. Valid subspecies: *M. l. syrius.*

Meriones arimalius Cheesman and Hinton, 1924. Ann. Mag. Nat. Hist. 14:554. Type from Jabrin, Saudi Arabia. Perhaps a valid subspecies. Extralimital.

Merionus syrius edithae Cheesman and Hinton, 1924. Ann. Mag. Nat. Hist. 14:555. Type from Khudud Spring, Hufuf (El Hufuf), Saudi Arabia. Synonym.

Merionus syrius evelynae Cheesman and Hinton, 1924. Ann. Mag. Nat. Hist. 14:555. Type from Khorasan Spring, Hufuf (El Hufuf), Saudi Arabia. Synonym.

Diagnosis.—The Libyan jird is a medium-sized jird (head and body length average less than 140 mm) with dark yellowish brown color. The ears are not pigmented. The claws are pigmented. The terminal tuft on the tail is better developed than in *M. crassus*. The accessory tympanum is present. The suprameatal triangle is closed or almost closed posteriorly. The auditory meatus is swollen to the level of the zygomatic process of the tympanum. *Meriones libycus* is difficult to separate from the northern African *Meriones shawi* (Rozet, 1833, Voyage dans la Regence d'Alger, 1:243, type from Oran, Algeria). Various characters were proposed to distinguish between the two species but some of these characters are variable or intermediate in Jordanian specimens (Table 1). The reasons for this are unclear, but it is possible that this is another case of character displacement that otherwise breaks down outside the areas of sympatry. The specimens from Jordan that I have examined all have dark claws. The specimens from the Holy Land are intermediate in many characters including the degree of inflation of the tympanic bulla. The structure of the tympanic bulla, although intermediate in morphology, more closely resembles that of typical *M. libycus* than *M. shawi.* Many of the skull characters approximate those of *M. libycus* rather

Table 1. Differences between *M. shawi* and *M. libycus* according to various authors.

	M. shawi	Holy Land	*M. libycus*
Claws [1,3]	Pale	Dark	Dark
Tail relative to head and body length [2]	Less than	Intermediate	More than
Ear color [3]	Distal one-third blackish	Pale	Pale
Exposure of infraorbital foramen in lateral view [3]	Exposed	Exposed	Not exposed
Length of bulla relative to occipito-condylar length [1]	Greater than one-third	Intermediate	Less than one-third
Suprameatal triangle [3]	Open	Open	Closed or almost closed by hamular process
Inflation of meatal lip of bulla [3]	Moderate	Conspicuous but less than *libycus*	Conspicuous, touching or almost touching zygomatic arch
Posterior margin of nasal relative to frontopremaxillary suture [3]	Anterior to	Anterior to	Level with
Bulla extending well beyond paraoccipital [3]	Yes	Intermediate	No
Accessory tympanum [3]	Absent	Present	Present

[1] Chaworth-Musters and Ellerman (1947)

[2] Aulagnier and Thevenot (1986)

[3] Osborn and Helmy (1980)

than those of *M. shawi*. Also, in countries where both *M. shawi* and *M. libycus* occur, the latter is more of an inland species. Most specimens from the Holy Land are from areas away from the coastal regions. These observations strongly suggest that the name for the examined specimens from the Holy Land should be *M. libycus*. I thus conclude that the specimens reported from Jordan as *M. shawi* and *M. libycus* are all referable to *M. libycus*. The name *syrius* described from Karyatein, Syria, applies to this species and probably represents a valid subspecies of *M. libycus*.

According to Aharoni (1932), in *M. libycus* ("*erythrourus*") the zygomatic arch lies on the auditory meatus, whereas in *M. sacramenti* it is spaced or barely touching. It may well be that the eastern Mediterranean *M. sacramenti*

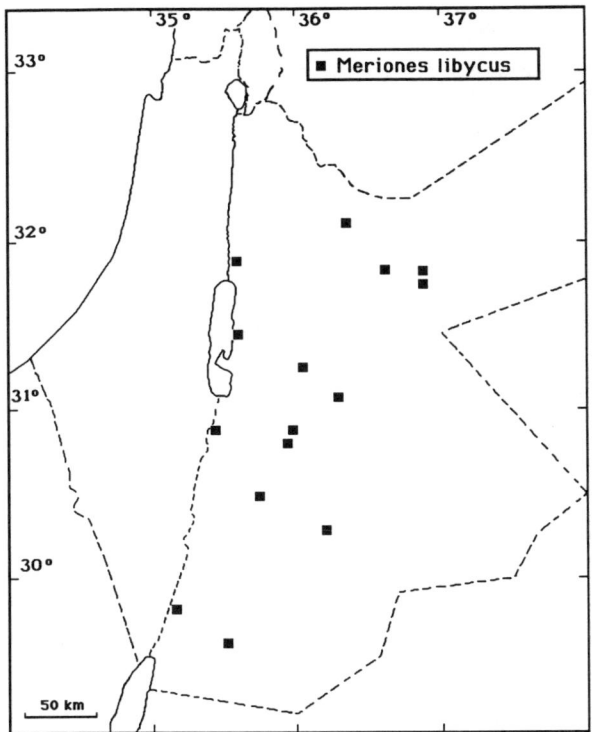

Fig. 92. Locality records of the Libyan jird (*Meriones libycus*).

and *M. syrius* correspond to the northern African *M. shawi* and *M. libycus*. This view is supported by many morphological characters and is also supported by distributional data: *M. shawi* occurs near the sea and *M. libycus* is found more inland. The form *arimalius* has been considered a valid pale-colored form of *M. libycus* (Harrison, 1972).

Range of species.—Egypt, Libya, Arabia, southwestern Asia to Azerbaijan SSR and Pakistan.

Local status.—The Libyan jird has so far been reported in the Holy Land only from areas east of the Rift Valley (Fig. 92). The species was first reported from Beersheba and Nahr Al Rubin (Zahr Al Rubin, near Jaffa) (Aharoni, 1932; Chaworth-Musters and Ellerman, 1947; Ellerman and Morrison-Scott, 1951). However, these records actually belong to *M. sacramenti* (Zahavi and Wahrman, 1957). More established records are reported (as *M. libycus* or *M. shawi*) from Azraq Shishan (Atallah, 1967*b*), El Jafr (Atallah, 1967*a*), 5 km N Al Hasa, 12 km N Al Hasa, 3–4 km ENE Qasr Amrah, 2–4 km S Shawmari Agricultural Research Station, 45 km N Ma'an (Atallah, 1978). Additional specimens are reported from Al Hallabat, Shawmari

Wildlife Reserve, Wadi el Khunayzir (Tributary of Wadi Araba) (Qumsiyeh et al., 1986), Azraq Shishan, King Hussein Bridge, Qatrana, Qasr Al Hallabat, Al Mahammadia, and El Jafr (Amr and Disi, 1988). A specimen from El Jafr is also available at the JNHM. The species was reported from northern Sinai at Bir Lehfan (14 km S Al Arish) (Wassif, 1954b).

Biology.—In the Holy Land, this species seems to occur in the lower parts of wadis or basins. According to Atallah (1978:16) these jirds "are locally abundant forming large colonies in wadis or along highways." They construct elaborate burrow systems with storage compartments and numerous entrances. In Jordan (Atallah, 1978) and northern Saudi Arabia (Lewis et al., 1965), this species appear to feed on the donkey melon (*Citrullus colocynthis*), as well as other vegetation. In Egypt, this species was found in sand mounds near *Nitraria retusa*, *Lycium* sp., and other plants (Osborn and Helmy, 1980).

Although mainly nocturnal, there are reports of diurnal activity in this species (Harrison, 1956; Lewis et al., 1965; Ranck, 1968; Osborn and Helmy, 1980). In Egypt, two to four young are born in April and May (Osborn and Helmy, 1980). A female collected 26 October in Iran contained four embryos (Lay, 1967).

Genetics.—2N = 44, AA = 74 was reported for specimens from Jordan (Qumsiyeh et al., 1986). Although we stated in that paper that the specimens from Jordan have a similar karyotype to *M. shawi*, this is not correct; the sex chromosomes and banded karyotype are more similar to *M. libycus* (M. B. Qumsiyeh, unpubl. data).

Human interactions.—This common jird in northern Africa and Arabia was named *M. libycus* in reference to its description first from Libya.

Meriones sacramenti   Buxton's jird, Negev jird   **K**

Meriones sacramenti Thomas, 1922. Ann. Mag. Nat. Hist. ser. 9, 10:552. Type from 10 mi S Bir As Seba (Beersheba), Palestine.
Meriones erythrourus legeri Aharoni, 1932. Z. Säugetierkd., 7:202. Type from Wadi el Abiad, southwest of Beersheba, Palestine. Synonym.

Diagnosis.—This is a large jird (head and body length 140–190 mm, tail length 110–170 mm, greatest length of skull 41–45.5 mm) with a dark cinnamon-brown color. The tail has a conspicuous black brush. The feet are pale with pale claws. The skull is robust and more angular and ridged than any of the other species in the region. The tympanic bulla is inflated and the accessory tympanum is absent. This species appears to be closely related to *M. shawi*, from which it differs in having a larger tail brush, a less inflated mastoid region, and a less exposed infraorbital foramen (Osborn and Helmy, 1980).

Range of species.—Endemic to the Negev and the coastal region of northern Sinai and the Holy Land.

Local status (Fig. 91).—This species was recorded from Beersheba, 16 km S Beersheba (Thomas, 1922; Chaworth-Musters and Ellerman, 1947; Petter, 1957), Jaffa, Wadi El Abiad (Aharoni, 1932), Holon, Rishon le Zion, Palmahim (Nabi Rubin), Nir Gallim, Nahal Lavan (= Wadi el Abiad), Beer Mashabim (Bir 'Asluj) (Zahavi and Wahrman, 1957), Zahr el Rubin, Beersheba, Ramleh, Rishon le Zion, and Rafah (Osborn and Helmy, 1980). Other localities represented by specimens at the HUJ museum and TAUM are from Nes Ziyona, Nir 'Oz, Sede Boqer, Revivim, and Agur Sand Dunes.

Biology.—This species lives in coastal dunes with limited vegetation cover. Zahavi and Wahrman (1957) suggested that this species is colonial.

Genetics.—The diploid chromosome number is 46 (Zahavi and Wahrman, 1957).

Meriones tristrami   Tristram's jird   O

Psammomys tamaricinus Pallas, 1778. Nova Spec. Quad. Glir. Ord., p. 322. Type from Saraitschikosk, near mouth of the Ural River, USSR. Used by Tristram, 1884, Fauna and Flora of Palestine, no. 51, p. 13 as *Gerbillus tamaricinus*. Forgotten name.

?*Gerbillus taeniurus*, Tristram, 1884. Fauna and Flora of Palestine, no. 46, p. 12.

Meriones tristrami Thomas, 1892. Ann. Mag. Nat. Hist. ser. 6, 9:148. Type from the Dead Sea region, Palestine.

Meriones blackleri Thomas, 1903. Ann. Mag. Nat. Hist., ser. 7, 12:189. Type from Smyrna, Turkey.

Meriones tamaricinus bodenheimeri Aharoni, 1932. Z. Säugetierkd., 7:197. Type from Kafrun, Nussarieh Mountains, Syria. Synonym.

Meriones tamaricinus karieteni Aharoni, 1932. Z. Säugetierkd., 7:200. Type from El Karjatein (Karyatein), Syria. Synonym.

Diagnosis.—Tristram's jird is a small *Meriones* with the soles of the hind feet naked at the heels. The head and body length is about 100–155 mm, hind foot length is 24–37 mm, and greatest length of skull of 32–40 mm. It is distinguished from other jirds of the region of similar size by its rather small bulla. The mastoid portions of the bulla never extend beyond the level of the occipital condyles. The chromosomes are also distinctive with 2N = 72, AA = 76-80 (Zahavi and Wahrman, 1957; Qumsiyeh et al., 1986). Harrison (1972) recognizes three subspecies in the region. *Meriones t. tristrami* is small, reddish brown, and occurs in the hills west of the Jordan

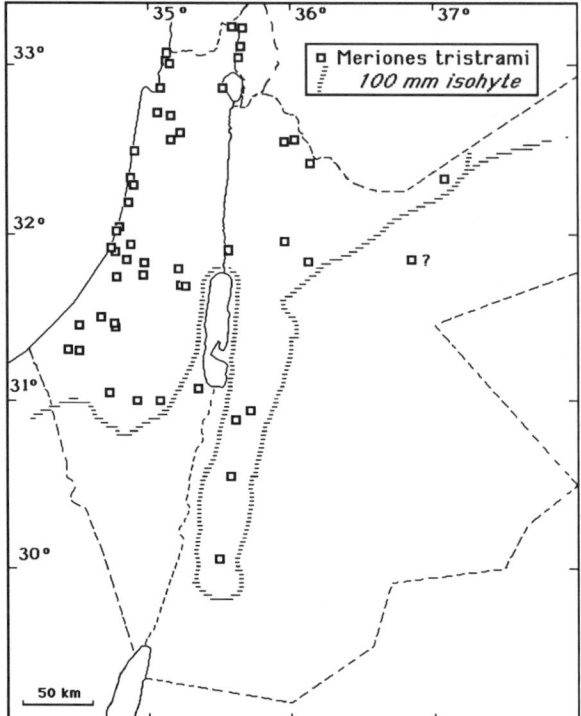

Fig. 93. Locality records of Tristram's jird (*Meriones tristrami*) and line of 100 mm average annual rainfall.

River. *Meriones t. bodenheimeri* is larger, darker, and more grayish and occurs in Lebanon, possibly entering northern Israel. *Meriones t. karieteni* is a small pallid form present in the Syrian Desert and in northern Jordan. Specimens I collected from Al Muwaqqar are referable to the latter form. An adult male from Al Muwaqqar was pale in color and had a greatest length of skull of 38.3 mm as compared with 32–37 mm found in specimens I collected from the Jordan Valley and near the Dead Sea. However, many specimens are intermediate in size and color (for example, those from Irbid). The clinal nature of this variation (influenced by humidity, elevation, and substrate) precludes formal recognitions of subspecies.

Range of species.—Turkey through the southern regions of the USSR south to northwestern Iran and west through Iraq, Jordan, Syria, Palestine, and marginally in Sinai.

Local status.—Tristram's jird is common in most areas of the Holy Land, avoiding only the extreme arid deserts of the south and east (Fig. 93). Its distribution seems to be influenced by rainfall and is limited to areas receiving more than 100 mm of rainfall annually. Tristram (1884) collected two specimens from the Dead Sea and Carmel Caves, which he referred to "*Gerbillus taeniurus*" and "*Psammomys tamaricinus.*" These two specimens were those on which Thomas (1892) based his description of *M. tristrami*. The southernmost localities known for this species in the western highlands include Revivim, Rekhme, Kurnub, and Ashalim. Reports from this area include Mount Carmel (Thomas, 1892), Jaffa (Nehring, 1901), Shobek, Tafila (Allen, 1915), Jaffa, Tabgha, Rehobot, Beersheba, Jezreel Plain (Aharoni, 1932), Mount Carmel, Beersheba, Ramleh (as *M. shawi tristrami*) (Chaworth-Musters and Ellerman, 1947; Ellerman, 1948), Plain of Esdraelon (Jezreel), and Jerusalem (Bodenheimer, 1949a). Zahavi and Wahrman (1957) reviewed these and other records and added the following important localities: Kefar Gileadi, Dan, Lahavot Habashan, Hulata, Rosh Haniqra (Ras en Naqura), Sa'ar, Kabri, Naharia, Emeq Zevulun, Allonim, Mishmar Ha'Emeq, Qeisariya, Natanya, Hadassim, Kfar Shemaryahu, Holon, Palmahim (Nabi Rubin), Rehovot, Har-el, Tsor'a, Arugot, Dorot, Beit Qama, Beeri, Magen, Urim, Revivim, Ashalim, Yeroham (Rekhme), and Mamshit (Kurnub). *Meriones tristrami* is also known from Amman (von Lehmann, 1966), Azraq ed Druz, Azraq Shishan (Atallah, 1967b), Al Muwaqqar, 10 km E Irbid, Ghor Nimrin (near King Hussein Bridge) (Qumsiyeh et al., 1986), Jawa (Searight, 1987b), Bethlehem, Beit Sahur, Deir Ibn Ubaid, Ekron, Rehobot, Tabgha, Jezreel Valley, 5–10 km S Ras en Negeb, 2 km N Azraq Shishan, Schniller Boarding School (Amman) (Atallah, 1978; and M. B. Qumsiyeh, personal collection), and Ramtha (specimen at JNHM). Additional specimens are available at the JUNHM from Jawa, Irbid, Surra Reserve Station, and Jebel Ghazaleh (Amr and Disi, 1988).

Biology.—This is one of the most common jirds in the hills of Mediterranean climates in the region. Its range appears to correspond well to areas with an annual rainfall of more than 100 mm (Fig. 93). However, exceptions to this are found in animals collected from areas in northeastern Jordan and eastern Syria in desert habitats. These pale desert forms are usually referred to as *M. t. karieteni*. Burrows are variable in size and complexity and can be as short as 50 cm and as long as several meters (Petter, 1961). Bodenheimer (1949a) reported breeding throughout the year (peaks between April and September) with a gestation period of 25–29 days and one to seven young (average 3.6). Atallah (1978) reported that this species appears to have a 24-day gestation period with six to eight young born in late spring.

Genetics.—2N = 72, AA = 76–80 was reported for specimens from the Holy Land (Zahavi and Wahrman, 1957; Qumsiyeh et al., 1986, 1988).

Human interactions.—The name of the species is appropriate; Harrison (1972:572) wrote "it seems most appropriate that this characteristic mammal of the Palestine countryside should bear the name of its discoverer, Canon Tristram, whose diligent researches and many discoveries in this region truly entitle him to be remembered as the father of Natural History in Palestine." In some areas of Palestine, this species can become a pest (Zahavi and Wahrman, 1957).

Genus *PSAMMOMYS*   Fat sand rats

The fat sand rats are heavily-built and superficially more resemble a rat than a gerbil. They have a relatively short tail and small, haired ears. The skull is robust and angular with strong ridges. They are adapted for colonial, diurnal life in arid and semiarid environments. There are two species of this peculiar genus, of which one, the fat sand rat (*Psammomys obesus*), occurs in the Holy Land.

Psammomys obesus   Fat sand rat   O

Psammomys obesus Cretzschmar, 1828. *In* Rüppell, Atlas Reise Nördl. Afr., Säugeth., p. 58, pl. 22, 23. Type from near Alexandria, Egypt (at the Senckenberg Museum).

Psammomys terraesanctae Thomas, 1902. Ann. Mag. Nat. Hist. ser. 7, 9:363. Type from the Dead Sea region, Palestine. Valid subspecies.

Diagnosis.—This is a large rodent (in *P. o. terraesanctae*, head and body length 125–185 mm, greatest skull length 43–49 mm) with a distinctive robust body and short, haired ears (Fig. 94A). The tail is thick and shorter (at 95–160 mm) than the head and body length. The skull is robust and is strongly angular. Unlike *Meriones*, the upper incisors are smooth anteriorly. From the nominate subspecies, *P. o. terraesanctae* is distinguished by its smaller size, paler color, flatter dorsal profile of the skull, and by having a broad stripe of clear, pale yellow color along the sides and by the shorter extent of the black hairs on the dorsal side of the tail (reaching 40–50% of tail) (Osborn and Helmy, 1980).

Range of species.—Occurs in northern Africa from Morocco to Egypt and Sudan and also in Saudi Arabia, the Holy Land, and Syria.

Local status.—The fat sand rat is common in desert and semiarid regions of the Holy Land (Fig. 95). It was first reported as "extremely abundant in sandy places about the Dead Sea, and also in the plains and uplands of Southern Judea" (Tristram, 1884:12). Records are available from Beersheba (Hart, 1891), Swemi (Nehring, 1901), Ain Abu Heran (north of 'Aqaba)

Fig. 94. The fat sand rat (*Psammomys obesus*). (Upper left) External appearance of a juvenile, (Upper right) *Nitraria retusa*, a staple in the diet of the sand rat, (Lower left) colony along a ridge at Al Halabat, (Lower right) child from Al Halabat with cutaneous leishmaniasis, a disease caused by a parasite whose reservoir host is the sand rat.

(Allen, 1915), Ain Gedi, Sedom (Aharoni, 1932), Rafah (Flower, 1932), Wadi Kelt, Kallia, Ain Gedi, Sedom, Tureiba Plain, Beeret Oded (Bir Jdeid), Nahal Mor (Wadi Murra), Nahal Nafha (Wadi Nafkh), Yotvata (Ain Ghidyan) (Zahavi and Wahrman, 1957), Bir Jdeid, 40 km on the Beersheba–Nizzana road, Wadi Kelt (Harrison, 1972), Jericho (Wahrman et al., 1988), El Jafr (Atallah, 1967*a*), Azraq Shishan, 2–4 km S Shawmari Agricultural Research Station (Atallah, 1967*b*), 5–10 km N Jiza, 45 km N

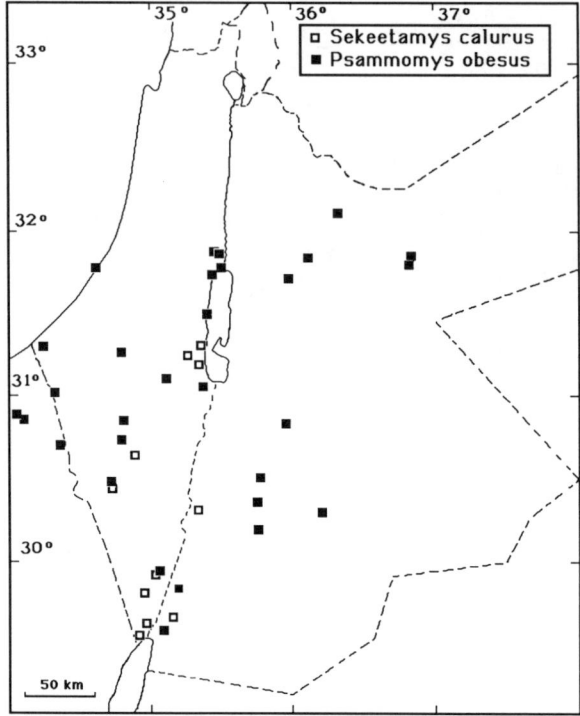

Fig. 95. Distributions
of the fat sand rat
(*Psammomys obesus*),
and the bushy-tailed
jird (*Sekeetamys calurus*).

Ma'an, 62 km N Ma'an, 26 km N Ma'an, Ma'an, 10 km N Al Hasa, near Qumran, Ain Gedi (Atallah, 1978), Al Muwaqqar (22 km E Amman), and Al Hallabat (Qumsiyeh et al., 1986). In the Sinai Peninsula, this species was collected from near Abu Aweigila, Khabra Abu Guzour, Awlad Ali, and Quseimah (Wassif, 1954*b*; Hoogstraal, 1963; Osborn and Helmy, 1980).

Biology.—This is a colonial rodent that lives in complex burrow systems. Its biology has been well characterized (Naggan et al., 1970; Daly and Daly, 1975; Atallah, 1977; Amr and Saliba, 1986). Unlike other gerbils, which are nocturnal and subsist primarily on high-energy seeds, the fat sand jird is diurnal and subsists on low-energy succulent plants. Reported food plants include members of the families Chenopodiaceae, Zygophyllaceae, Carophyllaceae, Cruciferae, and Graminae. Plants taken include *Suaeda mollis*, *Traganum nudatum*, *Salsola foetida* (Daly and Daly, 1973), *Anabasis articulata*, *Sueda monoica*, *Arthrocnomum glaucum* (Wassif and Soliman, 1979), *Arteplix halimus*, *Anabasis articulata*, *Zygophyllum dumosum*, *Z. simplex* (Abu-Dieyeh, 1988), and other halophilic plants primarily in the Chenopodiaceae (Osborn and Helmy, 1980). Fat sand rats are similar in habits to other diurnal colonial rodents, such as prairie dogs of the genus *Cynomys* in North America, in their characteristic posture, clearing the surrounding vegetation, and their reluctance to enter traps (Aharoni, 1932; Flower, 1932; Naggan et al., 1970; Atallah, 1978; Osborn and Helmy, 1980).

Although mostly restricted to open terrain of soil or sand, a dense population of *Psammomys* occurs at 5–10 km N Jiza in a very rocky area (Atallah, 1978).

Osborn and Helmy (1980) reported that in Egypt the breeding period is from September to May with a litter size of two to eight (average 3.2) young. In Algeria, the gestation period is 23–25 days (Daly and Daly, 1975). *Psammomys obesus* develops diabetes mellitus easily under laboratory conditions and is thus a useful laboratory animal for studying that disease.

Genetics.—2N = 48, AA = 74 or 75 was reported (Zahavi and Wahrman, 1957; Qumsiyeh et al., 1986). Chromosome banding studies and protein electrophoretic data indicate a close relationship to *Meriones* (Qumsiyeh and Chesser, 1988).

Human interactions.—The name *Psammomys obesus* derives from *psammos*, Gr., sand, *mus*, Gr., a mouse or rat, and *obesus*, L., fat (i.e., the fat sand rat). This rodent acts as a reservoir for cutaneous leishmaniasis throughout the Near East (Gunders et al., 1968; Naggan et al., 1970; Saliba et al., 1985) (Fig. 94).

Genus *SEKEETAMYS*   Bushy-tailed jirds

There is a single species in this genus, *S. calurus*. The study of the relationship of this species to other gerbils had a long and colorful history. The first specimens were collected in 1837 in Sinai and were housed at the British Museum for many years until described by Thomas (1892) as *Gerbillus calurus*. Although several authors kept this species within *Gerbillus* or its closely related genus *Dipodillus* (Anderson and de Winton, 1902), Thomas himself later referred to it as an aberrant form of *Meriones* (Thomas, 1919*a*) and this was followed by several authors (Flower, 1932; Allen, 1939; Ellerman, 1941). Chaworth-Musters and Ellerman (1947) in their revision created the name *Sekeetamys* as a subgenus of *Meriones*. However, other authors continued to discuss its similarity to *Gerbillus* (Wassif, 1954*a*). *Sekeetamys* was elevated to generic rank by many recent authors, which helped alleviate the need to discuss its affinities to either *Gerbillus* or *Meriones* (Atallah and Harrison, 1967; Harrison, 1972; Petter, 1971; Atallah, 1978). Recent chromosome banding studies clearly show that this genus is allied to *Meriones* (Qumsiyeh and Chesser, 1988).

Fig. 96. The bushy-tailed jird (*Sekeetamys calurus*). Photo: H. Mendelssohn.

Sekeetamys was named after a character in *Winged Pharaoh*, a book on ancient Egypt by Joan Grant.

Sekeetamys calurus   Bushy-tailed jird   R

Gerbillus calurus Thomas, 1892. Ann. Mag. Nat. Hist. ser. 6, 9:76. Type from Sinai, fixed to Tor, Sinai by Chaworth-Musters and Ellerman (1947)

Diagnosis.—This is a large jird that is superficially similar to a large *Gerbillus*. The head and body measures 105–130 mm, the tail length is 125–160 mm, and the skull length is 34–38.5 mm. The tail is bushy for more than half its length (Fig. 96). The bushy areas are black in color, occasionally with a distinct white tip (missing in some specimens). The fur is long and colored brown-yellow with black tips. The skull has an inflated tympanic bulla.

Range of species.—Eastern Egypt, southern Palestine and Jordan, Saudi Arabia.

Local status.—The bushy-tailed jird is found in rocky mountainous habitats of the Negev, the Dead Sea Basin, and Wadi Araba (Fig. 95). Records of the species are available from Masada, vicinity of Rosh Zohar (Ras Zuweira), Nahal Boqeq (Wadi Um Baghaq), Makhtesh Ramon (Wadi Ruman), Har 'Arif (Jebel Ureif), Sede Avraham, Nahal Timna (Wadi Meneya), Bir Ora (Bir Hindis), Nahal Shelomo (Wadi Masri) (Zahavi and Wahrman, 1957), and 34 km N 'Aqaba on the 'Aqaba–Ma'an highway (Atallah and Harrison, 1967).

Biology.—This species is restricted to rocky habitats throughout its range. It is well suited for this habitat and also for a relatively arid climate, as it was captured in areas that appear to be devoid of plant life (Wassif, 1954*a;* Zahavi and Wahrman, 1957). This is a nocturnal animal that feeds on *Zilla spinosa, Citrullus colocynthis, Zygophyllum coccineum, Aerva javanica, Cleome droserifolia,* and other desert plants and perhaps insects (Osborn and Helmy, 1980). There are no reports of reproductive habits from natural populations. Captive animals raised by Flower (1932) bred throughout the year with an average of 2.8 young per litter.

Genetics.—2N = 38, AA = 70 (Wahrman and Zahavi, 1955) and chromosome banding studies suggest a close relationship to the genus *Meriones* (Qumsiyeh and Chesser, 1988).

FAMILY SPALACIDAE   Blind mole-rats

Mole-rats have adapted to a largely subterranean life (fossorial habits). The external ears (pinnae) and the eyes have been reduced to vestigial forms (Fig. 97). The vestigial eyes are subcutaneuous and completely nonfunctional (hence the name blind mole-rat). The facial hairs are stiff bristles that give the head a broad wedge shape that is used to shovel dirt or pack the sides of tunnels. Mole-rats are found in southern Europe, southwestern Asia, northern Africa, and the southwestern parts of the USSR.

Genus *SPALAX*   Blind mole-rats

Spalax leucodon   Mole-rat   K

Spalax typhlus leucodon Nordmann, 1840. Observations sur la fauna Pontique, Demidoff Voy., 3:34. Type from near Odessa, Ukraine, USSR.

Spalax ehrenbergi Nehring, 1898. Schriften Berl. Ges. Naturf. Fr. Berl., (for 1897), p. 178, pl. 2. Type from Jaffa, Palestine. Valid subspecies.

Spalax kirgisorum Nehring, 1898. Schriften Berl. Ges. Naturf. Fr. Berl., (for 1897), p. 176, pl. 4. Type probably from northern Syria. Synonym.

Spalax berytensis Miller, 1903. Proc. Biol. Soc. Wash., 16:162. Type from Beyrout, Syria (now in Lebanon). Synonym.

Diagnosis.—Mole-rats are rodents that are adapted for fossorial life and thus are easily recognized among other mammals of the region by the "torpedo" or cylindrical shape, the rudimentary eyes, and the absence of visible external ears and tail. The fur is short and is different from other rodents of the region in that it is nondirectional. Mole rats use their strong incisors for loosening the soil and their "shovellike" heads for pushing the

Fig. 97. The mole-rat (*Spalax leucodon*). Photos: M. B. Qumsiyeh.

loose soil. Harrison (1972) was not able to separate *S. ehrenbergi* from *S. leucodon* by the two characters listed in Ellerman (1948). For example, the presence of two isolated islands in the adult upper third molar rather than one is not a reliable character. Nevertheless, Harrison (1972) and Harrison and Bates (1991) regarded *S. ehrenbergi* as valid based on its smaller size.

Range of species.—Greece, southern Ukraine, Turkey, the eastern Mediterranean region, Iraq, and the Mediterranean regions of Egypt and Libya.

Local status.—Records are numerous in the Holy Land. The characteristic dirt mounds made by this species can be seen in most cultivated areas in the region (Tristram, 1866*b*, 1884, as *S. typhlus*, Pall.). Available records for the western part of the Holy Land are too numerous to list. The distribution map shown in Figure 98 is a summary based on the work of Nevo and colleagues (Costa and Nevo, 1969; Nevo, 1969, 1982, 1985; Wahrman et al., 1985). Historically significant records and those from the eastern part of

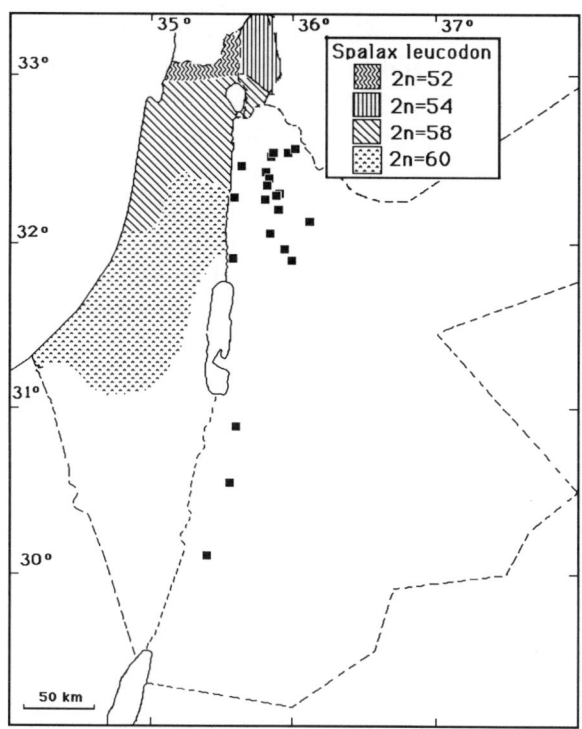

Fig. 98. Distribution of the mole-rat (*Spalax leucodon*). The distributions of the chromosomal forms west of the Rift Valley are from the work of Nevo and colleagues (Costa and Nevo, 1969; Nevo, 1969, 1982, 1985; Wahrman et al., 1985). East of the Jordan River, individual locality records are given (no published chromosome data are available).

the Holy Land are from the Jordan Valley and Safje (Nehring, 1901), Talkarem, Jericho, Djesi (60 km E Dead Sea), El Bassa (Ellerman, 1948), Nathania (Howells, 1956), and Shobek (Mountford, 1965). Specimens at JNHM are from Al Mazar, Hawfa, Jubeiha, and Wadi Sukhn. I have seen mounds in many parts of Jordan: Jordan University Campus, Suweilih, 7 and 9 km S Jerash, Jerash, Jerash Refugee Camp, Dibbine Forest, 2 km S Dibbine, 5 km W Amman, Sahab, 3 and 7 km S Irbid, 5 and 10 km E Irbid, Shobek, and also in the Ghor in alluvial soils and terra rosa-type soils such as those near Tabqat Fahl and Ghor Nimrin. I collected one specimen near the Jordan University Campus. Specimens are available at the JNUHM from Ramtha, Jubeiha, near Tafilah, Ibbin, and between Jerash and Suf (Amr and Disi, 1988), and Al Mashari'.

Biology.—Mole-rats are found throughout the country in all habitats with rainfall exceeding 100 mm annually (Nevo, 1961). The mole-rat is a fossorial rodent that rarely wanders outside of its elaborate tunnels. Such surface activity exposes them to predation, mainly by owls (Bate, 1945; Dor, 1947) but is essential for dispersal. These animals forage for bulbs (for example, those of *Asphodeles microcarpus* and *Leopoidia comosa* [Wassif and Soliman, 1979]), roots, rhizomes, tubers, corms, and fresh green plants in certain times of the year. They generally use their large incisors to dig in hard soil and use their shovel-shaped head to push and pack soft soil. I and others have caught them by plugging their retreat with a shovel when they attempt to plug an open burrow system. Burrowing systems were studied in Iraq (Reed, 1958).

In Lebanon, mole-rats give birth to two to three young in February at lower elevations and as late as April at higher elevations (Lewis et al., 1967). In northern parts of the Holy Land, these animals construct an elaborate breeding mound with an average size of 160 × 135 × 40 cm, which includes a nest, food, and sanitary chambers (Nevo, 1961, 1969). Copulation takes place in or around the breeding mound and a gestation period of 28 days is reported. Parturition is reported to occur from January to March.

Genetics.—Nevo and colleagues have published numerous papers that characterized the chromosome variation in this species (Nevo and Amir, 1964; Costa and Nevo, 1969; Nevo et al., 1977; Nevo, 1978, 1985, 1988). In summary, there appear to be four "karyotypic races," which Nevo believes are "semi-species" (Fig. 98). The diploid numbers noted were 52 (from the Upper Galilee Hills), 54 (from the Golan Heights and Mount Hermon), 58 (from north-central areas east of Haifa to Jaffa), and 60 (from the Negeb and the Judean and Samarian hills). Nevo argued in these papers that the chromosomal difference is adaptive and that each of these forms is specialized for its own environment. It is not clear which chromosomal forms exist in Jordan or Lebanon. Considering the extensive chromosome variability in *Spalax* from other areas of the world, it is not wise at present to recognize these forms in a formal fashion by assigning species or subspecies names to them. The north African form, *Spalax leucodon aegyptiacus*, has a diploid number of 60, similar to those from southern Palestine, a condition suggesting that this is the primitive chromosomal form (Lay and Nadler, 1972).

Human interactions.—*Spalax* is Greek (genitive *spalaktos*) for a mole. Farmers detest these rodents, which they call *khlund* or *khuld,* because mole-rats consume the roots of many commercial agricultural plants. I have observed a mole-rat in the early morning hours crossing a road in Beit Sahur. Paved roads can only be crossed if these animals leave their burrows. It can only be speculated as to how far mole-rats normally travel outside of their burrows. It is clear however from my own observations and others that dispersal can be accomplished in this fashion (especially for males seeking new a territory).

Family Arvicolidae   Voles, lemmings, and muskrats

Arvicolids have a broad face due to a shortened skull, especially in the nasal region. The cheekteeth are high crowned (hypsodont) with a prismatic pattern (Fig. 99). This family occurs in North America, Europe, Asia (excluding tropical regions), and marginally in northern Africa. There are 17 genera with at least 125 species.

Key to the Arvicolidae of the Holy Land

1. Large, head and body length more than 140 mm, greatest
   length of skull greater than 36 mm. . . . . *Arvicola terrestris*

   Small, head and body length less than 120 mm, greatest length
   of skull less than 31 mm. 2

2. Tail almost 50% of head and body length. Soles of hind feet
   with six tubercles. Third upper molar with two reentering folds.
   . . . . *Microtus nivalis*

   Tail less than 30% of head and body length. Soles of hind feet
   with five tubercles. Third upper molar with three reentering
   folds. . . . . *Microtus guentheri*

Genus *ARVICOLA*

Arvicola terrestris   Water vole   I

Mus terrestris Linnaeus, 1758. Syst. Nat., 10th ed., 1:61. Type from Uppsala, Sweden.
Arvicola terrestris hintoni Aharoni, 1932. Z. Säugetierkd., 7:209. Type from Island of Tel el Sultan, Antioch Lake, northern Syria (now in Turkey). Perhaps a valid subspecies.

Diagnosis.—This vole is distinguished from other microtines in the region by its robust size. The total length is greater than 230 mm, and the greatest length of the skull is 30–38 mm.

Range of species.—Most of the northern Palearctic region (Europe and northern Asia) south to Palestine and Iran.

Local status.—So far, the water vole has been observed or collected only near Hula Lake (Fig. 100). Tristram (1884) listed this species from Palestine and Aharoni (1932) lists a specimen at BZM from Tel el Sultan. Bodenheimer (1958) also included this species based on owl pellets reported from Yessod Hamaale by Dor (1947) and based on observation by Hurwitz from Melaha. Specimens from Hula and Melaha are at HUJ.

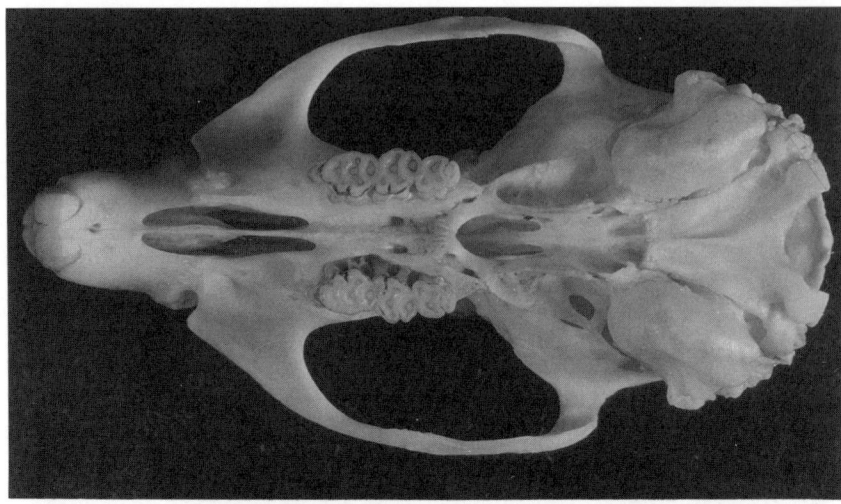

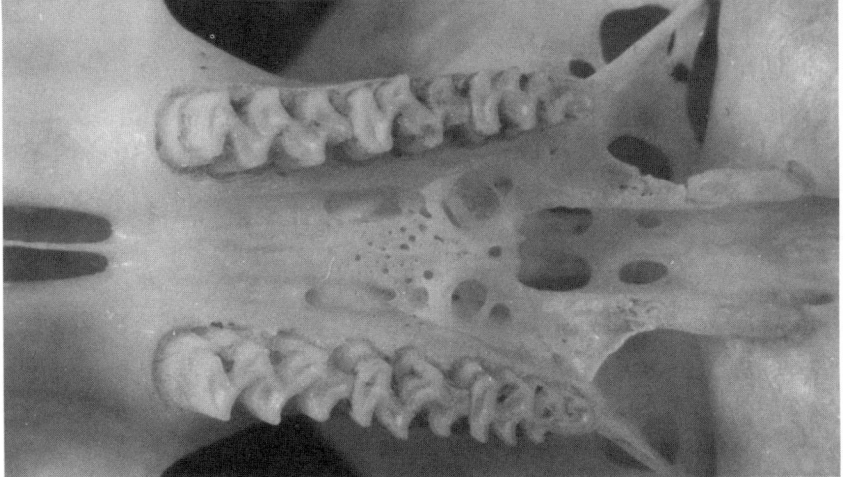

Fig. 99. Skull morphology in microtine rodents. (Upper) Ventral view of skull of *Arvicola terrestris*, (Lower) upper cheekteeth of *Microtus socialis*.

Biology.—The water vole, as the name indicates, is present around streams and irrigation ditches. It is mainly diurnal and feeds on both aquatic and terrestrial plants. In Iran, the ranges of *Nesokia indica* and the water vole did not overlap, suggesting potential competitive exclusion (Lay, 1967). Indeed, these two species have similar marshy habitats in the Holy Land (in the north for *Arvicola* and around the Dead Sea for *Nesokia*). In Europe, the water vole was noted feeding on foliage and roots of grasses as well as on

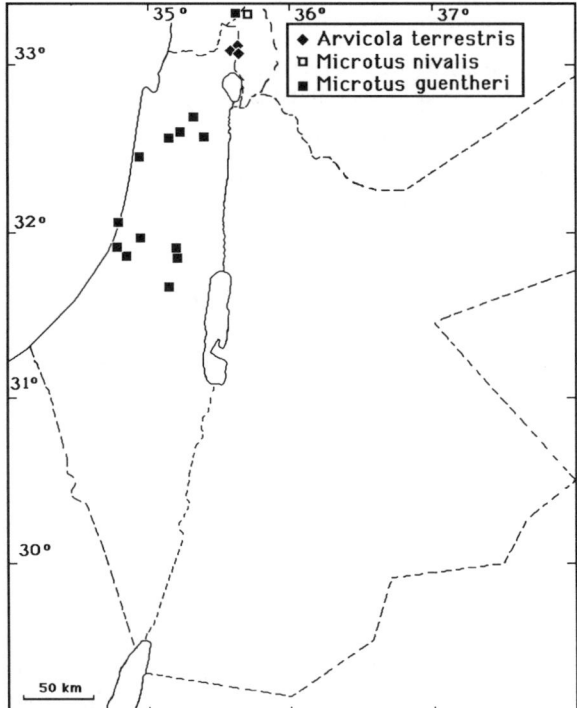

Fig. 100. Locality records of voles of the Holy Land. The water vole (*Arvicola terrestris*), the snow vole (*Microtus nivalis*), and the Mediterranean vole (*Microtus guentheri*).

sedges and other plants. Four to six young are born in the summer (Corbet and Ovenden, 1980).

Dor (1947) reported predation by barn owls on this species in northern Palestine. In Armenia, 3–14 young per litter are born in an extended breeding season covering spring, summer, and early fall (Dahl, 1954).

Genetics.—2N = 36, AA = 60–68 was reported for specimens from Europe (reviewed in Zima and Král, 1984). The X is submetacentric and the Y is a very small acrocentric or submetacentric.

Human Interactions.—The water vole was perhaps a common rodent in the Hula swamps until they were drained in 1957. There is a very limited area in the Hula Nature Reserve where this species still survives. *Arvicola* and Arvicolidae are words derived from *arvum*, L., ploughed land, and *colo*, L., I till, cultivate. The specific name *terrestris* is Latin for "of the ground." The derivations of these names indicate that these are animals that live in cultivated agricultural lands.

Genus *MICROTUS*

The social vole, *Microtus socialis* (*Mus socialis* Pallas, 1773, Reise Prov. Russ. Reichs., 2:705, type from near the Ural River), was reported from Lebanon (Bate, 1945; Bodenheimer, 1958). However, these are most likely misidentifications (Atallah, 1978). The name *Microtus* is derived from the Greek words *mikros* (small) and *otos* (ear).

Microtus guentheri   Mediterranean vole   **K**

Hypudaeus syriacus Brants, 1827. Het. Gesl. d. Muizen, 1827:92. Type from Syria. Synonym of *M. socialis* or *M. guentheri* (Atallah, 1977). Forgotten name.

Arvicola guentheri Danford and Alston, 1880. Proc. Zool. Soc. Lond., 1880:62. Type from Marash, Turkey.

Microtus philistinus Thomas, 1917. Ann. Mag. Nat. Hist. ser. 8, 19:450. Type from Ekron, Palestine. Synonym of *M. guentheri.*

Microtus irani Thomas, 1921. J. Bombay Nat. Hist. Soc., 27:41. Type from Bagh-i-Rezi (= vineyard), Shiraz, Iran. Valid subspecies.

Diagnosis.—See under *M. nivalis.* This is a small vole with a greatest length of skull of 26–30 mm. The problems of distinguishing *M. guentheri* from *M. socialis* were extensively debated. Some authors (Ognev, 1950; Lay, 1967; Hassinger, 1973) suggested that the characters used to distinguish them are unreliable and thus considered *M. guentheri* as a subspecies of *M. socialis.* However, according to other authors (Felten et al., 1971; Atallah, 1978; Kock and Nader, 1983), *M. socialis* is distinguished from *M. guentheri* by the lateral inflation of the mastoid portion of the tympanic bulla. This situation results in a partial hiding of the occipital condyles in lateral view. In ventral view the foramen tympanum is also hidden or partially hidden by the tympanic bulla. Similarly, Kock and Nader (1983) felt that *M. irani* is a separate species and they used that name for Syrian and Holy Land specimens. Harrison (1972) and Harrison and Bates (1991) considered the differences small and variable and placed *guentheri* as a synonym of *socialis.* Some authors (Atallah, 1965, 1978; Lewis et al., 1967) believed that there is no direct evidence for the presence of *M. socialis* in the Holy Land and Lebanon and suggested that all were closer to *M. guentheri.* I accept this view until more modern genetic data on this group becomes available.

Range of species.—South-central USSR to southeastern Europe, south to the Holy Land and Iran. An isolated population in northeastern Libya.

Local status.—The Mediterranean vole is known from many localities in the mountains and near the coastal regions of Palestine (Fig. 100). It has not yet been reported from Jordan. Records were first made by Tristram

(1884) from the plains of Gennesaret (= Nazareth); he stated that it probably occurs in southern Judea. Further records are from near the Sea of Galilee (Nehring, 1902*a*), Ekron, Jaffa, Ain Herod, Jezreel Plain, Ramallah (Aharoni, 1932), Ben Shemen, Hadera, and north of Jerusalem (Bodenheimer, 1935). This species was reported from localities west of Mount Hermon (Allen, 1915), 4 km NW Beit Fajjar, Ramallah, Jaffa, Ekron, Jezreel, and Mishmar Ha'Emeq (Atallah, 1978). Specimens at HUJ are from Ekron and Rehovot (labeled *philistinus*) Emeq, Kishon Lake, and Mishmar Ha'Emeq.

Biology.—This is a colonial species that occurs in moist areas, especially near irrigated or watered agricultural lands. It is a common species in the eastern Mediterranean region. Limited information on the Mediterranean vole is available (Atallah, 1965, 1978; Lewis et al., 1967). These voles occur near cultivated areas and in shrubby mountain regions where they apparently feed on succulent vegetation. This species is active almost 24 hours a day. Vole populations fluctuate dramatically and in some years they are considered to be dangerous agricultural pests in the Holy Land (Bodenheimer, 1949*b*, 1959; Atallah, 1978; Ilani and Bouskila, 1982). In Iran, these voles were noted to feed on alfalfa, grass, clover, *Canna*, and *Bromus* (Lay, 1967).

This is a highly prolific rodent with reproduction throughout the year and with an average of eight (and as many as 17) embryos per female (Atallah, 1978). Predators on these voles include owls, hawks, foxes, mustellids, and snakes. Tristram (1884) found an individual eaten by the snake *Caelopeltis lacertina*. Predators do not increase in number fast enough to control vole outbreaks. Thallium sulfate poisoning was used in Israel for vole control for many years (Bodenheimer, 1949*b;* Ilani and Bouskila, 1982). However, poisoning is ecologically harmful when the substances enter into the food chain (see Chapter 9, Conservation). For further discussion of reproduction and population structure in this species in the Holy Land, the reader is referred to Cohen-Shlagman et al. (1984*a*,1984*b*).

Genetics.—2N = 54, AA = 52 was reported for specimens from Bulgaria (Belcheva et al., 1980). The X and Y chromosomes are acrocentric.

Microtus nivalis   Snow vole   K

Chironomys nivalis syriacus Brants; 1827. Het. Gesl. d. Muizen, 1827:92. Type from Syria.

Arvicola nivalis Martins, 1842. Rev. Zool., p. 331. Type from Faulhorn, Bernese Oberland, Switzerland.

Microtus hermonis Miller, 1908. Ann. Mag. Nat. Hist. ser. 8, 1:103. Type from Mount Hermon, Palestine. Valid subspecies.

Diagnosis.—This is a small vole with a head and body length less than 126 mm and greatest length of the skull of 26.5–31 mm. From *Arvicola*, this vole is distinguished by its smaller size. From *Microtus guentheri*, the snow vole is distinguished by its relatively longer tail (about one-half the head and body length, Fig. 101) and the simple structure of the third upper molar. The third upper molar has only two inner reentrant folds. The soles on the hind feet have six tubercles (fleshy pads). *Microtus nivalis hermonis* is reported to differ from the nominate subspecies by paler color (dorsally "light clay brown," Harrison, 1972).

Range of species.—Mountainous regions of Europe, the eastern Mediterranean region, Turkey, Caucasus, and Iran.

Local status.—The name *M. hermonis* was given by Miller in 1908 to specimens collected by Tristram (1884) from Mount Hermon. In fact, this is the only locality where the snow vole was reported (Fig. 100). Many localities from the Lebanon mountains are available, including the western slopes of Mount Hermon (Felten et al., 1973; Atallah, 1977).

Biology.—Little is known of the biology of this species in the region. In Lebanon, it occurs at high altitudes (1300–2700 m, Atallah, 1977). In Iran, specimens were collected with *M. arvalis* but the latter was much less frequent (Lay, 1967). Lay (1967) reported that a female collected 7 August, had five embryos. In Lebanon, Atallah (1977) reported that females deliver young in May and early June. The animals are nocturnal in Lebanon. In Europe, however, these animals are active night and day and may even be observed "basking in the sunshine" on rocky slopes (Corbet and Ovenden, 1980).

Fig. 101. The snow vole (*Microtus nivalis*) from Mount Hermon. Photo: A. Shoob.

Genetics.—2N = 54, AA = 52 for specimens from Europe (Burgos et al., 1989).

Human interactions.—The name *nivalis* comes from *nix*, L., genitive *nivis*, meaning snow. Hence the English common name: snow vole.

FAMILY MURIDAE   Rats, mice

Holy Land murids are distinguished by the presence of distinct cusps on the cheekteeth (Fig. 83A; except in *Nesokia*) and by an external appearance that includes elongated snouts and a long tail that is never bushy. Murids have radiated into diverse habitats throughout Africa, Asia, Europe, and Australia. They are missing from some of the coldest regions of northern Asia and the Americas. However, the commensal forms (the house mouse, black rat, and brown rat) have been introduced by man to almost all areas of the world. There are 117 genera of murids with almost 500 species.

Key to the Genera of Muridae of the Holy Land

1. Dorsal pelage, except on head and neck, consisting of spines;
   bony palate extending far behind teeth, margin about halfway
   between teeth and bullae (Fig. 82B). . . . . *Acomys*

   Dorsal pelage not spiny or with some rather weak spines mixed
   with normal hair; palate not extending far behind teeth
   (Fig. 82A). 2

2. Larger: length of head and body usually in the range of 140–250
   mm, hind feet more than 25 mm long (usually over 30 mm),
   upper molar row longer than 6 mm. 3

   Smaller: length of head and body less than 140 mm, hind feet
   less than 30 mm long, upper molar row less than 6 mm long. . . . 4

3. Tail length less than body length. Ventral color not much
   lighter than dorsal color. Cheekteeth with transverse laminae
   and almost no trace of separate cusps. . . . . *Nesokia*

   Tail length greater than body length. Ventral color much lighter
   than dorsal color. Cheek-teeth with clearly defined cusps. . . . *Rattus*

4. M3 very small, not more than half length and width of M2;
   upper incisors usually appear notched when viewed from the

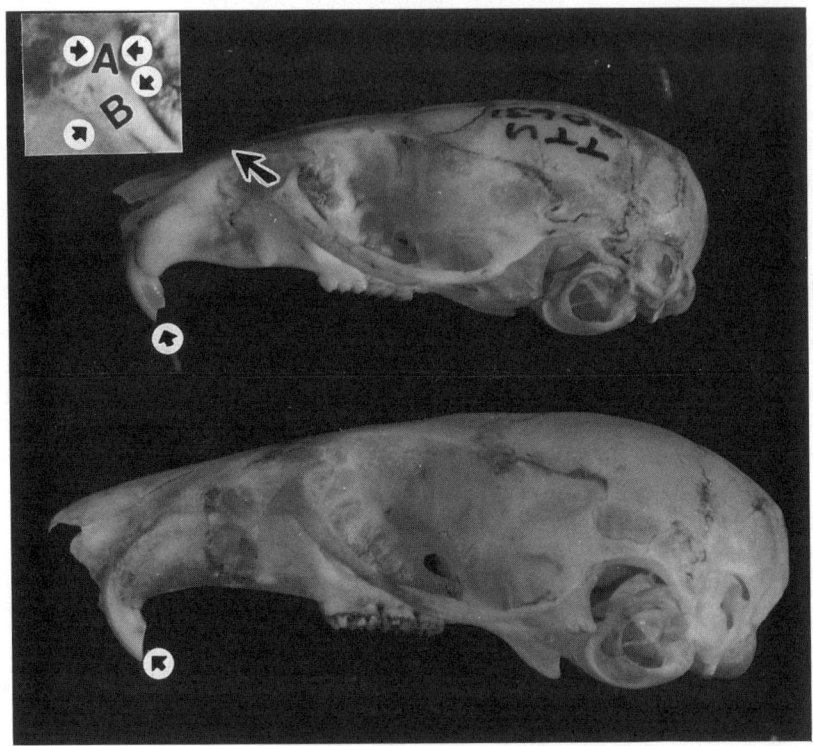

Fig. 102. Lateral views of skulls of *Mus musculus* (Upper) and *Apodemus mystacinus* (Lower). Note the notched upper incisor of *Mus*. Insert shows the measurements used to calculate the zygomatic index (ZI) in *Mus*: ZI = A/B (A = width of anterior part of malar, B = width of anterolateral part of zygomatic arch).

side (Fig. 102, upper); M1 with three roots (i.e., only one lingual root). . . . . *Mus*

M3 larger; upper incisors not notched (Fig. 102, lower); M1 with three or four roots. . . . . *Apodemus*

Genus *ACOMYS*   Spiny mice

Spiny mice are peculiar murids with unique adaptations, including the shape of their skull (Fig. 82) and their spiny pelage (Fig. 103). There are only two species of these unique rodents in the Holy Land. The name *Acomys* derives from *akoke*, Gr., sharp point, and *mus*, Gr., a mouse (a spiny mouse).

Key to the Species of *Acomys* of the Holy Land

1. Soles and palms black; tail distinctly shorter than head and body length; spiny pelage of back extending to back of head.
   . . . . *A. russatus*

   Soles and palms pale; tail at least as long as head and body length; spiny pelage of back does not reach back of head.
   . . . . *A. cahirinus*

Acomys cahirinus   Porcupine mouse, common spiny mouse   O

   Mus cahirinus Desmarest, 1819. Nouv. Dict. Hist. Nat. Paris, 29:70 (É. Geoffroy St.-Hilaire, 1803. Cat. Mamm. Mus. Nat. Hist. Nat. Paris, p. 195; nomen nudum, see Aharoni, 1932). Type from Cairo, Egypt.

   Mus dimidiatus Cretzschmar, 1826-1827. *In* Rüppell, Atlas Reise Nördl. Afr., Säugeth., pt. 3, p. 37, taf. 13, fig. a. Type from Sinai (now at Senckenberg Museum). Probably valid species.

   Mus hispidus Brants, 1827. Het. Geslachte der Muizen, Berl., p. 154. Type from Arabia. Synonym.

   Mus megalotis Lichtenstein, 1829. Darstell. neuer Säugeth., pl. 37, fig. 2. Type from Arabia. Synonym.

   Diagnosis.—This spiny mouse differs from the golden spiny mouse (*A. russatus*) in the following ways: footpads are pale rather than black, tail length usually exceeds 95 mm in adults, underside is light in color (most specimens have white underside) and easily demarcated from the darker upper color, spines are more restricted to the back and are not present over the nape or the lower sides, and lower diploid number (38 or 36 chromosomes). Earlier authors such as Tristram (1884) distinguished between *A. cahirinus* and *A. dimidiatus* as allopatric species in the region. Other authors do not corroborate the specific status of *A. dimidiatus*, which was considered a synonym of *A. cahirinus* and at best a subspecies (Atallah, 1977; Osborn and Helmy, 1980).

   Range of species.—Semiarid regions of northern Africa south in eastern Africa and east through Arabia and southwestern Asia to Sind.

   Local status.—The common spiny mouse lives up to its name "common." In areas where it occurs, it can be the dominant rodent. Its range is well established, at least west of the Rift Valley (Fig. 104). Tristram (1866*b*, 1884) reported these animals in the Dead Sea Basin and up the hillside to Der Mar Saba (as *cahirinus*) and in gravel and sandy areas in other regions of Palestine such as Ain Fashkhah (as *dimidiatus*). More recent authors found a clinal variation and thus no evidence for the validity of these forms as species

Fig. 103. The common spiny mouse (*Acomys cahirinus*). Photo: S. I. Atallah.

(Osborn and Helmy, 1980). The species has been reported from Ain Gedi (Nehring, 1901), Petra, Wadi Kerak, Tafileh (Allen, 1915), Tabgha, Jerusalem, Ghor es Safi, Hizma, Moab (Aharoni, 1932), Wadi Zarqa Ma'in, Jerusalem (Ellerman, 1948), Bir Hindis, (Petter, 1954), Wadi Masri (near Eilat), Ain Kerem, Nahal Amram, near Haifa, near Beit Oren, Wadi Arus (NW Hulata), Wadi Araba near Eilat, Carmel Caves, Tel el Qadi, Dan, Yotvata, Wadi Menayeh, Kfar Gileadi, E Dead Sea, Bethlehem, Beit Sahur (Harrison, 1972), Deir Mar Saba, Ain Fashkhah, 4 km N Beit Fajjar, Jericho, 34 km N 'Aqaba (Atallah, 1978; and M. B. Qumsiyeh pers. coll.), 'Aqaba (Qumsiyeh et al., 1986), Hurfeish, Beit Oren, Mitspe Ramon, and Timna (Nevo, 1985, 1989). I also collected specimens at Wadi Al Finan near Dhana and at 40 km N 'Aqaba in Wadi Araba on 18 December and 10 February, respectively. Specimens are available at the JUNHM from 'Aqaba, Gharandal, and Ghor as Safi (Amr and Disi, 1988). Other specimens were collected by Abu-Dieyeh (1988) from these and other localities in Wadi Araba. Specimens of the common spiny mice are available in many museums from many areas of the Holy Land. The records listed above are major important records from a historical perspective or because they

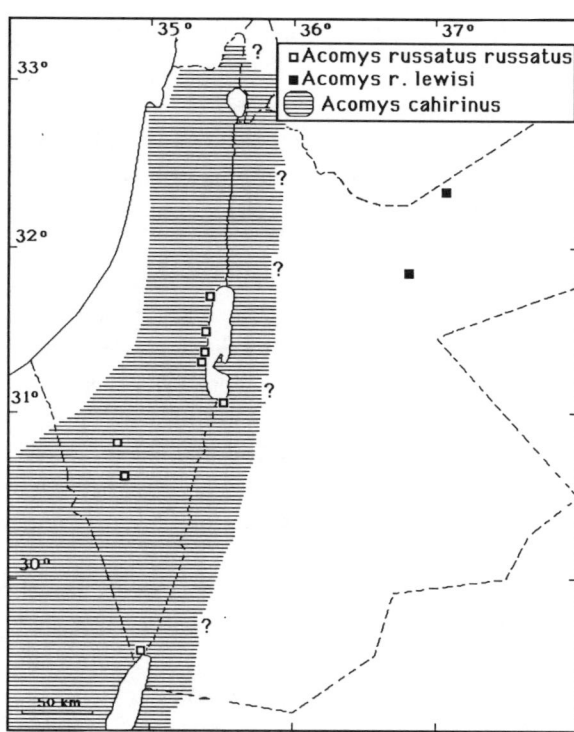

Fig. 104. Locality records of the common spiny mouse (*Acomys cahirinus*) and the golden spiny mouse (*Acomys russatus*). Eastern distributional limits for *A. cahirinus* are not well known.

represent marginal localities. The species seems to be absent from northern Sinai (Wassif, 1954*b*) and most of the coastal regions of the Holy Land.

Biology—This is a common species in hills, hilly soils, and rocky areas. The only limit to the range is that the species cannot be found at very high altitudes where temperatures fall significantly. These spiny mice are found in Mediterranean woodlands in the hills both east and west of the Jordan River. They are also found in the Araba Valley and in arid habitats of the Sinai, the Negev, and the Syrian deserts. They are primarily nocturnal and crepuscular in activity. Burrow systems in hilly soils of the Araba Valley appear to be simple structures of 30–90 cm length (Abu-Dieyeh, 1988). In rocky habitats the animals presumably use the spaces between rocks and natural crevices.

Atallah (1978) suggested, based on stomach contents and other field observations, that these animals are omnivorous and that a large portion of the diet consists of insects, snails, and other animals (including other spiny mice). The snail species they feed upon include *Helix* and *Leucochroa* (Bodenheimer, 1949*a*). In Egypt, spiny mice have been reported to feed on palm dates and in one area on feces of the Egyptian fruit bat (*Rousettus* *aegyptiacus*) (Maser, 1966; Osborn and Helmy, 1980). In Wadi Araba, food consists of such plants as *Mesembryanthemum forsskalii, Pteramus dichotomus, Fagonia mollis, Diplotaxis harra,* and *Gymnocarpus decandrum,* as well as snails (*Euporyphy* and *Rumina*) and insects (Abu-Dieyeh, 1988). In many areas of Egypt, the nominate subspecies is found to invade human habitations. In some instances, it even outnumbers the usual commensal species *Mus musculus* and *Rattus.*

Pregnant females were found with two to four embryos each in April in arid habitats of the Negev and Araba (Harrison, 1972; Abu-Dieyeh, 1988) and in more mesic habitats around Bethlehem (Sana Atallah's field notes). In Beit Sahur, I found pregnant females with four embryos each on 24 December, males with enlarged testes in early January, and suckling young on 1 May. Thus, it is possible that spiny mice breed all year as suggested by Bodenheimer (1949a). In Saudi Arabia, Al-Khalili and Delany (1986) gave a gestation period of 42–46 days, with a mean litter size of 4.5 (range one to six in the lab and four to six in the wild) and the main breeding period in January to July and in November. Al-Khalili and Delany (1986) also discussed the postnatal developmental events in this species.

Spiny mice in general are notorious for their delicate and loose skin, especially on the tail. Many collecters (for example, Harrison, 1972) have commented on the ease of separation of the tail skin from the body (pulls off as a sheath) and on how easily the skin breaks. The functional significance of this is yet to be determined; one suggestion is that it is a predator escape mechanism similar to that seen in geckos and other reptiles. The color in this species is quite variable and depends on the habitat. In the lava fields in Saudi Arabia, the color is very dark (Atallah and Harrison, 1967; Harrison, 1972).

Genetics.—Two karyotypic forms occur in the Holy Land: 2N = 36, AA = 68 in Sinai and 2N = 38, AA = 68 further north (Wahrman and Goitein, 1972; Qumsiyeh et al., 1986). Nevo (1985) showed that genic differences as measured by protein electrophoresis between the two forms were small. Janecek et al. (1991) demonstrated a close genic association between *A. cahirinus* and *A. ignitus.*

Human interactions.—In Egypt, common spiny mice have been reported to be "almost exclusively commensal" and in these situations they acquire a dark, almost melanistic color (Setzer, 1959). Bodenheimer (1935) also commented that these mice invade houses in Jerusalem in the winter, presumably to escape the cold weather. I have rarely collected these mice in old houses (but usually not inhabited by humans) especially in dry regions of the Holy Land. In the eastern desert of Egypt, 92 individuals were examined for various antigens: 29 animals were positive for the murine typhus *Rickettsia*, 5 were positive for the tick typhus *Rickettsia*, and 11 were positive for Q fever *Coxiella* (Hoogstraal et al., 1967). The name *cahirinus* refers to the Egyptian city of Al Cahira (meaning the conquerer), the Arabic name that was corrupted later to Cairo. These spiny mice are common in and around Cairo.

Acomys russatus   Golden spiny mouse   **R?**

> *Mus russatus* Wagner, 1840. Abh. Bayer. Akad. Wiss. Math.-Naturwiss. Kl., 3:195, pl. 3, fig. 2. Type from Sinai, probably Nohel (Osborn and Helmy, 1980).
>
> *Mus affinis* Gray, 1843. List Spec. Mamm. British Mus., 108. Nomen nudum.
>
> *Acomys russatus aegyptiacus* Bonhote, 1912. Abstr. Proc. Zool. Soc. Lond., no. 103, p. 3; Proc. Zool. Soc. Lond., 1912:320. Type from Wadi Hof, near Helwan, Egypt. Valid subspecies, extralimital.
>
> *Acomys lewisi* Atallah, 1967. J. Mammal., 48:258. Type from 3 km NW Azraq Shishan, Jordan. Valid subspecies.
>
> *Acomys russatus harrisoni* Atallah, 1970. Univ. Conn. Occas. Pap. Biol. Sci. Ser., 1(4):202. Type from ½ km S Qumran Caves, near Ain Faschka, West Bank of Jordan (Palestine). Perhaps a valid subspecies.

Diagnosis.—See under *A. cahirinus*. There are three forms described from the Holy Land: *A. r. russatus*, which is described from Sinai and also occurs in the Negev; *A. r. harrisoni*, which is thought to be smaller and paler than the nominate form and occurs around the Dead Sea (Atallah, 1970*b*); and *A. r. lewisi*, which is extremely distinctive in being melanistic and is isolated in the hammada and lava rocks of the Azraq (Atallah, 1967*c*). The continuous range and the few differences between *harrisoni* and *russatus* do not justify subspecies recognition. Thus, I consider only two valid subspecies in the Holy Land: *A. r. russatus* and *A. r. lewisi*.

Range of species.—Egypt, the Holy Land, Saudi Arabia, and Yemen.

Local status.—The golden spiny mouse (*A. russatus russatus*) occurs along the rocky hills of the Negev and around the Dead Sea, whereas the black form, *A. r. lewisi* (which maybe a valid species), occurs in the foothills of Jebel ed Druz in such areas as Azraq and Jawa (Fig. 104). The first records were by Tristram (1866*b*, 1884) from near Sebeh (Sebbe, Masada South) toward the southern end of the Dead Sea. Golden spiny mice were reported from Moab and Ain Gedi (Nehring, 1901), Avdat (Shkolnik and Borut, 1969), Moab, Ain Fashkhah, 3 and 4 km NW Azraq Shishan, Azraq Druz, Ghor as Safi (Atallah, 1978; Qumsiyeh et al., 1986), Jawa (Searight, 1987*b*), and Mitspe Ramon (Nevo, 1985, 1989). Other localities represented by specimens at HUJ are from Eilat, Ain Gedi, and Wadi Siyal. The species seems to be absent from most areas in northern Sinai (Wassif, 1954*b*). It is surprising that no records from Wadi Araba exist yet, where one would expect this species to be found.

Biology.—*Acomys russatus* is diurnal in areas where it is in competition with the common spiny mouse (*A. cahirinus*), being particularly active during morning and late afternoon hours. *Acomys russatus* can, however, resort to nocturnal activity in areas where *A. cahirinus* does not occur (Shkolnik, 1966). *Acomys russatus* tolerates higher ambient temperatures than the nocturnal *A. cahirinus* and is thus able to withstand day heat in the semiarid areas where it lives (Shkolnik and Borut, 1969). *Acomys russatus* also seems to be more tolerant of higher salt concentrations in its diet and lives in more arid environments than its congener. *Acomys russatus* is sensitive to colder temperatures, which would explain its absence from certain high elevations (Haim and Borut, 1975), whereas *cahirinus* tolerates colder temperatures better. I kept three *A. russatus* for 3 years in captivity until a cold spell (temperatures dropped to about 15-17°C) resulted in the death of all three animals. This species was found by Shkolnik and Borut (1969) to feed on *Artiplex halimus*, *Hammada scorpia*, and *Anabasis articulata*. The melanistic form, *A. r. lewisi*, was found to be active in the late afternoon as well as at night and to feed on *Diplotaxis harra*, *Erodium*, and *Alium* as well as various grasses. This species lives in rocky habitats and matches the substrate in color, clearly a protective advantage for a diurnal species.

Reproduction occurs in the late spring and early summer in the wild. In captivity, *A. r. lewisi* reproduced year-round with a gestation period of 44 days (Searight, 1987*a*). The earliest report on reproduction and growth in this species were observations made by Bonhote (1912) on a specimen brought to him alive from the Sinai. This female gave birth to three young, which were at an "advanced" stage of growth, being large and already covered with hairs and spines. Their color was grayish and the pelage later changed by molt to the adult color of golden yellow. I made similar observation on *A. r. lewisi* from Jordan. A female captured in May gave birth in early July to three young that were very dark in color. As they grew their color lightened somewhat but the overall color remained black-brownish.

Genetics.—The diploid number for specimens of *A. russatus* is 2N = 66, AA = 76 (Qumsiyeh et al., 1986), which is very different from *A. cahirinus*. Janecek et al. (1991) also demonstrated that *A. russatus* is closer to *A. percivali* and *A. wilsoni* than to the *cahirinus* group.

Genus *APODEMUS*   Field Mice

Apodemus (from *apode*, Gr., away from home, in the fields, and *mus*, Gr., a mouse = a field mouse) are very abundant in wooded areas having a Mediterranean climate. Habitat destruction by human activity, especially in the first half of the 20th century, resulted in these animals becoming less common and their distribution becoming patchy. Field mice are small animals with long thin tails, large eyes and ears, and pale-colored hind feet. Although easily collected, these mice are difficult to identify. The confusion about the status of species in the Holy Land is far from being resolved. I recognize two species here, pending further molecular and cytogenetic studies. The population from Mount Hermon was recognized as a distinct species, *Apodemus hermonensis* (Filippucci et al., 1989), but is here treated as a synonym of *A. flavicollis* (see discussion).

Key to the species of *Apodemus* of the Holy Land

1. Size large, skull length 28–33 mm. Dorsal color grayish.

. . . . *A. mystacinus*

Size smaller, skull length 22–29 mm. Dorsal color brownish.

. . . . *A. flavicollis*

Apodemus flavicollis   Wood mouse, common field mouse   I

Mus flavicollis Melchior, 1834. Danske Staat. Norg. Pattedyr., p. 99. Type from Sielland, Denmark.

Mus arianus Blanford, 1881. Ann. Mag. Nat. Hist., ser. ? 7:162. Type from Kohrud, northern Iran. Valid subspecies: *A. f. arianus*.

?*Mus sylvaticus tauricus* Barret-Hamilton, 1900. Proc. Zool. Soc. Lond., 1900:412. Type from Zebil, Taurus Mountains, Turkey.

Apodemus flavicollis argyropuloi Heptner, 1948. Dokl. Akad. Nauk. SSSR, 60:178. Type from Delizhan, Armenia. Synonym.

Apodemus hermonensis Filippucci, Simson and Nevo, 1989. Boll. Zool., 56:361. Type from Mount Hermon, Golan Heights. Synonym.

Diagnosis (Fig. 105).—See *Apodemus mystacinus*. Recent studies from Iran (Darviche et al., 1979) and the Holy Land (Filippucci et al., 1989) suggest that the smaller *Apodemus* of the Near East is biochemically more similar to *A. flavicollis* than to *A. sylvaticus*. *Apodemus flavicollis* is a small species with a head and body length of 75–100 mm and a skull length of 24–27.7 mm. The dorsal color of *A. sylvaticus* supposedly lacks the "both the intense fulvous tint" of *A. flavicollis* and "the greyish cast" of *A. mystacinus* (Harrison, 1972). The dorsal fur color of local specimens is pale buffy brown ranging to fulvous yellowish brown. The ventral color of *A. flavicollis* is whitish to yellowish gray with the bases of the hairs gray and occasionally with an orange fulvous or reddish brown collar (this collar is usually a very small light spot in the specimens from the Holy Land but can form a ring around the neck in specimens from southern and eastern Europe).

Range of species.—Most of the western Palearctic region including all of Europe.

Local status (Fig. 106).—The wood mouse was reported from the "plains of Palestine" (Tristram, 1884). No specific localities were given by Tristram or subsequently by Bodenheimer (1935, 1958). Specimens at HUJ are from Masada North, Moza, Sasa, and Horshat Ha'arbaim (Horshat Tel). Specimens at the MCZ were obtained from Shiba, Rasheya, and Ain Hersha in southern Lebanon (Allen, 1915). Filippucci et al. (1989) reported on populations in Galilee, Mount Hermon, and Tel Arad, and others in northern Holy Land.

Biology.—The field mouse is a nocturnal animal that feeds on acorns and seeds of shrubs (Lewis et al., 1967) as well as succulent grasses, buds of other plants, insect larvae, and other invertebrates. Field mice dig their own burrows, which have a nest chamber, usually locating the burrows under trees and shrubs. As in *A. mystacinus*, the limited available reproductive data suggest a continuous breeding season with more than one litter per reproductive season. *Apodemus flavicollis* is found in areas in northern Palestine where the dominant vegetation includes *Quercus calliprinos*, *Pistacia palaestina*, *Proterium*

Fig. 105. The wood mouse (*Apodemus flavicollis*). Animal from Mount Meron. Photo: H. Mendelssohn.

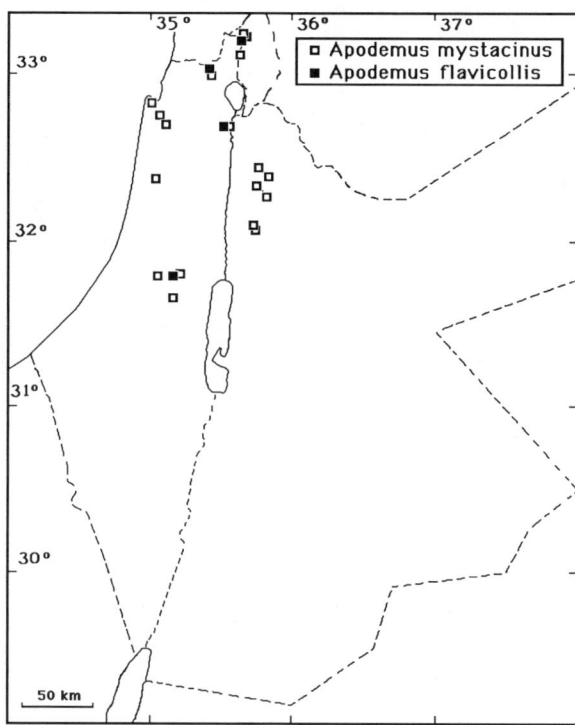

Fig. 106. Locality records of the broad-toothed field mouse (*Apodemus mystacinus*) and the wood mouse (*Apodemus flavicollis*).

spinosum, and *Cistus* sp. (Abramsky, 1981). In Afghanistan, five and six embryos were found in females collected on 17 July and 18 August (Hassinger, 1973).

Genetics.—2N = 48, AA = 46 was reported for specimens from Europe (reviewed in Zima and Král, 1984). Additional small chromosomes, called B chromosomes, have been reported in some European populations (Vujošević and Blagojević, 1994). The X and Y chromosomes are both acrocentrics.

The degree of genetic differentiation between "*A. hermonensis*" (described by Filippucci et al., 1989) and populations of *A. flavicollis* in the Holy Land (Nei's genetic distance of 0.056–0.058) does not justify specific recognition of *hermonensis*. Genetic distances between species of *Apodemus* are much higher and are usually associated with fixed allelic differences (Gemmeke, 1980). Morphologically, specimens from Mount Hermon are slightly smaller than those from other populations of *A. flavicollis*. However, such differences are expected in view of the habitat and elevation and do not represent major evolutionary changes.

Human interactions.—The name *flavicollis* derives from *flavus*, L., for yellow and *collum*, L., the neck (yellow collar on the neck).

Apodemus mystacinus   Broad-toothed field mouse   O

Mus mystacinus Danford and Alston, 1877. Proc. Zool. Soc. Lond., 1877:279. Type from Zebil, Bulgar Dagh, Turkey.

Apodemus (Sylvaemus) flavicollis pohlei Aharoni, 1932. Z. Säugetierkd., 7:183. Type from Kafrun, Nussarijeh Mountains, Syria. Synonym.

Diagnosis.—There is some confusion as to whether there are only two species of *Apodemus* in this region (*A. mystacinus* and *A. flavicollis*) or three (including *A. sylvaticus*). The three taxa are reportedly distinguished mainly by size with the largest being *A. mystacinus* and the smallest *A. sylvaticus*. The head and body length in *A. mystacinus* usually exceeds 95 mm (range 90–125 mm) and the skull length exceeds 28 mm (range 28–31 mm). A long (105–140 mm) thin tail is slightly longer than the length of the head and body. The dorsal color is dark gray with a faint brownish tinge on the sides that increases with age. The ventral color is dull white and is sharply delineated from the dorsal dark color. Aharoni (1932) described *pohlei* as a large form of *A. flavicollis*. However, *pohlei* is actually *A. mystacinus*.

Range of species.—Southeastern Europe, Turkey, the Holy Land, Iraq, Georgia SSR, Crete, Rhodes, and Aegean islands.

Local status.—The broad-toothed field mouse is common in areas with a Mediterranean climate in the northern parts of the Holy Land (Fig. 106). Allen (1915) first reported *A. flavicollis* from Ain Hersha at the base of Mount Hermon. Bodenheimer (1958) stated that M. Dor collected skulls from owl pellets on Mount Carmel. Harrison (1972) reexamined the two specimens of Allen (1915) and confirmed the designation "*flavicollis*" on one of them. Harrison (1972) also recorded specimens of *A. flavicollis* housed at Tel Aviv University from Mount Hermon and Masada North (Golan Heights) and further suggested that specimens from the Hula Lake area reported by von Lehmann (1965) are probably also referable to *A. flavicollis*. Allen (1915) reported a series of 14 skins from Mount Hermon that he referred to *A. mystacinus*. *Apodemus mystacinus* was also reported from Jebel Jarmak, Mount Carmel (Ellerman, 1948), Jerusalem (Bodenheimer, 1958), Tel el Qadi (Dan), Beit Meir, Tivon, Haifa, and St. John's Monastery (near Jerusalem) (Harrison, 1972), 4 km NW Beit Fajjar, Dibbine Ranger Station (Atallah, 1978), and Ajloun (Harrison and Bates, 1991). A specimen is also available at JUNHM from Ibbin (Amr and Disi, 1988). Many other localities from the western uplands are represented by specimens at the HUJ, including Jalama, Sataf, and Tel Al Qadi. I also collected specimens from Salt, Zubiya Forest, Dibbine Forest, and Zayy (all in May 1990).

Biology.—This field mouse is common in most forested or well-vegetated parts of the Holy Land having Mediterranean climates. It feeds primarily on grains, pine seeds, acorns, and carob pods (Lewis et al., 1967; Atallah, 1978). Other food items taken include land snails and insects. It is a strictly nocturnal species. In May 1990 in Jordan, in some 400 trap nights in several localities, this was the only rodent obtained. It appears to be very common in the habitat dominated by *Quercus calliprinos,* the acorns of which form its main diet (acorns were seen at burrow entrances). *Apodemus mystacinus* is found in a wider variety of habitats than the smaller *A. flavicollis.*

Atallah (1978) reported females pregnant with five embryos in September; young in April, May, and October; and lactating females in June to August. A specimen collected near Jerusalem gave birth to four young when collected on 21 April (Harrison, 1972). These data suggest a rather continuous breeding season with the possibility of two litters per season.

Genetics.—2N = 48, AA = 50 was reported for specimens from Europe (reviewed in Zima and Král, 1984). The X and Y chromosomes are both acrocentrics. I obtained similar data on Jordanian specimens.

Genus *MUS*   Common mice

Key to the Species of *Mus* of the Holy Land

1. Zygomatic index (width of anterior part of malar relative to width of anterolateral part of zygomatic arch) 0.2–0.7. Tail longer (about 50% of head and body length). . . . . . *M. musculus*

   Zygomatic index 0.6–1.2. Tail shorter (about 45% of head and body length). . . . . *M. macedonicus*

Mus macedonicus   Wild mouse, short-tailed mouse   **O**

Mus macedonicus Petrov and Ruzic, 1983. Proc. Fauna SR Serbia, Serbian Acad. Sci. Arts, Belgrade, 2:177. Type from Valandova, Macedonia, Yugoslavia.

Diagnosis.—This species is very similar in characters to *M. musculus.* The overall size is smaller with a relatively shorter tail. The best skull character is the zygomatic index (ZI; the ratio of the width of the anterior part of the malar process to the width of the anterolateral part of the zygomatic arch [Auffray et al., 1988]). In the inset of Figure 102 this would be the ratio of the distance A to B. In *M. musculus*, the ZI averages 0.47–0.52 (range 0.2–0.7) whereas in *M. macedonicus* it is 0.74–0.80 (range 0.6–1.2) (Orsini et al., 1983; Auffray et al., 1988). In *M. musculus*, the ratio of the head and body length to tail length averages 1.03, whereas in the short-tailed mouse it averages 1.28. There are more consistent biochemical differences between the two species. According to the biochemical classification of European *Mus*, the short-tailed form in the Holy Land belongs to group 4A, the Latin name for which is still being reevaluated (*spretoides, spretus, spicilegus, macedonicus*, or *abbotti*) (Marshall and Sage, 1981; Bonhomme, 1986; Kratochvil, 1986; Auffray et al., 1988, 1990; Nadachowski et al., 1990). I use the name *macedonicus* following Harrison and Bates (1991).

Range of species.—Europe and the Near East.

Local status.—Reported from a few localities west (Auffray et al., 1990) and east (Bates and Harrison, 1989) of the Jordan River. Further collecting and biochemical studies are needed to document its distribution in the Holy Land. I could identify two specimens from near the Dead Sea as this species. Other specimens collected by Atallah are all *M. musculus*.

Biology.—In sympatric areas, *M. musculus* occurs mainly in association with human habitations, whereas *M. spretoides* occurs in the wild. In areas where *M. macedonicus* is absent, the house mouse may be found outdoors (Auffray et al., 1990).

Genetics.—Biochemical differences allow the diagnosis of this species in the Holy Land as corresponding to the type 4A reported from Europe (Auffray et al., 1988). More extensive protein electrophoretic data from the Holy Land are needed to evaluate the status of the forms of *Mus*.

Mus musculus   House mouse   O

Mus musculus Linnaeus, 1758. Syst. Nat., 10th ed., 1:62. Type from Upsala, Sweden.

Mus orientalis Cretzschmar, 1826. *In* Rüppell, Atlas Reise Nördl. Afr., Säugeth., p. 76, pl. 30, fig. a. Synonym.

Mus praetextus Brants, 1827. Het. Geslachte der Muizen, Berl., p. 125. Type from Sakhara, Syria. Synonym.

Mus gentilis Brants, 1827. Het. Geslachte der Muizen, Berl., p. 126. Type from southern Egypt. Synonym.

Diagnosis.—Members of the genus *Mus* in the Holy Land are small mice with a head and body length of 74–90 mm and a skull length of 21–24 mm. The tail is about the same length as the head and body. The color is brown to gray with occasional melanistic or albino individuals seen. They are distinguished from other murids in the region by their small size, smooth uniform pelage lacking contrasting markings, and the presence of a distinct notch on the upper incisors (Fig. 102). Also, the first upper cheektooth has a crown area larger than that of the second or third cheekteeth. The two species of *Mus* in the Holy Land are very similar; characters to distinguish them are discussed under *M. macedonicus.*

Range of species.—Cosmopolitan.

Local status.—As stated as early as 1884 (p. 11) by Tristram "the European house mouse is common in all the towns." The house mouse was reported from diverse towns in areas such as Jerusalem, Moab, Ghor es Safi (Aharoni, 1932), 'Aqaba, Shobek (Allen, 1915), Jericho (Ellerman, 1948), Amman (von Lehmann, 1965), Hulata, Haifa, Sa'ar, near Eilat, Wadi Araba, Jerusalem, and Rehobot (Harrison, 1972). Specimens are also reported from Bethlehem, Beit Sahur, Givat Yeshayahu (near Beisan), near Qumran N end of Dead Sea, 4 km NW Beit Fajjar, Azraq Druz, Azraq Shishan, El Jafr, and 10 km NE Amman (Atallah, 1967*b*, 1978; and M. B. Qumsiyeh pers. coll.). The species was also recorded from northern Sinai from Al Arish (Wassif, 1954*b*). Allen (1915) reported four pale-colored animals from 'Aqaba, and from Rasheya, Hasbeiya, and Shiba in southern Lebanon, which he assigned to *M. m. orientalis,* and others from Shobek, Wadi Kerak, and El Karak, which he assigned to *M. m. gentilis.* The extensive color and size variation in the Holy Land does not warrant subspecific recognition and I tentatively assign all these animals to *M. m. musculus.*

Biology.—Although this species was so successful because of its commensal relationship with humans, it easily adapted to local habitats and in many areas can be trapped far from human habitation. It is found in almost every habitat in the area. There is extensive color variation and specimens from some desert areas are light colored, suggesting early introduction with subsequent evolution of the lighter color. The diet of *M. musculus* is varied; these animals are primarily opportunistic feeders, mainly on plant material. There is no evidence of a limited breeding season. Young (up to eight per litter) are born naked and blind after a gestation period of slightly more than 21 to 23 days.

Genetics.—The genus *Mus* in Europe shows extreme variability in diploid numbers due to numerous Robertsonian translocations. The primitive karyotype appears to be 2N = 40 with all acrocentric autosomes (reviewed in Jotterand-Bellomo, 1984; Zima and Král, 1984). Specimens collected in the Holy Land exhibit the primitive karyotype (M. B. Qumsiyeh, unpublished data).

Human interactions.—This species is of an obvious economic importance because it lives and feeds in homes, storehouses, and other human structures. The wild *M. musculus* is the ancestor of the laboratory mouse, of which many varieties have been selected. The lab mouse is now used extensively in almost all branches of biology and medicine. Mus is Greek for mouse, *culus* is the diminutive suffix in Latin (hence the name *Mus musculus* means mouse, little mouse).

Genus *NESOKIA*   Bandicoot rats

There is only a single species of this genus in the region. These rats are superficially similar to *Rattus*. However, the body is overall more adapted to fossorial life. The skull is particularly wide and has a short rostrum.

Nesokia indica   Short-tailed bandicoot rat   **R**

Arvicola indica Gray and Hardwicke, 1832. Illustr. Indian Zool., 1:pl. xi. Type from India.

Nesokia bacheri Nehring, 1897. Zool. Anz., 547:503. Type from Ghor es Safi, Holy Land. Valid subspecies.

Diagnosis.—This is a large (head and body length 180–260 mm, skull length 45–47 mm) brown rat with a short tail (110–145 mm). The tail is always much shorter (50–70%) than the head and body length. The size and relative length of the tail as well as the uniform color (with little distinction between dorsal and ventral surfaces) distinguish this species from the other murids of the region. The only other rat with which it may be confused is the brown rat (*Rattus norvegicus*). However, the latter has a clear distinction between the dark color of the back and the very light underside. The skull is robust, short, and broad with strong temporal and parietal ridges. The mandible is well developed compared to the genus *Rattus*. The cheekteeth have cusps that are completely fused to form transverse lamina. The Palestine bandicoot rat, *N. i. bacheri* is larger than other subspecies. In the Holy Land, many individuals are seen with a ventral white spot between the forelegs.

Range of species.—Steppes and cultivated areas from northwestern India and Sinkiang through southwestern Asia and the northern Arabian Peninsula to Egypt.

Local status.—The short-tailed bandicoot rat is common around oases and springs near the Dead Sea and in Wadi Araba (Fig. 107). Tristram (1884) first obtained a single specimen from Palestine but did not state the locality. Later specimens from Ghor es Safi, which were collected by F.

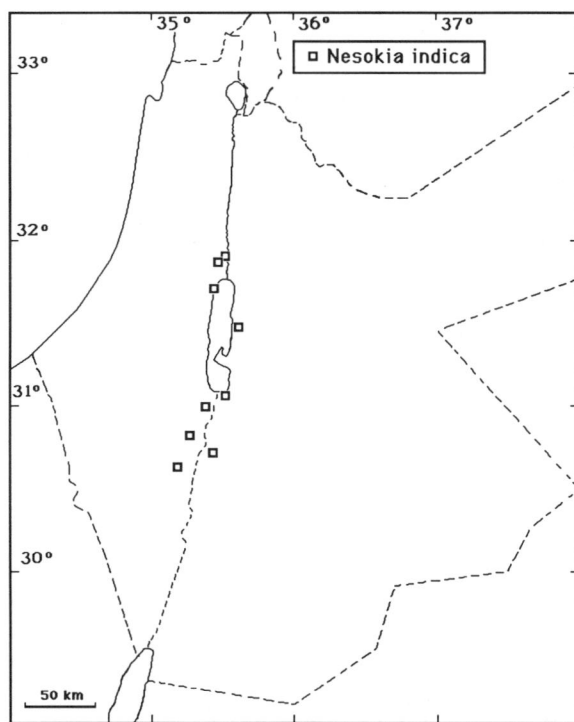

Fig. 107. Locality records of the short-tailed bandicoot rat (*Nesokia indica*).

Bacher (hence the name *bacheri*), were described by Nehring (1897). I. Aharoni (1930) and B. Aharoni (1932) repeated this record and added "Moab" (no exact locality). I obtained several specimens from Ain Fashkhah and Abu-Dieyeh (1988) reported a specimen from Wadi Al Fidan. Specimens are also available at HUJ from Allenby Bridge, Hatseva (Ain Husab), 4 km E Musa El Alami (near Jericho), Jericho, Ain Yahav, and Neot Hakikar. This species could have had a much wider range in the past than it does today, as illustrated by the scattered areas where it occurs now. Because it is adapted to moist conditions, the expanding desert conditions have left isolated pockets in certain areas. This also appears to be the case in Egypt (Osborn and Helmy, 1980).

Biology.—Bandicoot rats live in moist soils near water sources in areas of dense vegetation. The specimens I obtained from Ain Fashkhah were collected at midday where one could watch these animals foraging in the lush vegetation of the spring. Bandicoot rats were found on trips from March to August. In Pakistan, bandicoot rats construct elaborate tunnels close to the surface; these may serve as feeding tunnels (Roberts, 1977). In Egypt, they were reported to feed on fleshy roots of *Alhagia* and *Typha* (Hoogstraal, 1963; Osborn and Helmy, 1980).

One female obtained at Ain Fashkha on 18 August had two embryos. In Iran, this species appears to breed year-round with two to eight embryos recorded (Lay, 1967). Bandicoot rats are preyed upon by the Palestine jungle cat (*Felis chaus*), snakes, and owls (Al Jumaily et al., 1975).

Genetics.—2N = 42, AA = 66 was reported for specimens from Iran (Kamali, 1975).

Genus *RATTUS*   Common rats

Key to the Species of *Rattus* of the Holy Land

1. Tail longer than head and body. Ear relatively larger (length more than one-half of hind foot length). First upper molar with anterior lateral cusp. Temporoparietal ridges curving laterally. Skull more elongate. . . . . *R. rattus*

   Tail shorter than head and body. Ear relatively smaller (length less than one-half of hind foot length). First upper molar without anterior lateral cusp. Temporoparietal ridges almost parallel. Skull more oblong. . . . . *R. norvegicus*

Rattus norvegicus   Brown rat, Norway rat   o

Mus norvegicus Berkenhaut, 1769. Outlines of the natural history of Great Britain and Ireland, 1:5. Type from England.

Mus maniculatus Wagner, 1848. Archiv. Naturgesch. 14:186. Type from Egypt.

Diagnosis.—See *Rattus rattus*. From the black rat, the brown rat is distinguished by color, a shorter and thicker tail, and a more robust body. Also the muzzle, ears, and legs are all shorter in the brown rat. The head and body length is 200–250 mm, the tail length is 170–210 mm, and the skull length is 45–54 mm. The first upper molar is without an anterior lateral cusp.

Range of species.—Originally from southern Asia, brown rats are now cosmopolitan but mainly concentrated in coastal regions. They are commensal with humans.

Local status.—A cosmopolitan pest that "has found its way to Palestine, and is as common there as elsewhere" (Tristram, 1884). This species was recorded from Haifa, Jaffa, Tel Aviv (Bodenheimer, 1935, 1958), and Wadi

Zarqa, S Jerash (Bates and Harrison, 1989). A specimen at JNHM is from Jordan University Campus (Jubeiha). I have specimens in my personal collection from Beit Sahur, Bethlehem, and Auja (northern Jericho).

Biology.—In the Holy Land, the brown rat is more commensal than its relative the black rat but is relatively less widespread. It occurs in human habitations in the port cities and has managed to penetrate into some cities in the interior. It seems to prefer more mesic areas than *R. rattus*. Burrows can be found along stream banks (including sewage streams) in many areas. Little is known of the reproduction of the brown rat in this area. In other parts of its range female brown rats reach sexual maturity at a few months of age and give birth to several litters per year, each having up to 10 young.

Genetics.—2N = 42, AA = 62 was reported for specimens from many areas in Europe (reviewed in Zima and Král, 1984).

Human interactions.—See comments on the black rat (*R. rattus*). The name *norvegicus* is Latin for "of or belonging to Norway" where this species was first described. Brown rats are the ancestors of the laboratory white rat, which is used extensively for research, teaching, and even as pets in many parts of the world.

Rattus rattus   Ship rat, black rat, house rat   O

Mus rattus Linnaeus, 1758. Syst. Nat., 10th ed., 1:61. Type from Upsala, Sweden.

Mus alexandrinus É. Geoffroy St.-Hilaire, 1803. Cat. Mamm. Mus. Nat. Hist. Nat. Paris, 1803:192 (also Desmarest, 1819, Nouv. Dict. Hist. Nat. Paris, 29:47, 192). Type from Alexandria, Egypt. Synonym.

Mus flaviventris Brants, 1827. Het. Geslachte der Muizen, Berl. 1827:108. Type from Arabia. Synonym

Musculus frugivorus Rafinesque, 1814. Précis des Découv. et Travaux. Simiol., p. 13. Type from Sicily. Synonym.

Diagnosis.—Compared with *R. norvegicus*, the black rat has a relatively longer tail (longer than the head and body length), relatively larger ear (length more than one-half of hind foot length), and more slender and elongate body. The head and body measures 145–200 mm, the tail 190–250 mm, and the skull 38–45 mm. The temporal ridges on the braincase are curved in *R. rattus* and almost parallel in *R. norvegicus*. The skull of *R. rattus* is also more elongate than that of *R. norvegicus*. The first upper molar has an anterior lateral cusp.

There are several color forms seen in this region, which have been given subspecific names by some authors (synonyms listed above): blackish brown (*rattus*), dark grizzled brown (*alexandrinus*), and light grizzled brown with whitish or yellowish underside (*frugivorous* when upper and undersides sharply delimited or *flaviventris* otherwise) (Aharoni, 1932).

Range of species.—This rat is originally from southwestern Asia. It spread as a commensal form with humans in warmer parts of the world.

Local status.—The black rat was reported as a common commensal species in Palestine (Bodenheimer, 1935, 1958). It is also recorded from Rehobot, Jerusalem, Ghor es Safi (Aharoni, 1932), Wadi Kelt, Bir Salem, Haifa (Harrison, 1972), and Beit Sahur (Atallah, 1978). The black rat is also reported from Al Arish in northern Sinai (Hoogstraal, 1963).

Biology.—Black rats are commensal, opportunists, omnivorous, and are active during the day or night. Females reach reproductive age in only 3 months. They breed at any time during the year, producing up to 10 young in each litter. Because they can have several large litters each year, black rats can be very prolific and cause much destruction to agricultural products.

Genetics.—2N = 38, AA = 58 was reported for specimens from Europe (reviewed in Zima and Král, 1984). However, other variants were reported for Asian specimens.

Human interactions.—In the Holy Land, most rats (both brown and black) live inside houses and storage areas and are completely dependent on man. In some areas, rats have adapted to local wild habitats and can be considered feral. Rats are economic pests because they feed on grains and other human foods. They are despised by the locals (known to the Arabs as *jardoun*, pl. *jaradin* and in Hebrew as *Halda*) and are commonly trapped. Up until a few years ago, there was not a single market place in any city in the Holy Land where rat traps were not sold. The black rat was one of the main carriers of bubonic plague and outbreaks in the Holy Land have been reported as recently as 1947 in Haifa (Gratz, 1957). Black rats act as carriers for other diseases such as scrub typhus, trichinosis, and even rabies. *Rattus* is the new Latin word for rat.

FAMILY GLIRIDAE   Dormice

The Gliridae (from *glis*, L., genitive *gliris*, a dormouse) or dormice are small to medium-sized rodents that externally resemble a small squirrel. The tail is heavily furred and the pelage is thick and soft. The main diagnostic characters for this family are in skull morphology. The infraorbital foramen is large and the mandible has angular processes bent laterally to the side. There are 13 species known in areas of Europe, Africa, southwestern Asia,

Fig. 108. The forest dormouse (*Dryomys nitedula*). Animal from Mount Meron. Photo: H. Mendelssohn.

and Japan, of which two are found in the Holy Land: *Eliomys melanurus* and *Dryomys nitedula*. Reports of *Glis glis* by Tristram (1866*a* and 1884) from the Jordan Valley were probably an error (Atallah, 1978).

Key to the Species of Gliridae of the Holy Land

1. Tail with short hairs at base and a more bushy tuft at the end; larger animal with skull length more than 30 mm.

    . . . . *Eliomys melanurus*

    Tail uniformly bushy; smaller animal with skull length less than 30 mm. . . . . *Dryomys nitedula*

Genus *DRYOMYS*   Forest dormice

Dryomys nitedula   Forest dormouse   I

Mus nitedula Pallas, 1779. Nova Spec. Quad. Glir. Ord., p. 88. Type from Lower Volga, USSR.

Dryomys nitedula phrygius Thomas, 1907. Ann. Mag. Nat. Hist., ser. ? 20:407. Type from Murad Dagi, 40 km NE Usak, Usak Province, Turkey. Named after the ancient area of of Phrygia. Perhaps a valid subspecies.

Diagnosis.—From *Eliomys melanurus* (the only other dormouse of the region), this species differs in its smaller size (head and body length 75–105 mm, tail length 65–105 mm, skull length 24–29 mm), overall brown color, and the rather uniform-colored and bushy tail (Fig. 108). The long bushy tail and the gray-brown color that matches tree bark are presumed adaptations for the arboreal life of this species. The skull is slender and has small

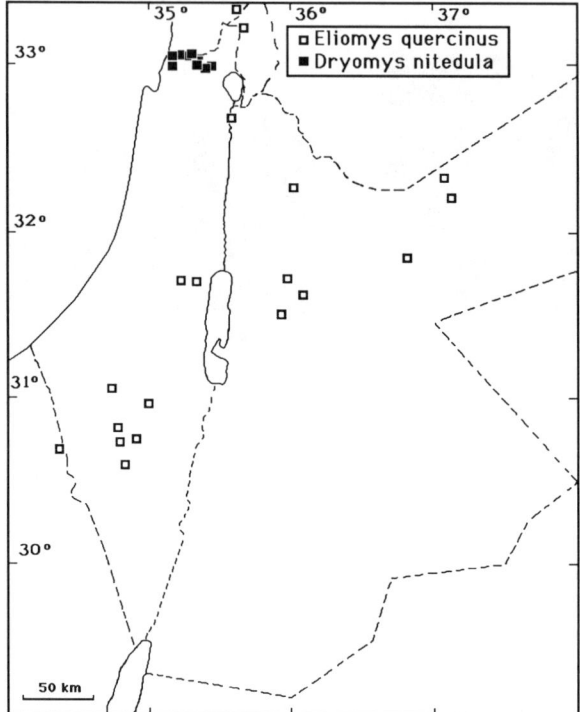

Fig. 109. Locality
records of the Sinai
dormouse (*Eliomys
melanurus*) and the
forest dormouse
(*Dryomys nitedula*).

bullae, much smaller than in *Eliomys melanurus* of the region, even when the
relative skull size is taken into consideration. *Dryomys nitedula phrygius* was
named for the Turkish populations because it had "brighter and yellower
color" than other forms. Specimens from the Holy Land do not have such
coloration but if the Turkish form proves valid, they may be referable to it
on geographic grounds.

Range of species.—In deciduous forests of eastern Europe and the western
USSR south to Turkey and Iran. Isolated populations (perhaps from the last
Pleistocene glacial) in the eastern Alps, the Holy Land, and other areas.

Local status.—The forest dormouse is known only from forests in the
northern Galilee (Fig. 109). Bodenheimer (1958) reported a specimen from
Hurfeish, and stated that Tristram's *Myoxus dryas* Schreber may be referable
to this species. A discussion of the biology and a list of other localities is
provided by Nevo and Amir (1961, 1964): Mount Meiron, W Meiron,
Mount Zeved, Nahal Afaim (Wadi Abu Ali), Nahal Manba (Wadi Manba),
Har Zevul (Jebel Sabalan), SE Hurfeish, Hurvat Bar Zeit (Khirbet Barza),
Nahal Kziv (Wadi Qarn), SW Montfort, Nahal Ga'aton, S Eilon near
Hanita, Eilon, Har Ushman (Tel Marda), and Nahal Ukam (Wadi Suad).

Biology.—Forest dormice are omnivorous, nocturnal, and arboreal. The following data are primarily from Nevo and Amir (1964) and Roberts (1977). The species occurs in Mediterranean forests at high elevations. *Quercus calliprinos* and *Pistacia palestina* are the dominant trees in these habitats. In addition to acorns of many trees, the forest dormouse feeds on fruits of *Rosa*, *Arbutus*, and *Styrax*, as well as animal matter (bird eggs and insects). Unlike European forest dormice, the forest dormice in the Holy Land enter into hibernation for only very short periods (hours) and are active throughout the year.

Both males and female participate in building the nest but the male abandons the female shortly before parturition. The nest is constructed about 3 m above ground. Nests may occur in groups but populations are very sparse and individuals may be separated by tens to hundreds of meters. One to four young are born from March to December with two to three litters each year. The gestation period is about 5 weeks. In Turkey, four to five embryos were found in June, and very young individuals were found in nests in August (Felten et al., 1973).

Genetics.—2N = 48, AA = 92 was reported for specimens from Europe (reviewed in Zima and Král, 1984). The X chromosome is metacentric and the Y is a small acrocentric.

Human interactions.—*Dryomys nitedula* is derived from *drus*, Gr., genitive *druos*, the oak; *mus*, Gr., a mouse; and *nitedula*, L., for a small mouse. Hence, *Dryomys nitedula* is "the little mouse that lives on oaks."

Genus *ELIOMYS*   Garden dormice

Eliomys melanurus
Sinai dormouse; southwest Asian garden dormouse   I

[*Mus quercinus* Linnaeus, 1766. Syst. Nat., 12th ed., p. 84. Type from Germany. Used by numerous authors for Holy Land *Eliomys*.]

Myoxus melanurus Wagner, 1839. Gelehrte Anz. I. K. Bayer. Akad. Wiss., München, 8(37):299. (Further details in Wagner, 1840, Abh. Bayer. Akad. Wiss. Math.-Natur-wiss. Kl., 3(1):176, fig. 3 and 4, pl. 3, fig. 1.) Type from the Sinai Peninsula (collected by Von Schubert) and restricted to Mount Sinai by Nader et al. (1983).

Diagnosis.—The overall appearance of the dormouse is like that of a small squirrel. Dormice are distinguished from other rodents of the region by their coat color and appearance (Fig. 110). A black stripe around and extending back from the eye gives a masked appearance. In *Eliomys*, the tail is bushy only on the distal region, which is also blackish; this distinguishes it from

Fig. 110. Sinai dormouse (*Eliomys melanurus*). Photo: Y. Werner.

the other dormouse in the region, *Dryomys nitedula*. Also, *Eliomys* is larger with a head and body length of 105–140 mm and a skull length of 32–37 mm.

The form *melanurus* was considered to be distinct from *quercinus* by the very large tympanic bulla in *melanurus* (almost one-third of the occipitocondylar length) (Ellerman, 1948). Preliminary genetic data support the separation of a Middle Eastern/northern African *E. melanurus* from a European *E. quercinus* (Fillippucci et al., 1988*a*). The situation is more complex chromosomally, with several chromosome forms (2N = 48–54) reported in Europe (Delibes et al., 1980; Zima and Král, 1984; Filippucci et al., 1988*a*). Because no chromosomal phylogeny is available based on G-band studies, it remains unclear what the relationships between the 2N = 48 form from the Holy Land (*E. melanurus*) are to forms from Europe (*E. quercinus*) with the same diploid number. Further study of this problem at the genetic level is required before a formal reevaluation of the systematics is undertaken (Filippucci et al., 1988*b*). Color is not helpful in this regard, because it shows marked polymorphism. As an example, two specimens from the H5 region at JUNHM are distinct from each other, with one being almost melanistic.

Range of species.—In most countries surrounding the Mediterranean and east and north through wooded regions of eastern and northeastern Europe. Also in Arabia and southwestern Asia.

Local status.—The Sinai dormouse was found both in the northern regions as well as in the southern arid regions of the Holy Land (Fig. 109). Tristram (1877) obtained two specimens, from Um Rasas and the Roman city of Jiza, both in Moab. Bodenheimer (1935, 1958) stated that the species is common in Palestine and southern Jordan without giving details. Three specimens at the MCZ were collected at Ain Hersha and Rasheya at the base of Mount Hermon in southern Lebanon (Allen, 1915). This species was reported from Azraq Druz (Atallah, 1966, 1967*a*), Wadi Naphekh (Negev), Dan (Harrison, 1972), Mount Hermon, Bab el Haoua (= Hawa), Masada North, Avdat, Makhtesh Ramon, Wadi Nafkha (Nahal Nafha) (Nader et al., 1983), and Wadi Rajil (near Jawa) (Searight, 1987*b*). Specimens are also known from Dab'ah (Amr and Disi, 1988), and H5 (all at JUNHM). This species was also collected from El Quseimah in Sinai (Flower, 1932). Kahmann (1981) summarized the available knowledge of this species in the Holy Land and Arabia and listed the following localities based on museum specimens and field observations: Mitspe Ramon, Avdat, Sede Boqer, Wadi Nafha, Revivim, Maktash haGadol, Birket Bab El Haoua (=Hawa), Masada North, Mount Hermon, Ain Quneiya, Dan, Beit Sahur, Mar Saba, Dab'ah, Amman, Um Rasas, Ziza, and Azraq.

Biology.—The Sinai dormouse is nocturnal and is found in forested or rocky habitats or in small valleys near rocky or limestone hills in varied climates in this region. The reason for this diversity in habitats in a species that evolved as an arboreal animal is not clear. It was speculated that the expansion of the desert and deforestation provided a strong selective force for adaptations to the nonforested habitat having a harsh climate (Harrison, 1972). At Nahal Nafha (Wadi Nafha), a small colony was found in rocky outcrops on a hill overlooking a few pistachio trees (Harrison, 1972).

Stomachs examined by Atallah (1978) contained more animal (including a gecko, *Ptyodactylus*) than vegetable matter. Similarly, Kahmann (1981) and Nader et al. (1983) found fragments of insects, centipedes, snails, and mammalian remains in the stomachs of this rodent. Kahmann (1981) reported a pregnant female in April and one lactating on 4 May in the Negev. In the laboratory, there is an average of 2.8 young per litter (Kahmann, 1986). Kahmann (1986) details growth and development of these animals (including the pattern of molting) under laboratory conditions. Animals from the Near East ranged in weight from 47 to 67 g (Nader et al., 1983).

Genetics.—See under diagnosis.

Human interactions.—*Eliomys* is derived from *eleios*, the Greek word for this species of dormouse. The name *quercinus* is from *quercus,* L., for oak and *inus,* L., suffix meaning pertaining to. The English name (dormouse) was probably derived from the Latin *dormio* (= I sleep), an allusion to the length of hibernation in these rodents.

FAMILY DIPODIDAE   Jerboas, birch mice, jumping mice

Jerboas are distinguished from other rodents of the region by the long hind legs and long tail, which allow a ricochetal mode of locomotion (Fig. 111). The family name Dipodidae comes from this mode of locomotion (*di* is from *dis,* Gr., two; *pous,* Gr., genitive *podos,* the foot; *idae,* L., suffix for family names). The infraorbital foramen is greatly enlarged and no zygomatic plate is developed. The sides of the mandibles may show perforations in some species. Fourteen genera with about 44 species are found in areas of North America, eastern Europe, Asia, Arabia, and northern and eastern Africa.

Key to the Species of Dipodidae of the Holy Land

1. Ears long (more than one-half hind foot length). Bulla not as below. Four upper cheekteeth. Five digits on the hind feet.
. . . . *Allactaga euphratica*

   Ear short (less than one-half hind foot length). Bulla highly inflated with mastoid regions clearly visible from the dorsal aspect. Three upper cheekteeth. Three digits on the hind feet.   (*Jaculus*) 2

2. Size large, head and body length greater than 130 mm in adults. Hind foot length more than 70 mm. . . . . *Jaculus orientalis*

   Size smaller, head and body length less than 120 mm in adults. Hind foot length less than 70 m. . . . . *Jaculus jaculus*

Genus *ALLACTAGA*   Five-toed jerboas

Allactaga euphratica   Euphrates jerboa, five-toed jerboa   V

Allactaga euphratica Thomas, 1881. Ann. Mag. Nat. Hist. ser. 5, 8:15. Type from Iraq.

Diagnosis.—This is a medium-sized jerboa with long (30–45 mm) slender ears (Fig. 111C). It is distinguished from *Jaculus* by the presence of all five digits on the hind feet, three of which are functional. The tail is long and has a black banner with a contrasting white tip.

Range of species.—Semiarid regions in Turkey, the Syrian Desert, northern Saudi Arabia, Iraq, the Caucasus, Iran, and Afghanistan.

Fig. 111. Three dipodids from the Holy Land. (Upper) Lesser Egyptian jerboa (*Jaculus jaculus*). Photo: S. I. Atallah. (Middle) Greater Egyptian jerboa (*Jaculus orientalis*). Photo: A. Shoob. (Lower) Five-toed jerboa (*Allactaga euphratica*). Animal from Jawa. Photo: A. Searight. Not to the same scale.

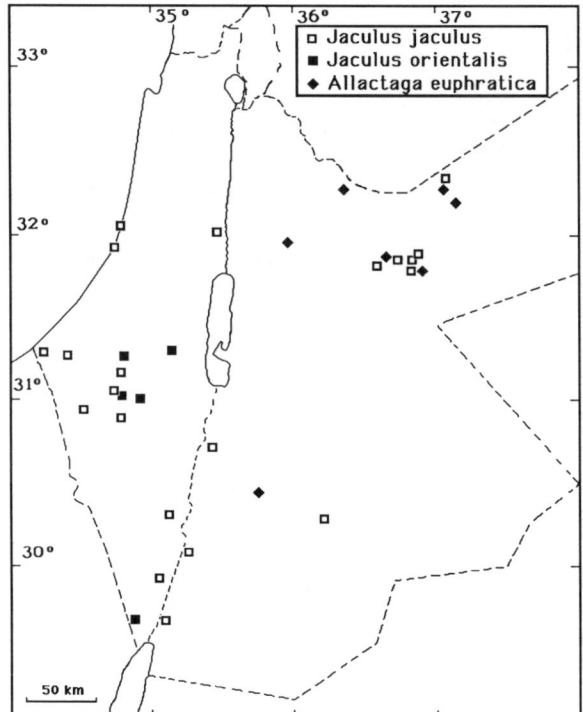

Fig. 112. Locality records of the lesser Egyptian jerboa (*Jaculus jaculus*), the greater Egyptian jerboa (*Jaculus orientalis*), and the five-toed jerboa (*Allactaga euphratica*).

Local status.—The five-toed jerboa in the Holy Land is recorded only from arid regions of the Syrian Desert (Fig. 112). The distribution was stated to be "east of a line joining Amman and Ma'an" (Atallah and Harrison, 1968). Localities in the region include 5–10 km NE Qasr Amrah, 4 km ENE Qasr Amrah, 30 km E Mafraq, 45 km N Ma'an (Atallah, 1978), Amman (Harrison, 1972), Jawa (Searight, 1987*b*), Shawmari (Amr and Disi, 1988), and H 5 (specimen at JNHM).

Biology.—Five-toed jerboas live in extremely arid regions of the Syrian Desert and share their habitat with other rodents including the lesser Egyptian jerboa (*Jaculus jaculus*). However, where found together, *Allactaga* is always less common than *Jaculus*. Five-toes jerboas seem to prefer foothills, especially near wadi systems and in the grassy areas of the hammada deserts. The burrow system has been described and is similar to that of *Jaculus* (Atallah, 1978). Two females collected by Atallah (1978) in April had six and nine embryos. Misonne (1957) stated that the young are born with the hind legs short and similar to the front legs but that a few days later the tail and the hind legs grow disproportionately faster than the rest of the body.

The five-toed jerboa molts its hair once a year about in July (Kadhim et al., 1979).

Human interactions.—The Mongolian name for this animal is *alak-dagha,* which is the origin of the genus name *Allactaga.* The name *euphratica* refers to the Euphrates River (this jerboa was first caught in Iraq). See also under *Jaculus jaculus* for other comments.

Genus JACULUS   Three-toed jerboas

Jaculus jaculus   Lesser Egyptian jerboa, three-toed jerboa   V

    Mus jaculus Linnaeus, 1758. Syst. Nat., 10th ed., 1:63. Type from Giza Pyramids, Egypt.

    Dipus macrotarsus Wagner, 1843. Abh. K. Bayer. Akad. Wiss. München, Math.-Phys., 3:214, pl. 4, fig. 2. Type from Wadi Feiran, Mount Sinai, Egypt. Synonym?

    Jaculus jaculus schlueteri Nehring, 1901. Schriften Berl. Ges. Naturf. Fr. Berl., p. 163. Type from the coastal region south of Jaffa, Palestine. Valid subspecies.

    Jaculus loftusi vocator Thomas, 1921. Ann. Mag. Nat. Hist., ser. ? 8:441. Type from near Muscat, Yemen (now Oman). Valid subspecies: *J. j. vocator* in the Syrian Desert.

    Jaculus jaculus syrius Thomas, 1922. Ann. Mag. Nat. Hist., ser. ? 9:296. Type from Karyatein, Syria. Synonym of *J. j. vocator.*

Diagnosis.—The lesser Egyptian jerboa (Fig. 111A) is distinguished from *Allactaga* by the much smaller ears (21–30 mm) and the three digits of the hind feet. From *Jaculus orientalis*, it is distinguished by the smaller size, especially in the ear and hind foot measurements, and the geographic distribution. The head and body length measures 95–130 mm, the tail length is 150–190 mm, the hind foot length is 50–78 mm, the greatest skull length is 32–36 mm, and the weight is 45–75 g. See also discussion under *J. orientalis.* There appear to be two subspecies in this area that can be distinguished by body size and hind foot length: *Jaculus j. vocator* in the Syrian Desert and southern Jordan, and the larger *J. j. schlueteri* in the coastal plains of Israel. According to Bodenheimer (1958), *J. j. schlueteri* has larger tympanic bullae that "touch at their inner margin." However, the two subspecies may intergrade in the Negev Desert (Harrison, 1972). This problem can only be solved by additional systematic studies in the area, preferably using genetic techniques.

Range of species.—Sahara Desert in northern Africa, Arabian Desert to Syrian Desert and the Iranian Desert.

Local status.—The lesser Egyptian jerboa is found in the sandy and hard soil deserts of the south and east (Fig. 112). The species was reported by

Tristram (1884) as *Dipus hirtipes*. Specific records are available from El Jafr (Atallah, 1967*a*), 2–3 km NE Azraq Shishan, Azraq Shishan, 5 km W Azraq Shishan, near Qasr Amrah (Atallah, 1967*b*), 50 km E Beersheba, Bir Melikha (Wadi Araba) (Harrison, 1972), Jaffa (Atallah, 1978), Wadi Finan (Amr and Disi, 1988), and Jawa (Searight, 1987*b*). Specimens at JNHM are from Wadi Rajil (near Azraq) and W Shawmari Wildlife Reserve. Specimens at TAUM are from many localities in Israel and the occupied territories including Revivim, Sede Boqer, Wadi Rubin, Ain Radian, Auja–Rafah road, Wadi El Abiad, 15 km S Beersheba, and 22 km N Jericho. The species was also reported from northern Sinai near Al Arish (Wassif, 1954*b*).

Biology.—These animals are adapted to nocturnal life in the desert. Their saltatorial locomotion allows them to move easily between one foraging bush and the next and to perform leaping maneuvers to escape their enemies (which include the three species of foxes in the same habitat, the jackal, and the desert cat). Jerboas cover their burrow entrances during the day and emerge shortly after dark (Atallah, 1978). Atallah (1978) also discussed their complex tunnel systems, suggesting that they prefer to build their burrow systems in flat areas, whereas *J. orientalis* prefers a burrow entrance at the foot of hills. However, Wassif (1954*b*) reported that in northern Sinai, hills and hillocks with loose sands are their prefered areas. Lesser Egyptian jerboas appear to wander as solitary individuals, sometimes for long distances, and can be caught with a hand net as they "freeze" under a spotlight. Osborn and Helmy (1980) reported that jerboas in Egypt feed on such plants as *Cleome droserifolia, Aerva javanica, Salsola baryosma, Zygophyllum album, Cornulaca monocantha*, and various grasses (monocots).

The gestation period appears to be 25 days with three to four young born without a particular defined breeding season (Lewis et al., 1965; Osborn and Helmy, 1980). Lesser Egyptian jerboas in Iraq molt once a year around September–October (Eissa et al., 1975; Kadhim et al., 1979).

Genetics.—2N = 48, AA = 92 was reported for specimens from the Volga, USSR (Vorontsov and Malygina, 1973). A similar karyotype was reported for specimens from Saudi Arabia (Al-Saleh and Khan, 1984).

Human interactions.—The name *Jaculus* is from *jaculor* (*iaculor*), L., meaning to throw a javelin, an allusion to the jumping behavior of these rodents. Jerboas are familiar to all who have spent some time observing the beauty of the desert night, as those little jumping animals that are seen in the dim night light. They are considered to be edible by several tribes of Bedouin in Jordan and the Negev (called *yerboua* or *jerboua* in Arabic).

Fig. 113. Greater Egyptian jerboa (*Jaculus orientalis*). Photo by A. Shoob.

Jaculus orientalis   Greater Egyptian jerboa   V

Jaculus orientalis Erxleben, 1777. Syst. Regn. Anim., p. 404. Type from Sinai, Egypt.

Diagnosis.—This is the largest jerboa in the region (Figs. 111B, 113). Its head and body length measures 135–180 mm, the tail length is 210–240 mm, and the greatest skull length is 38–42 mm. The hind feet have only three toes. The soles are covered with hair. These modifications make this species well adapted for a ricochetal mode of locomotion (jumping kangaroo-style) in sandy desert areas. This species differs from *J. jaculus* mainly in having a larger body size and in the V-shaped posterior margins of the nasal bones (Osborn and Helmy, 1980).

Range of species.—Northern Africa, Sinai, and the coastal region of the Holy Land.

Local status.—The greater Egyptian jerboa is known to penetrate into the Negev Desert from Sinai (Fig. 112). It was reported from Tel Arad (Tel Shoket) (Harrison, 1972). Specimens at the HUJ are from Mashabei Sade and Yeroham (Rekhme) and those at TAUM are from Beersheba, 13 km S Beersheba, Tel Arad, 10 km SW Tel Arad, and NW Eilat. This species was also recorded from El Kuntila, Sinai, near the border with the Holy Land (Haim and Tchernov, 1974).

Biology.—The greater Egyptian jerboa occurs in the seashore areas of Palestine and northern Sinai in sandy or marshy habitats. This species appears to be more sociable and docile than *J. jaculus* (Osborn and Helmy, 1980). Its biology and adaptation to desert environments were studied by Kirmiz (1965). The animals seem to feed on seeds, buds, and roots of various plants and to obtain all of their water from these food sources. The greater Egyptian jerboa builds its nest some 120 cm below ground in a burrow system that may exceed 250 cm in length (Wassif, 1954*b*). Apparently two to three young are born in a breeding season that extends from November to February.

Human interactions.—See under *J. jaculus.*

FAMILY HYSTRICIDAE   Porcupines

The members of the family Hystricidae (from *hustrix,* Gr., genitive *hustrikhos,* a porcupine) are distinctive among many rodents of the region in their large size and heavy built and in the covering of spines on the body. The skull characters are also distinctive with a large infraorbital foramen, prominent occipital region, and a rather smooth and rounded dorsal profile. The cheekteeth have deep reentrant folds, which in adults isolate four "lakes" (areas surrounded by ridges) for each tooth. Three genera are known with 11 species found in Africa and southwestern and southern Asia to southern Sulawesi (Celebes).

Genus *HYSTRIX*   Old World porcupines

Hystrix indica   Indian crested porcupine   **K**

Hystrix cristata var. *indica* Kerr, 1792. Animal kingdom, p. 213. Type from India.

Hystrix hirsutirostris aharonii Müller, 1911. Sitzungsber. Ges. Naturf. Fr. Berl., 1911:123. Type from Emmaus, Palestine. Synonym.

Hystrix hirsutirostris schmidtzi Müller, 1911. Sitzungsber. Ges. Naturf. Fr. Berl. 1911:126. Type from Ain Dschuheijir, northwestern Dead Sea, Palestine. Synonym.

Diagnosis.—The porcupine is the largest rodent in the region with a total length of almost 1 m and a weight of 10–17 kg. The body is massively built and is well adapted for digging. The dorsal surface is covered with quills that can attain lengths of 30–40 cm (Fig. 114). The skull is very unique (Fig. 115). The cranium is similar to other porcupines in having a very massive infraorbital foramen and expanded nasal and frontal regions (making the skull appear elevated anteriorly). The cheekteeth are strongly hypsodont. Several characters can be used to distinguish between the Indian crested

Fig. 114. Indian crested porcupine (*Hystrix indica*). Photo by D. M.Shafi.

porcupine (*Hystrix indica*) and the north African porcupine (*Hystrix cristata*) (Corbet and Jones, 1965). The ratio of the width of premaxilla to width of nasals is less than 36% in *H. cristata* and more than 44% in *H. indica*. The frontal to nasal ratio is less than 38% in *H. cristata* and more than 45% in *H. indica*. In an adult male from Lahav at the USNM, these ratios were 100% and 75%, respectively. This clearly indicates that the porcupine found in the Holy Land is *Hystrix indica*. Specimens from Sinai need to be examined to determine their identity.

Range of species.—Turkey and the eastern Mediterranean region through Turkestan and Iran to India and Nepal.

Local Status.—"The porcupine is common in all the rocky districts and mountain glens, though from its nocturnal habits seldom seen" (Tristram, 1884:10). This statement is still true and I have seen evidence of porcupines on every field trip in the Holy Land. Because of their presence in all major habitats, a distribution map is not necessary.

Müller (1911) recorded the species early in the century and named two forms from Palestine (listed above) but these were later considered to be synonyms of *H. indica* (Ellerman, 1948). Specimens were also recorded from Ain Terabeh and Wadi Um Baghak (Tristram, 1866*a*), Ain Dcheier (specimen collected by Schmitz and named by Müller [1911]), 'Asqelon, Kafr,

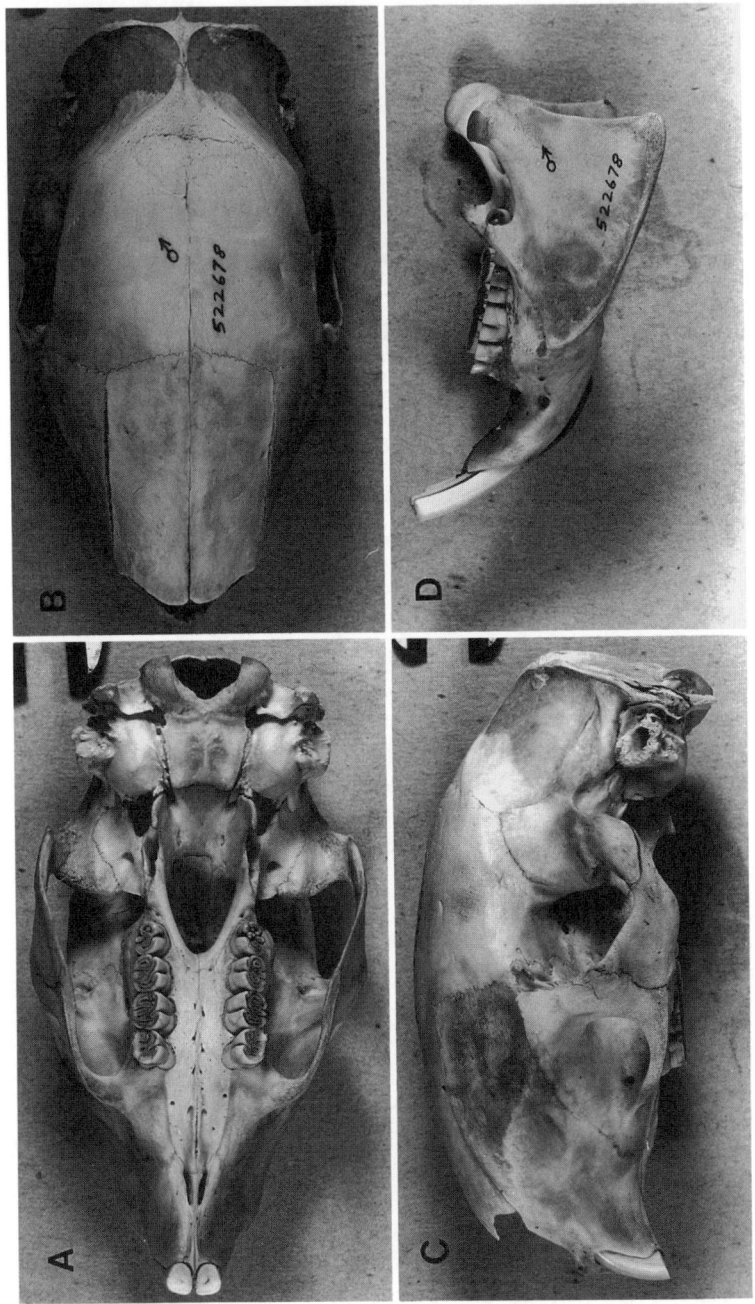

Fig. 115. Skull of the Indian crested porcupine (*Hystrix indica*). (A) Ventral (B) dorsal, and (C) lateral view of cranium; and (D) lateral view of mandible.

Qatana, Hammam Zarah, Jericho, Wadi Swenit, Ain Fawwar (Wadi Kelt), Silwan, Gaza (Müller, 1920), Jordan Valley, near Tiberias (Bodenheimer, 1935), Emmaus, Ain Dschuheijir, Rehobot, Jabal Abdul Aziz, Wadi Kelt, Silwan, and Wadi Swenit (Atallah, 1978). The species was reported from the Irbid area (Amr et al., 1987), and from Aqraba and Tafilah; and spines were reported from Wadi Fidan, Fuhays, King Hussein Bridge (=Allenby Bridge), and Wadi Shu'ayb (Amr and Disi, 1988). I found quills of porcupines in many areas in Palestine including Beit Sahur (an animal was also collected), Ain Fashkhah, Mount Qarantal, Deir Mar Saba, Beit Guvrin, and Bitan Aharon (a skull was found). Specimens at JNHM are from the Yarmouk River area and from Tafila. Specimens at TAUM and HUJ are available from many localities in Israel and the occupied territories. The species was reported from Ain Gudeirat, 90 km SE El Arish in northern Sinai (Wassif, 1954b).

Biology.—Porcupines are nocturnal animals that retreat to their burrows in small caverns or on the sides of hills during the daytime. These burrows are dug in earth, usually along small hills or elevations. They feed on roots, bulbs, bark, and other plant material. Up to four young are born in the spring in the Holy Land. Porcupines are preyed upon by leopards, hyenas, and perhaps wolves.

Genetics.—2N = 60, AA = 114 was reported for a specimen from Europe (Renzoni, 1967).

Human interactions.—Locals believe that the porcupine can shoot its quills at attackers like darts; however, this is a false impression. Alarmed porcupines rattle their quills to make a warning sound. During this process some of the older quills may fall off, which may have given rise to this belief. Porcupines are known to charge backwards toward their enemies, which can result in impaling quills deep in the body of the attacker (Harrison, 1972). These animals are considered to be agricultural pests by the locals, who also use them for food (near Beit Sahur, I obtained a skull from scraps left from preparing a meal).

Al Mneisah, a village in northern Jordan, derives its name from the common name of the crested porcupine. Darwish Al Shafi informed me that the locals related to him that porcupines were very abundant in this area two decades ago but are now much rarer. The porcupine is a threatened species in the Holy Land. It is hunted for food and also because it is assumed to be an agricultural pest. The specific name *indica* refers to India, where the species was first described.

ORDER LAGOMORPHA   Rabbits, hares, and pikas

The Lagomorpha (from *lagos,* Gr., a hare, and *morphe,* Gr., the form, shape) includes rabbits, hares, and pikas. These mammals were formerly included with the rodents in the group called Glires. It is now well established that lagomorphs are a distinct order that may not even be closely related to the order Rodentia. Lagomorphs are distinguished quickly from that group by the presence of four upper incisors instead of two. The small pair of upper incisors is hidden behind the large functional ones. Many other characters distinguish lagomorphs including a short tail, characteristic posture, foot structure, closable nostrils (by skin folds), lobes of upper and lower lips that can meet behind incisors, and lateral jaw movement. The pelage is soft and varies in color depending on substrate.

FAMILY LEPORIDAE   Rabbits and hares

Rabbits and hares (family Leporidae) are distinguished from pikas (family Ochotonidae) by long movable ears (longer than wide). Leporids are present on every continent except Australia, where they have been introduced by European settlers.

Genus *LEPUS*

The taxonomic status of the hares of the Holy Land is entangled by the many named forms. Tristram (1866*b,* 1884) lists five forms of the hare in Palestine. Modern research documents only a single species of wild hare (*Lepus capensis*) in the Holy Land. The domestic rabbit (*Oryctolagus cuniculus*) is raised extensively in the Holy Land for meat and pelts. The domestic rabbit can be distinguished from *Lepus capensis* by the rabbit's shorter ears that lack black tips and by its more robust size with shorter legs. The dental formula in *Lepus* is i 2/1, c 0/0, p 3/2, m 3/3 = 28.

Lepus capensis   Cape hare   O

Lepus capensis Linnaeus, 1758. Syst. Nat., 10th ed., 1:58. Type from Cape of Good Hope, South Africa.

Lepus syriacus Hemprich and Ehrenberg, 1833. Symb. Phys. Mamm. 2:sig. U, t. 15. Type from Mount Lebanon. Synonym of *L. c. arabicus.*

Lepus judeae Gray, 1867. Ann. Mag. Nat. Hist., ser. 7, 20:222. Type from Ain Fashkhah, Palestine. Synonym of *L. c. arabicus.*

Lepus arabicus Hemprich and Ehrenberg, 1833. Symb. Phys. Mamm. 2:sig. r, t. 15. Type from Qunfidha, S of Mecca, Saudi Arabia. Valid subspecies: *L. c. arabicus.*

Lepus sinaiticus Ehrenberg in Hemprich and Ehrenberg, 1833. Symb. Phys. Mamm. 2:sig. t, pl. 14, fig. 1. Type from Jebel Musa, near Mount Sinai, Sinai Peninsula. Perhaps a valid subspecies in Sinai and Negev deserts (Osborn and Helmy, 1980).

Lepus aegyptius Desmarest, 1822. Encyclop. Methodique, Mammalogie, suppl., p. 350. Type from between Luxor and Karnak in Qena Province, Egypt. Valid subspecies for the southern Egyptian specimens.

Diagnosis.—See under order, family, and genus. Yom-Tov (1967) examined specimens from 62 localities in Israel and demonstrated a clinal variation between a larger, northern form with a relatively smaller bulla and a southern form with smaller skull and relatively large bulla (*L. arabicus*). The only locality listed by Yom-Tov for *arabicus* in Israel (pre-1967 borders) is Yotvata. Atallah (1977) recognizes two subspecies in the eastern Mediterranean region: *Lepus capensis syriacus* in brush and forested areas, and *L. c. arabicus* in desert habitats. In doing so, he reassigned the localities listed by Yom-Tov to these two subspecies. The hares of the region are quite polymorphic, and although Yom-Tov (1967) and Harrison (1972) considered the variation to be clinal, much more work remains to be done to elucidate variation. The earliest available name (including page priority in Ehrenberg's monograph) for hares in the Holy Land is *Lepus capensis arabicus*.

Range of species.—Most of Europe through Russia to China, Mongolia, and Siberia; Afghanistan, Pakistan, Iran, and most of the Arabian Peninsula. Widespread in northern, eastern, and southern Africa.

Local status.—The cape hare was reported to be common in Palestine by early travelers (Hart, 1891; Aharoni, 1930). Indeed, localities abound for this species throughout the area under study and thus no distribution map is necessary. Yom-Tov (1967) listed 62 localities in Israel. According to Atallah (1977), some specimens from Yom-Tov's localities referrable to *L. c. syriacus*, whereas others refer to *L. c. arabicus* (with *L. c. sinaiticus* and *L. c. judeae* as synonyms). The localities listed by Yom-Tov for the smaller race (*L. c. arabicus*) are: Ain Gedi, Neot Hakikar, Tureibe, Nahal Amatsya, Wadi Hatira, Sede Boker (= Boqer), Nahal Nizzana, Ain Hatzefa, Ain Yahav, Wadi Ramon, Nahal Nikrot, Beer Menuha, Nahal Karkom, Yotvata, Wadi Asluj–Wadi Abiad, and Eilat. Harrison (1972) recorded specimens from between Jaffa and Jerusalem, 15 mi W Jerusalem, and Berurim (S of Gedera) (as *L. c. syriacus*); and 56 km E Beersheba, near Ain Gedi, Yotvata, Jebel Usdum, and Auja (as *L. c. sinaiticus*).

East of the Jordan River, the cape hare is reported from Azraq, between Azraq Shishan and Qasr Amrah, Jebel Aseikhem (Atallah, 1967*b*, 1977), and El Jafr (Atallah, 1967*a*). Haim Hovel kindly sent me specimens from

Qatrana and near Beer Sheba). Searight (1987*b*) reported seeing them at Jawa. Specimens are available at the JUNHM from Wadi Fidan, Ghor as Safi, and Ma'an (Amr and Disi, 1988), and at JNHM from Ramtha. Cape hares were also reported from northern Sinai in wadis S Al Arish (Wassif, 1954*b*).

Biology.—The wild cape hares of the Holy Land are remarkable in the degree of their morphologic variations and their presence in almost every habitat from desert to forest and from the lowlands of the Ghor to Mount Hermon. This remarkable variability in *Lepus capensis* is similar to that found in *Oryctolagus cuniculus* (the domestic rabbit), which allowed the selection of a large number of races for human consumption or to be sold as fancy pets. In the Holy Land, the cape hare is mainly nocturnal and feeds on a large variety of grasses and shrubs. These hares also seem to be very versatile in their choice of dens, as they have been noted to sleep in hollow trees, crevices, small caves, burrows of other animals, ruins, between rocks, and so on. In most cases, these hares shelter in small depressions dug in the soil; their color acts to camouflage them. In Armenia, there are three to five young in a litter and peaks of reproduction were noted at least four times a year (Dahl, 1954).

Genetics.—2N = 48, AA = 90 was reported for specimens from Europe (reviewed by Zima and Král, 1984).

Human interactions.—*Lepus* is the Latin word (genitive *leporis)* for hare. The ending *ensis,* L., in *capensis* is a suffix meaning belonging to (first described from the Cape of South Africa).

11

Introduced and
Domesticated Mammals

Domestication of many animals occurred in the Near East in areas such as Egypt, the eastern Mediterranean, and Mesopotamia. The Holy Land represents the western horn of the so-called fertile crescent. The fertile crescent was an area of ancient civilizations stretching from southern Iraq north through Syria and then south through Lebanon and the Holy Land. This fertile territory was probably the area in which several species were domesticated. Certainly many domesticated forms were used by early humans of the Holy Land several millennia ago. Domestication carried with it unique modifications (morphologic and behavioral) of the wild species that were tamed. These changes are sometimes enough for us to recognize the domestic forms as separate species from their wild ancestors (for example, the dog from the wolf). Sites in the Holy Land where bones of potential domestic animals were uncovered include Jericho (7000 BC; cattle, goats, pigs, dogs), and Beida (7000–5000 BC; goats).

There is an argument about whether domestic animals should be given the names of the wild progenitors or should be given distinct names. We have already mentioned three rodents that are used in medical research and occasionally kept as pets: the house mouse (*Mus musculus*), the black rat (*Rattus norvegicus*), and the golden hamster (*Mesocricetus auratus*).

Myocastor coypus (Molina, 1782)   Nutria, coypu

The name nutria is given to this aquatic animal but is also used to designate its rich fur that has been of significant economic importance. The coypu (Fig. 116) is a large aquatic rodent native to South America. Nutrias are easily distinguished from other rodents in the region by size. The nutria has a head and body length of 40–55 cm, a tail of 30–45 cm, a skull length

Fig. 116. The nutria (*Myocastor coypus*). Specimen at JNHM.

of 11–14 cm, and a weight of 7–10 kg. All other rodents in the Holy Land are smaller in size with the exception of porcupines (which have spines). Coypus feed on plants, snails, molluscs, and other invertebrates and have been known to live in captivity for 6 years. Coypus are prolific animals. They give birth to five to seven (up to nine) young after a gestation period of about 130 days, with young usually born in the spring.

Nutrias were introduced by fish farmers in the areas of Kfar Masaryk and Kfar Ruppin some 50 years ago for the economic benefit from the fur (Bodenheimer, 1958; Ehrlich, 1959). The nutria has since spread from a semicaptive state near these fish ponds to other suitable aquatic habitats in the area (Atallah, 1978). Specimens of the coypu at HUJ are from Lahav and Hadera; those at JNHM are from 7 km S Jerash, Jerash reservoires ("sil Jerash"), and the Yarmouk River near Aqraba village.

Oryctolagus cuniculus Linnaeus, 1758   Rabbits

Rabbits most likely were domesticated from their wild ancestors by early Romans, probably in the Iberian Peninsula. In the Holy Land, as in many other parts of the world, they are used for both food and pelts.

Cavia porcellus Linnaeus, 1758   Guinea pig

Guinea pigs are South American rodents that were domesticated for meat and pelts at least 3000 years ago. Unlike the nutria, the time since domestication and the way guinea pigs are reared in captivity would probably prevent any establishment of feral populations. Guinea pigs give birth to one to four young following a gestation period of about 2 months. They have been known to live 8 years in captivity. Domestic guinea pigs have been used for research since the 18th century.

Felis silvestris (= *Felis catus*) Schreber 1777   Domestic cat

The taxonomy of domestic and wild cats in Europe, Asia, and Africa has been controversial. Recent morphologic (Ragni and Randi, 1986) and genetic studies (Randi and Ragni, 1991) suggest that all belong to a single variable species, *Felis silvestris*. The divergence of the European (*Felis s. silvestris*) and north African forms (*Felis s. libyca*) probably occurred some 20,000 years ago. Domestication was most likely first established in northern Africa (?Egypt) some 5000–10,000 years ago. In Egypt, cats were considered sacred and their mummies have been discovered entombed with kings and other royalties.

Domestic cats, called *Felis catus* by Linnaeus in 1758, most probably originated from the north African *Felis silvestris libyca* (Kratochvil and Kratochvil, 1976; Randi and Ragni, 1991). Genetic evidence suggests that divergence time was about 5000 years ago. Because of the many wild habits of domestic cats, including the ability to return to and survive in the wild, many do not believe that cats were ever truly "domesticated." As is true of many other places, feral cats are known from the Holy Land.

Canis familiaris Linnaeus, 1758   Domestic dog

People befriended dog ancestors in the Near East at least 12,000 years ago (Davis and Valla, 1978). In Egypt, hounds were probably used to hunt gazelles at least 3000 years ago. If one considers the dog as an extension of its ancestor, the wolf, then perhaps this association with humans is much more ancient. We still marvel at the early events associated with transforming the feared competitor of humans to a trusted ally. Wolf pups can be raised as "tame" and friendly animals; this is not unusual among Alaskan Eskimos. The actual domestication process is difficult to trace and probably took place independently in many parts of the world where humans and wolves coexisted during the last ice age.

Feral and domestic dogs are present around human habitations throughout the Holy Land. There appear to be several breeds, of which the true pariah dog is still half-wild (Bodenheimer, 1958; Tristram, 1866*b*). Several breeds (Greyhound, saluki, and tazi) were known to be used by Bedouin for gazelle hunting. These highly inbred strains are still the common breeds seen around the camps of the Bedouin (Fig. 117). In villages, the sheep dog is more often used. According to Bodenheimer, these breeds are all descended from the pariah dog.

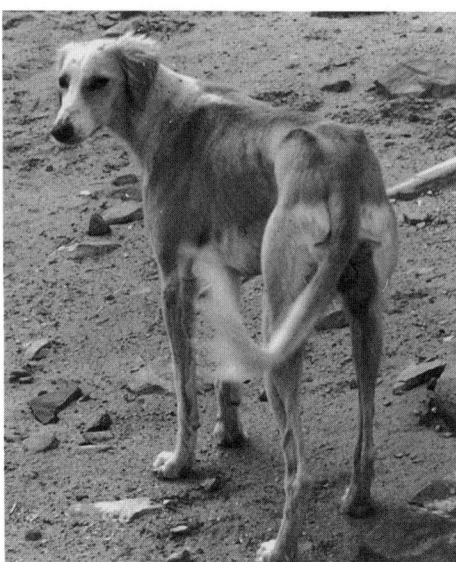

Fig. 117. The saluki breed of dog is common near desert camps.

Feral dogs are common in the Holy Land and may interbreed with wolves. Attacks on domestic animals by groups of feral dogs are sometimes ascribed to wolves by the locals. Distinguishing features between the dog and the wolf are given under the account of the latter. Domestic dog skulls generally have a shorter rostrum, more elevated braincase, and smaller bullae.

Equus caballus Linnaeus, 1758   Horse

The ancestors of the horse have all gone the path of extinction. Only one wild form remains in the Mongolian steppes, Przewalski's horse (*Equus caballus przewalskii*). Tristram (1866*b*) mentions that east of the Jordan River he and others with him saw only purebred Arabian horses. West of the river, such horses were only in the possession of "sheikhs and wealthy men." Today, as then, the horse is more a symbol of wealth and sport rather than a practical beast of burden in the Holy Land. *Equus* is Latin for horse and *caballus* is Latin for a pack horse or domestic horse.

Equus asinus Linnaeus, 1758   Ass, donkey

The north African wild ass (*Equus africanus*, synonym *E. asinus*) is the ancestor of the domestic ass or donkey. Domestic and wild asses will interbreed and produce fertile hybrids. Although domestic asses can be seen everywhere in the Holy Land, there is no evidence that the wild form was present there. Domestication probably occurred in northern Africa (perhaps Egypt). As Tristram (1866*b*) observed, this is the most common beast of burden in the Holy Land. Even today with motor vehicles common every-where, asses still are used in remote areas for transport, especially along difficult mountain passes.

In captivity, domestic horses and asses interbreed but their hybrid is sterile. A mule is a hybrid produced by a male ass (jackass) and a female horse (mare). A hinny is the offspring of a female ass (jenny) and a male horse (stallion). The hybrids are useful because they combine characteristics of both species. Hybrids were depicted in Egyptian tomb paintings at about 1400 BC (Clutton-Brock, 1981). The word *asinus* is Latin for an ass.

Sus scrofa (= *Sus domesticus*) Linnaeus, 1758   Domestic pig

The domestic pig is still very close to its wild ancestor and as such does not justify the use of the Latin name *Sus domesticus* Erxleben, 1777. In the Holy Land, the wild pig is still common; its status was discussed earlier. There is evidence of domestication or at least interference by humans in excavations at Jericho dated at 8000–7000 years BC (Clutton-Brock, 1981). A few domestic pigs are raised in the Holy Land for meat. They are raised in villages and towns where Christian populations predominated (for exam-ple, Nazereth, Ramallah, Bethlehem, Beit Sahur, Beit Jala).

Camelus dromedarius Linnaeus, 1758   Dromedary or Arabian camel

The dromedary camel (order Artiodactyla, family Camelidae; Fig. 118) no longer exists in the wild anywhere. The oldest records of wild camels in the Holy Land come from the Upper Paleolithic, some 35,000 years ago. They were probably domesticated some 3500 years ago in areas outside the Holy Land. The camel, weighing more than 700 kg, is used as the main beast of burden in deserts from Africa to China. Camels can travel with a heavy load for up to 40 km a day and have been known to live in captivity for 25 to 30 years. Camels give birth to one (unusually two) young following a gestation period of 11 months. In the Holy Land camels are encountered commonly in the deserts of the Negev, Wadi Araba, and east Jordan where their population is probably around 30,000 according to FAO yearbook data for 1976. *Camelus* is Latin for camel. The word *dromedarius* is derived from the Greek *dromas*, meaning running.

Bos taurus Linnaeus, 1758 and *Bos indicus* Linnaeus, 1758
Cattle, cows

The aurochs (*Bos primigenius* Bojanus, 1827), was the progenitor of domestic cattle in many parts of the world. Even though the aurochs is known in prehistoric and historic sites in the Holy Land, it apparently was

Fig. 118. The camel is man's best friend in the desert.

not domesticated here. Rather, domesticated cattle were probably first imported to the Holy Land from the east. Domestic cattle were known in the Holy Land for thousands of years. Cattle were especially plentiful near the coast. For example, the Bible states that "Shitrai the Sharonite had charge of the cattle which were grazing in Sharon" (1 Chronicles, 27:29). Cattle were still reared extensively, especially in the southern regions of Palestine and in eastern Jordan, at the time of Tristram (1866*b*).

Bubalus bubalis Linnaeus, 1758   Buffalo, water buffalo

The domestic water buffalo is descended from the wild Indian water buffalo (*Bubalus arnee* Kerr, 1792). The latter is now rare in the wild and found only in localized areas of Nepal and Southeast Asia (Clutton-Brock, 1981). Water buffalo were reared extensively in the Ghor and the Jordan Valley at the time of Tristram (1866*b*), where they replaced cattle as the work and food animal. Tristam also indicated that buffalo were used by the Bedouins of Beni Sakhr and others in the forested regions of Bashan. With increased aridity and drainage of swamps, the numbers of water buffalo declined. The remnant populations in the Hula Basin were decimated in the 1950s. Other beasts of burden and then mechanized agriculture replaced this large animal. The scientific name is based on the Greek *boubalos*, a buffalo.

Ovis aries Linnaeus, 1758   Domestic sheep

The ancestor of the domestic sheep was probably the Asiatic mouflon, *Ovis orientalis* Gmelin, 1774. Tristram (1866*b*) mentions the two varieties of sheep raised in Palestine: the bigger one raised in the north and a smaller, broad-tailed southern form (which he called *laticaudata*). *Ovis* is Latin for a sheep and *aries* is Latin for a ram.

Capra hircus Linnaeus, 1758   Domestic goats

Goats were thought to be first domesticated in Persia some 10,000 years ago. Wild goats are hardy animals adapted to harsh conditions on mountain slopes and hills. One such species, *Capra aegagrus* Erxleben, 1777, was thus easily transformed to a useful domestic animal having little maintenance cost. The wild goat or ibex, *Capra ibex*, has been discussed earlier as a successful inhabitant of mountains in the Holy Land. Tristram (1866*b*) mentions the the abundance of domestic goats as a food source in Palestine. The goat variety used is the black Syrian breed (which Tristram calls *Capra mambrica* L.). The goat gives birth to one to three (unusually more) young following a gestation period of 5–6 months. Goats may live to 10–15 years in captivity.

Glossary

AA. See autosomal arms.

aestivation. Declined activity and torpor during summer months, as opposed to hibernation, which is torpor during winter months.

Allen's rule. This rule states that animals from warmer regions have longer extremities (greater surface area-to-volume ratio) than those from cooler regions.

allopatric. Having disjunct (nonoverlapping) distributions.

altricial. Newborn animals that are completely dependent on their parents and helpless. Some rodents, for example, give birth to blind and naked young.

alveolus. A socket or cavity in the jaw that surrounds the root of a tooth.

angular process. The posterior projection of the lower jaw bone ventral to the condyle.

annular ring. Rings or ridges around the horns.

anterior. Toward the front end.

antitragus. A small projection of the margin of the ear pinna in some bats.

antler. A bony head ornament on some ungulates that is shed annually and is usually branched.

aquatic. Pertaining to water. Aquatic mammals are those adapted to live in water.

arboreal. Pertaining to trees. Arboreal mammals are those adapted to life in trees (e.g., the Syrian squirrel).

arcuate. Curved.

auditory bulla. A capsule of bone that protects the middle ear. If it is derived from the tympanic bone, it may be called the tympanic bulla.

autosomal arms (AA). The number of arms of all chromosomes except the sex chromsomes. Chromosomes with the centromere not at one end have two arms; those with the centromere very close to the end (called acrocentric chromsomes) are considered to have one arm.

axilla. Arm or leg pits.

baculum. A bony and/or cartilaginous structure in the penis of certain mammals. If bony, it is called os penis.

basisphenoid. A bone in the base of the skull located anterior to the basioccipital.

Bergmann's rule. This is the idea formulated by C. Bergmann that animals from warmer regions are generally smaller than their counterparts from cooler regions.

bicuspid. A tooth with two cusps.

brachydont. Pertaining to teeth with low crowns.

bulla. See auditory bulla.

bunodont. Pertaining to teeth with conelike cusps.

calcar. A cartilage or bone that extends from the heel or ankle of some bats and helps support the tail membrane.

canine. The first tooth in the maxilla bone (behind the incisors). Canines are usually conical with a single, sharp cusp.

carnassials. Modified cheekteeth in carnivores having a scissors-like action between the jaws. Usually they are derived from the primitive fourth upper premolar and first lower molar.

carnivore. An animal that consumes meat as the primary source of food.

carpal, carpi. Pertaining to the bones of the wrist joint.

caudal. Pertaining to the tail or toward the tail.

cheekteeth. The premolars and molars collectively.

chromosome. Literally meaning colored bodies. These are the threadlike structures in the nucleus of the cell that are composed of the genetic material (DNA) and associated proteins.

cingulum. Enamel shelf that covers the margins of an upper tooth. Called cingulid for lower teeth.

cline. A character gradient (gradual change in its value over a distance).

commensal. Pertaining to animals that are dependent on or live close to man. For example, the black rat.

coronoid process. A bone extension of the lower jaw that supports the jaw muscles.

crepuscular. Pertaining to animals that are active at the twilight periods of dusk and dawn.

crown. Tooth portion above the gum (i.e., excluding the root).

cursorial. Pertaining to running and running adaptations.

cusp. A projection on the tooth surface.

dental formula. A simplified statement of the structure of teeth in mammals. The numbers of teeth on one side are stated in the order of incisors, canines, premolars, and molars, with the denominater indicating mandibular teeth. For example, i 1/2, c 1/1, p 2/3, m 3/3 = 32 indicates that the upper teeth consist of 1 incisor, 1 canine, 2 premolars, and 3 molars on each side and that the lower teeth consist of 2 incisors, 1 canine, 3 premolars, and 3 molars on each side, with a total of 32 teeth.

diastema. A space in the toothrow such as that between the incisors and the cheekteeth in rodents and hares.

dimorphism. Having two forms or types.

diploid number (2N). The normal chromosome number in a mammal, which is derived by union of male and female gametes each with one set of chromosomes (thus the term diploid, which signifies two sets).

diurnal. Pertaining to daylight time. Animals that are active during the day and hide at night. Opposite of nocturnal.

dorsal. Pertaining to the back or upper surface.

enamel. The hard outer layer of the tooth.

Eocene. The epoch in geologic time starting at about 50 million years ago.

extant. Still in existence, as opposed to extinct.

extinct. An extinct species or population is one that no longer exists.

extirpated. A population decimated in the particular area (i.e., locally extinct).

feral. Introduced by man and now occurring in a wild state in an area the species otherwise does not inhabit. Usually an animal that reverts from a domesticated to a wild state.

foramen (pl. foramina). Any opening or perforation, especially in bone.

foramen magnum. The large opening in the back of the skull where the spinal cord emerges.

fossorial. Pertaining to the earth or soil. Fossorial mammals are those that live a large portion of their lives underground and are well adapted for digging.

frontals. Two bones in the anterior part of the cranium between the nasals and the parietals.

gestation period. Period of embryonic development (between fertilization and delivery or birth of young).

granivore. An animal that feeds on seeds.

guard hairs. Coarse hairs used for protection or sensory purposes. Distinct from the soft underfur that provides insulation.

habitat. The environment in which a species lives.

hamular process. A posteriorly oriented extension of the squamosal bone over the tympanic bulla.

herbivore. An animal that consumes plant material as the primary component of its diet.

hibernation. State of torpor in winter.

Holarctic region. The combined Nearctic and Palearctic faunal regions.

hoof (pl. hoofs or hooves). The keratinized covering of the digits in ungulates.

hypsodont. Teeth that have high crowns.

incisor. The anteriormost of the four basic kinds of teeth. These are usually the cutting teeth.

incus. One of the middle ear bones located between the stapes and the malleus.

infraorbital canal. A canal that runs through the anterior part of the zygomatic arch below the orbit.

infraorbital foramen. The anterior opening of the infraorbital canal.

insectivorous. Feeds primarily on insects and related groups.

interfemoral membrane. In bats, the skin stretching between the hindlegs and sometimes including the tail.

karyotype. An arrangement of the chromosomes cut from a photograph of a dividing cell. This allows counting of the chromosomes and chromosomal arms, and a study of chromosome morphology.

keratinized. Made of keratin (a fibrous protein).

lambdoidal crest. See occipital crest.

lamina (pl. laminae). A transverse ridge on the teeth, with or without cusps.

lancet. On the nose leaf of the horseshoe bats, this is the posteriormost and dorsal triangular swelling.

lateral. Pertaining to the sides, away from the midline.

litter. Young born in one birth to a mammal usually having more than one young at a time.

lophodont. These are teeth in which the cusps have fused into ridges.

malleus. One of the middle ear bones located more lateral than the stapes and the incus.

mamma (pl. mammae). The organ that secretes milk.

mandible. The lower jaw bone.

mandibular condyle. The portion of the lower jaw bone that articulates with the skull.

masseter. The main muscle of mastication in mammals running from the sides of the cranium, through the zygomatic arch, and to the lower jaw.

maxilla (pl. maxillae). The large bone of the upper jaw. Carries the upper teeth except the incisors.

maxillary. Pertaining to maxilla.

metacarpals. The elongated bones in the hand that connect the carpal bones to the phalanges.

mist net. A net of fine mesh used to capture bats.

molar. The most posterior of the cheekteeth in mammals. Unlike the other types of teeth, molars are not preceded by deciduous teeth.

nasals. The paired, anteriormost of the bones of the roof of the cranium. They cover the nasal cavity.

Nearctic region. The faunal region that includes North America north of central Mexico.

Neotropical region. The faunal region that includes South America and northern to central Mexico.

niche. A term intended to include all aspects of the environment for which a species is adapted (food, shelter, prey, predators, etc.).

nocturnal. Pertaining to the night. Animals that are active at night and hide during the daylight. Opposite of diurnal.

nomen nudum. A scientific name that does not meet the conditions of the International Code of Zoological Nomenclature.

nose leaf. A fleshy structure on the nose region of some bats believed to aid in directing echolocation waves.

nuchal. Toward the nape of the neck.

occipital bone. The bone on the back of the skull surrounding the foramen magnum.

occipital condyles. Two knoblike structures on either side of the foramen magnum (projections of the occipital bone).

occipital crest. Ridges of bone that extend laterally on the side of the braincase. Medially they meet on the sagittal crest.

omnivorous. Feeding on both animal and plant matter. For example, humans, *Acomys*.

orbit. The eye socket. The cavity in the skull where the eye is situated.

Oriental region. One of the faunal regions of the world. Includes the tropical and semitropical regions of Asia.

os penis. See baculum.

ossicle. Any small bone.

pad. A fleshy structure on the ventral sides of the feet and hands of mammals.

palate. The roof of the mouth between the teeth of the upper jaws, the bony part of which (the hard palate) is formed from the maxilla and premaxilla.

Palearctic region. One of the faunal regions of the world. Includes Europe, North Africa, Arabia, and most of Asia north of the Himalayas.

parietals. The paired bones at the roof of the braincase between the frontals and the occipital bone.

patagium. A fold of skin; the membranous skin forming the wing of a bat.

pelage. All the hairs on a mammal (the coat).

phalanx (pl. phalanges). The distal portion of the digits.

pinna. External or projecting part of the ear.

plagiopatagium. The wing membrane (skin) that extends between the elongated digits and the side of the body in bats.

Pleistocene. An epoch in geologic times starting at about one million years ago and characterized by numerous glaciations.

precocial. Pertaining to young that are capable of moving about and feeding without assistance immediately after birth.

premaxilla. The anteriormost paired bones on the upper jaw that form part of the ventral region of the nasal passage.

psammophile. Pertaining to animals adapted for sandy habitats.

reentrant angle and fold. The angles and folds formed on the sides of the cheekteeth of some mammals.

ricochetal. A bipedal movement involving jumping by using both hind limbs and landing simultaneously on both hind feet. Examples are kangaroos and jerboas.

root. The portion of the tooth below the gum line.

rostrum. The part of the cranium anterior to the orbits.

rut. The mating time for a species when both males and females are reproductively active.

sagittal crest. A dorsal ridge of bone on the braincase.

Saharo-Sindian region. The desert and semidesert region of the Palearctic extending from northwestern Africa to central Asia.

saltatorial. Adapted for leaping.

scats. Mammal feces or droppings.

selenodont. Pertaining to cheekteeth with crescent shaped ridges.

sella. A small structure projecting above the nares and below the lancet in the nose leaf of horseshoe bats.

semi-. Partially. For Example, semifossorial, semiarid, semiaquatic.

spines. Thick guard hairs of some mammals that are used for protection.

squamosal. This is a bone on the lateral aspect of the braincase that forms the posterior portion of the zygomatic arch. Derived from the temporal bone.

stapes. One of the three middle ear ossicles. Situated most interior to the incus and malleus.

subspecies. A geographic race of a species that is defined by morphologic or preferably genetic differences from other such races.

suprameatal triangle. An opening that is triangular and is formed above the tympanic bulla in certain rodents. This triangle is bordered ventrally by the hamular process of the squamosal.

sympatry. Occurring together; populations or species that are not geographically separated.

synonym. Each of two or more names applied to the same taxon.

tarsus (pl. tarsi). The bones of the ankle joints.

taxon (pl. taxa). Any group of organisms that is sufficiently distinct from others and is given a systematic name.

terrestrial. An organism that inhabits the land.

tine. Spike or point on antlers.

tragus. The projection in the medial portion of the ears of some mammals, especially bats.

tympanic bone. The bone forming the ring that supports the ear drum.

tympanic bulla. See auditory bulla.

type. A specimen or taxon designated specifically to be associated with the name of a particular systematic category. The type locality is the area at which the type specimen was collected or assumed to have been collected.

ungulate. A hoofed mammal of the orders Artiodactyla or Perissodactyla.

unicuspid. A tooth (usually an incisor) having a single cusp and thus is simple and cone-shaped.

uropatagium. Membraneous skin that stretches between the hind legs of bats and that may include the tail.

ventral. Pertaining to the lower surface or the underside.

vibrissae. Whiskers or long and usually sensory hairs on the upper lip of many mammals.

volant. Capable of flight

xerophylic. A plant or an animal with capacity to live in drought or arid conditions (in xeric habitats).

zoogeography. The field of study concerned with patterns and dynamics of animal distributions.

zygomatic arch. An arch of bone that encloses the orbit (forming the cheekbones). Formed by the jugal bone and processes of the maxilla and squamosal.

References

Abel, P. F.-M. 1938. Géographie de la Palestine. Vol. II. Géographie Politique Les Villes. Libraire Lecoffre, Paris. 538 pp.

Abramsky, Z. 1981. Habitat relationships and competition in two Mediterranean *Apodemus* spp. Oikos, 36:219–225.

———. 1984. Population biology of *Gerbillus allenbyi* in northern Israel. Mammalia, 48:197–206.

Abu-Dieyeh, M. H. 1988. The ecology of some rodents in Wadi Araba with a special reference to *Acomys cahirinus*. M.S. Thesis, Jordan University, Ammani, Jordan. 216 pp.

Aharoni, B. 1932. Die Muriden von Palästina und Syrien. (Translated by J. H. Lewis, No. 142, USNMRU-3, Cairo, Egypt.) Z. Säugetierkd., 7:166–240.

———. 1944. An Egyptian bat, new to Palestine. Bull. Zool. Soc. Egypt, Cairo, 6:26.

Aharoni, I. 1917. Zum Vorkommen der Saugetier in Palaestina und Syrien. Z. Mitt. Dt. Palast. Ver., 40:235–242.

———. 1930. Die Säugetiere Palästinas. Z. Säugetierkd., 5:327–343.

Akhtar, S. A. 1945. On the habits of the marbled polecat, *Vormela peregusna*. J. Bombay Nat. Hist. Soc., 45:142.

Aldridge, H. D. J. N., M. Obrist, H. G. Merriam, and M. B. Fenton. 1990. Roosting, vocalization, and foraging by the African bat, *Nycteris thebaica*. J. Mammal., 71:242–246.

Al Jumaily, M. M., I. A. Nader, and M. K. Bunni. 1975. Natural history study of the bandicoot rat *Nesokia indica* (Gray and Hardwicke). Bull. Nat. Hist. Res. Center, Univ. Baghdad, 6:55–59.

Al-Khalili, A. D., and M. J. Delany. 1986. The post-embryonic development and reproductive strategies of two species of rodents in south-west Saudi Arabia. Cimbebasia, 8:175–185.

Allen, G. M. 1915. Mammals obtained by the Phillips Palestine Expedition. Bull. Mus. Comp. Zool. Harv. Univ., 59:3–14.

———. 1939. A checklist of African mammals. Bull. Mus. Comp. Zool. Harv. Univ., 83:1–763.

Al-Robaae, K. 1966. Untersuchungen der Lebensweise irakischer Fledermäuse. Säugetierkd. Mitt., 14:177–211.

———. 1968. Notes on the biology of the tomb bat, *Taphozous nudiventris magnus* v. Wettstein 1913, in Iraq. Säugetierkd. Mitt., 16:21–26.

Al-Saleh, A. A., and M. A. Khan. 1984. Cytological studies of certain desert mammals of Saudi Arabia. 1. The karyotype of *Jaculus jaculus*. J. Coll. Sci. King Saud. Univ., 15:163–168.

———. 1985. Cytological studies of certain desert mammals of Saudi Arabia 3. The karyotype of *Paraechinus aethiopicus*. Cytologia (Tokyo), 50:507–512.

Amr, Z. S., and A. M. Disi. 1988. Jordanian mammals acquired by the Jordan University Natural History Museum. Dirasat Nat. Sci. (Amman), 15:3–32.

Amr, Z., and M. B. Qumsiyeh. 1993. Records of bat flies from Jordan, Libya, and Algeria. Entom. News, 104:43–46.

Amr, Z. S., and E. K. Saliba. 1986. Ecological observations on the fat jird, *Psammomys obesus dianae,* in the Mowaqqar area of Jordan. Dirasat Nat. Sci. (Amman), 13:155–161.

Amr, Z. S., S. Woodbury, and A. M. Disi. 1987. On a collection of mammals from Jordan. Dirasat Nat. Sci. (Amman), 14:131–136.

Anciaux de Faveaux, F. 1953. Chiropteres de Palestine. Bull. Inf. Fédér. Spéléol. Belg., 4:32–33.

Andersen, K. and Matschi. P. 1904. S. B. Gesch. Naturf. Fr. Berlin, p. 80 (not seen, see DeBlase, 1972).

Anderson, J., and W. E. de Winton. 1902. Zoology of Egypt: Mammalia. Revised and completed by W. E. de Winton. Hugh Rees Ltd., London, xvii + 374 pp.

Anonymous. 1945. Further notes on Palestinian mammals. Bull. Jerusalem Nat. Club, 12:1.

———. 1946a. Wolf reports. Bull. Jerusalem Nat. Club, 30:3.

———. 1946b. The Schmitz collection of mammals. Bull. Jerusalem Nat. Club, 23:1–2.

———. 1975. Reservate Zu beiden Seiten des Jordans. Vögel Heimat, 46:21.

———. 1976. Where, when, and how to see hyrax. Isr. Land Nat., 2:69.

———. 1978. Arabian oryx returning to desert homeland. Int. Zoo News, 25:27.

———. 1982. Tyche calling on 151.730,000 megaherz. Isr. Land Nat., 8:38–39.

———. 1983a. Leopards in the Negev Mountains. Isr. Land Nat., 8:171.

———. 1983b. Wolves in the Negev. Isr. Land Nat., 8:171.

———. 1988. Leopards in northern Israel. Isr. Land Nat., 14:246.

Asdell, S. 1964. Patterns of mammalian reproduction. Cornell Univ. Press, Ithaca, New York, 670 pp.

Atallah, S. I. 1965. Species of the subfamily Microtinae (Rodentia) in Lebanon. M.S. thesis, American University of Beirut, 32 pp.

———. 1966. Mammalogy. *In* International Jordan Expedition, 1966 (J. Boyd, ed.). Nature (Lond.), 212:664–666.

———. 1967a. A collection of mammals from El-Jafr, southern Jordan. Z. Säugetierkd., 32:307–309.

———. 1967b. Mammalogy, with a list of amphibians and reptiles. Pp. 56–63 *in* International Jordan Expedition, 1966 (J. Boyd, ed.), International Biological Program/CT Section, London, 340 pp.

———. 1967c. A new species of spiny mouse (*Acomys*) from Jordan. J. Mammal., 48:255–261.

———. 1970a. A new subspecies of the golden spiny mouse, *Acomys russatus* (Wagner) from Jordan. Univ. Conn. Occas. Pap. Biol. Sci. Ser., 1:201–204.

———. 1970b. Bats of the genus *Myotis* (family Vespertilionidae) from Lebanon. Univ. Conn. Occas. Pap. Biol. Sci. Ser., 1:205–212.

————. 1977. The mammals of the eastern Mediterranean region: their ecology, systematics and zoogeographical relationships (part 1). Säugetierkd. Mitt., 25:241–320.

————. 1978. The mammals of the eastern Mediterranean region: their ecology, systematics and zoogeographical relationships (part 2). Säugetierkd. Mitt., 26:1–50.

Atallah, S. I., and D. L. Harrison. 1967. New records of rodents, bats and Insectivora from the Arabian Peninsula. J. Zool. (Lond.), 153:311–319.

————. 1968. On the conspecificity of *Allactaga euphratica* Thomas, 1881 and *Allactaga williamsi* Thomas, 1897 (Rodentia: Dipodidae) with a complete list of subspecies. Mammalia, 32:628–638.

Auffray, J.-C., E. Tchernov, F. Bonhomme, G. Heth, S. Simon, and E. Nevo. 1990. Presence and ecological distribution of *Mus "spretoides"* and *Mus musculus domesticus* in Israel: circum-Mediterranean vicariance in the genus *Mus*. Z. Säugetierkd., 55:1–10.

Auffray, J.-C., E. Tchernov, and E. Nevo. 1988. Origin du commensalisme de la souris domestique (*Mus musculus domesticus*) vis-a-vis de l'homme. C. R. Acad. Sci. Paris, 307:517–522.

Aulagnier, S., and M. Thevenot. 1986. Catalogue des Mammiferes sauvages du Maroc. Travaux de l'Institut Scientifique, Rabat, Morocco, 164 pp.

Badr, R. M., and R. L. Asker. 1980. Prevalence of non-Robertsonian polymorphism in the gerbil *Gerbillus cheesmani* from Kuwait. Genetica (The Hague), 52/53:17–22.

Baharav, D. 1976. The 1500 gazelles of Ramot Yissakhar. Isr. Land Nat., 2:15–21.

————. 1981. Food habits of the mountain gazelle in semi-arid habitats of eastern Lower Galilee, Israel. J. Arid Environ., 4:63–69.

————. 1983. Observations on the ecology of the mountain gazelle in the upper Galilee, Israel. Mammalia, 47:59–70.

Baker, R. J., B. L. Davis, R. G. Jordan, and A. Binous. 1975. Karyotypic and morphometric studies of Tunisian mammals: bats. Mammalia, 38:695–710.

Baker R. J., M. B. Qumsiyeh, and C. S. Hood. 1987. Role of chromosomal banding patterns in understanding mammalian evolution. Pp. 67–96, *in* Current mammalogy (H. H. Genoways, ed.). Plenum Press, New York, xx + 519.

Barak, Y., and Y. Yom-Tov. 1991. The mating system of *Pipistrellus kuhli* (Microchiroptera) in Israel. Mammalia, 55:285–292.

Bar-Yosef, O., and E. Tchernov. 1966. Archeological finds and the fossil faunas of the Natufian and Microlithic industries at Hayonim cave (Western Galilee, Israel). Isr. J. Zool., 15:104–140.

Bate, D. M. A. 1904. The mammals of Cyprus. Proc. Zool. Soc. Lond., 2 (for 1903):341–348.

————. 1919. A fossil wart-hog from Palestine. Ann. Mag. Nat. Hist., ser. 10, 13:120.

————. 1937. The fossil fauna of the Wadi El Mughara caves. Pp. 140–142, *in* The Stone Age of Mount Carmel (D. A. E. Garrod and D. M. A. Bate, eds.). Clarendon Press, Oxford.

———. 1942. Pleistocene Murinae from Palestine. Ann. Mag. Nat. Hist., ser. 11, 9:465–486.

———. 1943*a*. Fossil mammals of Shukbah. Proc. Prehist. Soc. Cambridge, 8:1–20.

———. 1943*b*. Pleistocene Cricetinae from Palestine. Ann. Mag. Nat. Hist., ser. 11, 10:813–838.

———. 1945. Notes on small mammals from the Lebanon mountains, Syria. Ann. Mag. Nat. Hist., ser. 11, 12:141–158.

———. 1952. The Pleistocene mammal faunas of Palestine and East Africa. Proc. Pan-Afr. Congr. Prehist., Oxford, 1947:38–39.

Bates, P. J. J., and D. L. Harrison. 1989. New records of small mammals from Jordan. Bonn. Zool. Beitr., 40:223–226.

Belcheva, R. G., T. H. Peshev, and D. T. Peshev. 1980. Chromosome C- and G-banding patterns in a Bulgarian population of *Microtus guentheri* Danford and Alston (Microtinae, Rodentia). Genetica (The Hague), 52/53:45–48.

Belon, P. 1553. Les observations de plusieurs Singularitez et choses mémorables trouués en Gréce, Asie, Judée, Egypte, Arabie et autres pays estranges. Paris (not seen, see Kumerloeve, 1975).

Benazzou, T., E. Viegas-Pequignot, M. ProdHomme, M. Lombard, F. Petter, and B. Dutrillaux. 1984. Phylogenie chromosomique des Gerbillidae. III. Etude d'especes des genres *Tatera, Taterillus, Psammomys* et *Pachyuromys.* Ann. Genet., 27:17–26.

Ben-Yaacov, R., and Y. Yom-Tov. 1983. On the biology of the Egyptian mongoose, *Herpestes ichneumon* in Israel. Z. Säugetierkd., 48:34–45.

Bhatnager, A. N., and T. F. El-Azawi. 1978. A karyotype study of chromosomes of two species species of hedghogs, *Hemiechinus auritus* and *Paraechinus aethiopicus* (Insectivora: Mammalia). Cytologia (Tokyo), 43:53–59.

Bird, F. W. 1946. The hyaena in Palestine. Bull. Jerusalem Nat. Club, 21:1.

Blaine, G. 1913. On the relationship of *Gazella isabella* to *Gazella dorcas*, with a description of a new species and subspecies. Ann. Mag. Nat. Hist., ser. 8, 11:291–296.

Blake, I. 1966. A leopard in the wilderness of Judea, Jordan. IUCN Bull., 18:7.

Blanford, W. T. 1888–1891. The fauna of British India, including Ceylon and Burma. Mammalia. Taylor and Francis Publ., London, 617 pp.

Bodenheimer, F. S. 1935. Animal life in Palestine. L. Mayer, Jerusalem, xiii + 506 pp.

———. 1937. Prodromus Fauna Palestinae. Mem. Inst. Egyp., 33:47–51.

———. 1949*a*. Ecological and physiological studies on some rodents. Phys. Comp. Oecol. Haag., 1:376–389.

———. 1949*b*. Problems of vole populations in the Middle East. Report on the population dynamics of the Levant vole (*Microtus guentheri* D et A). Bull. Res. Counc. Isr., report, 77 pp.

———. 1958. The present taxonomic status of the terrestrial mammals of Palestine. Bull. Res. Counc. Isr. Sect. B, Zool., 7:165–191.

———. 1959. A remark on vole outbreaks. Bull. Res. Counc. Isr., Sect. B Zool., 8:97–98.

————. 1960. Animal and man in Bible Lands. E. J. Brill Publ., Leiden, 232 pp.

Boessneck, J., and A. Von den Driesch. 1977. Hirschnachweise aus frühgeschichlicher Zeit von Hesbon, Jordanien. Saügetierkd. Mitt., 25:48–57.

Bogdanov, O. P. 1953. Rukokrylyje (Chiroptera). Fauna Uzbekskoj T. III. 2 ed. Academy of Sciences, Tashkent, Uzbekistan SSR, 158 pp.

Bonhomme, F. 1986. Evolutionary relationships in the genus *Mus*. Pp. 19–34, *in* Current topics in microbiology and immunology: the wild mouse in immunology (M. Potter, J. H. Nadeau and M. P. Cancro, eds.). Springer Verlag, Berlin, 395 pp.

Bonhote, J. L. 1912. On a further collection of mammals from Egypt and Sinai. Proc. Zool. Soc. Lond., 1912:224–231.

Bouskila, Y. 1985. A closer look at the striped hyena. Isr. Land Nat., 10:50–56.

Briscoe, M. S. 1956. Kinds and distribution of wild rodents and their ectoparasites in Egypt. Am. Midl. Nat., 55:393–408.

Bromage, T. N. 1954. Wolves in the Middle East. Field, 22 April 1954:703.

Brooke, V. 1873. On the antelopes of the genus *Gazella* and their distribution. Proc. Zool. Soc. Lond., 1873:535–554.

Brosset, A. 1962. The bats of central and western India. J. Bombay Nat. Hist. Soc., 59:1–78.

————. 1966. La Biologie des Chiroptéres. Masson and Co., Paris, vii + 240 pp.

Burgos, M., R. Jiménez, and R. Diaz de la Guardia. 1989. Comparative study of G- and C-banded chromosomes of five species of Microtidae. Genetica (The Hague), 78:3–12.

Burns, T. W. 1956. Endocrine factors in the water metabolism of the desert mammal *G. gerbillus*. Endocrinology, 58:243–254.

Buxton, H. M. B., C. E. B. Buxton, and T. B. Buxton. 1895. On either side of the Red Sea. Edward Stanford, London, viii + 163 pp.

Buxton, P. A. 1923. Animal life in deserts. Edward Arnold Ltd., London, xv + 176 pp.

Canaani, G. 1976. Wild boars in Galilee. Isr. Land Nat., 2:68–71.

Capanna, E., M. V. Civitelli, and L. Conti. 1967. I chromosomi somatici del pipistrello "ferro di carvallo minore" (Mammalia—Chiroptera). Rend. Acc. Naz. Linçei (Rome), 42(8):125–128.

Carruthers, D. 1909. Big game of Syria, Palestine, and Sinai. The Field, London, 114:1135.

————. 1935. Arabian Adventure. H. F. and G. Witherby, London, 200 pp.

Catzeflis, F. 1983. Relations génétique entre trois especes du genre *Crocidura* (Mammalia: Soricidae) en Europe. Mammalia, 47:229–236.

Catzeflis, F., T. Maddalena, S. Hellwig, and P. Vogel. 1985. Unexpected findings on the taxonomic status of east Mediterranean *Crocidura russula* auct. (Mammalia, Insectivora). Z. Säugetierkd., 50:185–201.

Chaworth-Musters, J. L., and J. R. Ellerman. 1947. A revision of the genus *Meriones*. Proc. Zool. Soc. Lond., 117:478–504.

Cheesman, R. E. 1926. In unknown Arabia. MacMillan Publ., London, 447 pp.

Cheesman, R. E., and M. A. C. Hinton. 1924. On the mammals collected in the desert of central Arabia by Major R. E. Cheesman. Ann. Mag. Nat. Hist., (ser. 9), 14:548–558.

Clarke, J. E. 1977. 1979. Some personal experiences of conservation in dry land. Black Lechwe, 13:6–9.

Cloudsley-Thompson, J. L. 1977. Man and the biology of arid zones. Edward Arnold, London, ix + 182pp.

Clutton-Brock, J. 1971. The primary food animals of the Jericho Tell from the Proto-Neolithic to the Byzantine Period. Levant (Brit. School Archeol., Jerusalem), 4:41–55.

———. 1979. The mammalian remains from the Jericho Tell. Proc. Prehist. Soc., Cambridge, 45:135–157.

———. 1981. Domesticated animals from early times. Univ. of Texas Press and British Museum (Natural History), Austin and London, 208 pp.

Clutton-Brock, J., G. B. Corbet, and M. Hills. 1976. A review of the family Canidae, with a classification by numerical methods. Bull. Br. Mus. (Nat. Hist.), Zool. Ser. 29:119–199.

Cockrum, E. L., T. C. Vaughan, and P. J. Vaughan. 1976. *Gerbillus andersoni* de Winton, a species new to Tunisia. Mammalia, 40:467–473.

Cohen-Shlagman, L., Y. Yom-Tov, and S. Hellwig. 1984a. The biology of the Levant vole, *Microtus guentheri* in Israel: I. Population dynamics in the field. Z. Säugetierkd., 49:135–147.

———. 1984b. The biology of the Levant vole, *Microtus guentheri* in Israel: II. The reproduction and growth in captivity. Z. Säugetierkd., 49:148–156.

Corbet, G. B. 1978. The mammals of the Palaearctic region: a taxonomic review. British Museum (Natural History) and Cornell Univ. Press, London and Ithaca, viii + 314 pp.

———. 1988. The family Erinaceidae: a synthesis of its taxonomy, phylogeny, ecology and zoogeography. Mammal Rev., 18:117–172.

Corbet, G. B., and L. A. Jones. 1965. The specific characters of the crested porcupines, subgenus *Hystrix*. Proc. Zool. Soc. Lond., 144:285–300.

Corbet, G. B., and D. Ovenden. 1980. The mammals of Britain and Europe. William Collins Sons & Co. Ltd., London, 253 pp.

Costa, M., and E. Nevo. 1969. Nidicolous arthropod associated with different chromosomal types of *Spalax ehrenbergi* Nehring. Zool. J. Linn. Soc., 48:199–215.

Cribiu, E. P., V. Durand, J. F. Asmode, A. Greth, and S. Anagariyah. 1989. The G-banding and C-banding karyotype of the Arabian oryx (*Oryx leucoryx*). Ann. Génét., 32:200–203.

Dahl, S. K. 1954. The animal world of the Armenian S.S.R., Vol. I. Vertebrates. Part II. Mammals. Zoological Institute, Yeravan, 415 pp.

Daly, M., and S. Daly. 1973. On the feeding ecology of *Psammomys obesus* (Rodentia, Gerbillidae) in the Wadi Saoura, Algeria. Mammalia, 37:545–561.

————. 1975. Behavior of *Psammomys obesus* (Rodentia: Gerbillinae) in the Algerian Sahara. Z. Tierpsychol., 37:298–321.

Dar, S. 1975. The wild pigs of the Alexander River. Isr. Land Nat., 1:58–59.

Darviche, D., F. Benmehdi, J. Britton-Davidian, and L. Thaler. 1979. Données préliminaires sur la systématique biochimique des genres *Mus* et *Apodemus* en Iran. Mammalia, 43:427–430.

Davis, S. 1977. Size variation in the fox, *Vulpes vulpes* in the Palaearctic region today, and in Israel during the quaternary. J. Zool. (Lond.), 182:343–351.

Davis, S., and F. R. Valla. 1978. Evidence for domestication of the dog 12,000 years ago in the Natufian of Israel. Nature (Lond.), 276:608–610.

DeBlase, A. F. 1972. *Rhinolophus euryale* and *R. mehelyi* (Chiroptera: Rhinolophidae) in Egypt and Southwest Asia. Israel J. Zool., 21:1–12.

————. 1980. The bats of Iran: systematics, distribution, ecology. Fieldiana Zool., ns, 4:1–424.

De Hondt, H. A. 1972. Karyological studies of two insectivores of Egypt. Proc. Egypt. Acad. Sci., 25:171–174.

Delibes, M., F. Hiraldo, J. J. Arroyo, and C. Rodriguez-Murcia. 1980. Disagreement between morphotypes and karyotypes in *Eliomys* (Rodentia, Gliridae): the chromosomes of the central Morocco garden dormouse. Säugetierkd. Mitt., 40:289–292.

Dobroruska, L. J. 1959. *Meriones vinogradovi* Heptner, 1931 in Palästina. Säugetierkd. Mitt., 7:173.

Dobson, G. E. 1878. Catalogue of the Chiroptera in the collection of the British Museum. Reprinted 1966 and published by Verlag J. Cramer, Germany, xlii + 567 pp.

Dollman, J. G. 1927. A new race of Arabian gazelle. Proc. Zool. Soc. Lond., 1927:1005.

Dollman, J. G., and J. B. Burlace. 1935. Rowland Ward's records of big game, African and Asiatic sections. Rowland Ward Publications, London, 408 pp.

Dor, M. 1947. Observations sur les micromammiferes trouves dans les pelotes de la Chouette Effraye (*Tyto alba*) en Palestine. Mammalia, 11:49–54.

Doughty, C. M. 1888. Travels in Arabia Deserta. Cambridge Univ. Press. Published 1979, Dover Publ., Inc., New York, Vol. I, 674 pp.

Ducos, P. 1968. L'origine des animaux domestiques en Palestine. Publ. Inst. Prehist. Bordeaux, 6:1–187.

Dulíc, B., and M. Mrakovcíc. 1980. Chromosomes of European freetailed bat, *Tadarida teniotis teniotis* (Rafinesque, 1814. Mammalia, Chiroptera, Molossidae). Biosistematika, 6:109–112.

Dulíc, B., and F. A. Mutere. 1973. Comparative study of the chromosomes of some molossid bats from East Africa. Period. Biol., 75:61–65.

————. 1974. The chromosomes of two bats from East Africa: *Rhinolophus clivosus* Cretzschmar 1928 and *Hipposideros caffer* (Sundevall, 1946). Period. Biol., 76:31–34.

East, R. 1992. Conservation status of antelopes in Asia and the Middle East, part 1. Species, 19:23–25.

Ehrlich, S. 1959. Nutria breeding in fish ponds. Bamidgeh, 11:44–47.

Eissa, S. M., S. M. El-Ziyadi, and M. M. Ibrahim. 1975. Autecology of the jerboa, *Jaculus jaculus* inhabiting Al-Jalia desert area, Kuwait. J. Univ. Kuwait (Sci.), 2:111–120.

Ellerman, J. R. 1941. The families and genera of living rodents. British Museum (Natural History), London, xii + 690 pp.

———. 1948. Key to the rodents of southwest Asia in the British Museum collection. Proc. Zool. Soc. Lond., 118:765–816.

Ellerman, J. R., and T. C. S. Morrison-Scott. 1951. Checklist of Palearctic and Indian mammals, 1758 to 1946. British Museum, (Natural History), London, vi + 810 pp.

Fairon, J. 1980. Deux novelles especes de cheiropteres pour la fauna du Massif de l'Air (Niger): *Otonycteris hemprichi* Peters, 1859, et *Pipistrellus nanus* (Peters, 1852). Bull. Inst. R. Sci. Nat. Belg. Biol., 52:1–7.

Fedyk, S., and A. L. Ruprecht. 1976. Karyotypes of *Pipistrellus pipistrellus* and *Pipistrellus nathusii* (Chiroptera, Vespertilionidae). Caryologia, 29:283–289.

Felten, H. 1956. Fledermause fressen Skorrpione. Nat. Volk (Frankf.), 86:53–57.

Felten, H., F. Spitzenberger, and G. Storch. 1971. Zur Kleinsäugerfauna West-Anatoliens. Teil I. Senckenb. Biol., 52:393–424.

———. 1973. Zur Kleinsäugerfauna West-Anatoliens. Teil II. Senckenb. Biol., 54:227–290.

———. 1977. Zur Kleinsäugerfauna West-Anatoliens. Teil IIIa. Senckenb. Biol., 58:1–44.

Ferguson, W. W. 1981*a*. The systematic position of *Canis aureus lupaster* (Carnivora: Canidae) and the occurence of *Canis lupus* in North Africa, Egypt, and Sinai. Mammalia, 45:459–465.

———. 1981*b*. The systematic position of *Gazella dorcas* (Artiodactyla: Bovidae) in Israel and Sinai. Mammalia, 45:453–457.

Ferguson, W. W., Y. Porath, and S. Paley. 1985. Late Bronze period yields first evidence of *Dama dama* (Artiodactyla: Cervidae) from Israel and Arabia. Mammalia, 49:209–214.

Filippucci, M. G., E. Rodino, E. Nevo, and E. Capanna. 1988*a*. Evolutionary genetics and systematics of the garden dormouse, *Eliomys melanurus* Wagner, 1840. 2. Allozyme diversity and differentiation of chromosomal races. Boll. Zool., 55:47–54.

Filippucci, M. G., S. Simson, and E. Nevo. 1989. Evolutionary biology of the genus *Apodemus* Kaup, 1829 in Israel: allozymic and biometric analyses with description of a new species: *Apodemus hermonensis* new species (Rodentia, Muridae). Boll. Zool., 56:361–376.

Filippucci, M. G., S. Simson, E. Nevo, and E. Capanna. 1988*b*. The chromosomes of the Israeli garden dormouse, *Eliomys melanurus* Wagner, 1849 (Rodentia, Gliridae). Boll. Zool., 55:31–33.

Fitter, R. 1984. Operation Oryx—the success continues. Oryx, 17:136.

Fletcher, T. J. 1984. Other deer. Pp. 452–459, *in* Evolution of domesticated animals (I. L. Mason, ed.). Longman Group Ltd., New York. 489 pp.

Flower, S. S. 1932. Notes on recent mammals of Egypt, with a list of species recorded from that kingdom. Proc. Zool. Soc. Lond., 101:369–450.

Gaisler, J., G. Madkour, and J. Pelikan. 1972. On the bats (Chiroptera) of Egypt. Acta Sci. Nat. Acad. Sci. Bohemoslov. Brno, 6:1–40.

Gasperetti, J. 1978*a*. A note on *Procavia capensis*. J. Saudi Arabia Nat. Hist. Soc., 22:6–10.

———. 1978*b*. Notes on wild goats and sheep in Arabia. J. Saudi Arabia Nat. Hist. Soc., 23:10–18.

Gasperetti, J., D. L. Harrison, and W. Buttiker. 1985. The Carnivora of Arabia. Fauna Saudi Arabia, 7:397–461.

Geffen, E., R. Hefner, D. W. MacDonald, and M. Ucko. 1992. Diet and foraging behavior of Blanford's foxes, *Vulpes cana*, in Israel. J. Mammal., 73(2):395–402.

———. 1993. Biotope and distribution of Blanford's fox. Oryx, 27:104–108.

Gemmeke, H. 1980. Proteinvariation und Taxonomie in der Gattung *Apodemus* (Mammalia, Rodentia). Z. Säugetierkd., 45:348–365.

Genoud, M. 1988. Energetic strategies of shrews: ecological constraints and evolutionary implications. Mamm. Rev., 18:173–193.

Ghandour, A. M., G. M. El, and A. A. Banaja. 1983. Mammals of Saudi Arabia: a review of the one-humped camel or dromedary (*Camelus dromedarius*) in the western region of Saudi Arabia. Fauna Saudi Arabia, 5: 673–682.

Gharaibeh, B. M., and M. B. Qumsiyeh. 1995. *Otonycteris hemprichii*. Mamm. Species, No. 514:1-4.

Ghobrial, L. I. 1976. Observations on the intake of sea water by the dorcas gazelle. Mammalia, 40:489–494.

Ghobrial, L. I., and J. L. Cloudsley-Thompson. 1966. Effect of deprivation of water on the dorcas gazelle. Nature (Lond.), 212:306.

Golani, I., and A. Keller. 1975. A longitudinal field study of the behavior of a pair of golden jackals. Pp. 303–335, *in* The wild canids; their systematics, behavioral ecology and evolution (M. W. Fox, ed.). Van-Nostrand Reinhold, New York and London, 508 pp

Gratz, A. 1957. A rodent ectoparasite survey of Haifa Port. J. Parasitol., 43:328.

Green, A. A. 1986. Status of large mammals of northern Saudi Arabia. Mammalia, 50:485–493.

Green, M. J. B., and G. R. F. Drucker (with assistance by S. I. Day). 1991. Current status of protected areas and threatened mammal species in the Sahara-Gobian region. Pp. 5–69, *in* Mammals in the Palaearctic Desert: status and trends in the Sahara–Gobian region (J. A. McNeely and V. M. Neronov, eds.). The Russian Academy of Sciences, the Russian Committee for the UNESCO Programme on Man and the Biosphere (MAB), Moscow, 298 pp.

Greth, A., P. Sunnucks, M. Vassart, and H. Stanley. 1992. Genetic management of an Arabian oryx (*Oryx leucoryx*) population without known pedigree. Pp. 77–83, *in* Proceedings of the International Symposium "Ongulés/Ungulates 91" (F. Spitz, G. Janeau, G. Gonzalez, and S. Aulagnier, eds.). S.F.E.R.M.-I.R.G.M., Toulouse, Paris.

Gridler, R. W., and T. C. Southern. 1987. Structure and evolution of the northern Red Sea. Nature (Lond.), 330:716–721.

Gromov, I. M., D. I. Bibikov, N. I. Kalabukhov, and M. N. Meir. 1928. [Fauna of the USSR, Vol. 1. Pt. 2; Chiroptera.] U.S.S.R. Acad. Sci., Moscow, 320 pp.

Groves, C. P. 1969. On the smaller gazelles of the genus *Gazella* de Blainville, 1816. Z. Säugetierkd., 34:39–60.

———. 1983. Notes on the gazelles. IV. The Arabian gazelles collected by Hemprich and Ehrenberg. Z. Säugetierkd., 48:371–381.

———. 1985. An introduction to the gazelles. Chinkara, 1:4–10.

Groves, C. P., and D. L. Harrison. 1967. The taxonomy of the gazelles (genus *Gazella*) of Arabia. J. Zool. (Lond.), 152:381–387.

Groves, C. P., and V. Mazak. 1967. On some taxonomic problems of the Asiatic wild asses: with the description of a new subspecies (Perissodactyla, Equidae). Z. Säugetierkd., 32:321–355.

Gunders, S. A., R. Lidror, B. Montilio, and P. Amitai. 1968. Isolation of *Leishmania* species from *Psammomys obesus* in Judea. Trans. R. Soc. Trop. Med. Hyg., 62:465.

Günther, A., and H. J. Gebauer. 1982. The chromosomes of the otter, *Lutra lutra* (Carnivora: Mustelidae: Lutrinae). Mamm. Chrom. Newsl., 23:97–101.

Hahn, H. 1934. Familie der Procavidae. Z. Säugetierkd., 9:207.

Haim, A., and A. Borut. 1975. Size and activity of a cold resistant population of the garden spiny mouse (*Acomys russatus*: Muridae). Mammalia, 39:605–612.

Haim, A., and E. Tchernov. 1974. The distribution of myomorph rodents in the Sinai Peninsula. Mammalia, 38:202–223.

Haltenorth, T. 1953. Die Wildkatzen der Alten Welt. Geest and Portig Publ., Leipzig, 117 pp.

———. 1959. Beitrag zur Kenntnis des Mesopotamischen Damhirsches—*Cervus* (*Dama*) *mesopotamicus* Brooke, 1875 und zur Stammes—und Verbreitungsgeschichte der Damhirsche allgemein. Säugetierkd. Mitt., 8:1–89.

Haltenorth, T., and H. Diller. 1980. A field guide to the mammals of Africa. (Translated by R. W. Hyman.) Reprinted 1984 by Collins, London, 400 pp.

Happold, D. C. D. 1968. Observations on *Gerbillus pyramidum* (Gerbillinae, Rodentia) at Khartoum, Sudan. Mammalia, 32:44–53.

Harding, G. L. 1954. Desert kites. Antiquity, 28:165–167.

Hardy, E. 1947*a*. The biblical coney. Zoo Life, 2:62.

———. 1947*b*. The Palestine leopard. Soc. Preserv. Fauna Empire, 55:16–20.

Harrison, D. L. 1955. On a collection of mammals from Oman, Arabia, with a description of two new bats. Ann. Mag. Nat. Hist., ser. 12, 8:897–910.

———. 1956. Mammals from Kurdistan, Iraq, with description of a new bat. J. Mammal., 37:257–263.

———. 1957. Some systematic notes on the trident bat (*Asellia tridens* E. Geoffroy) of Arabia. Mammalia, 21:1–8.

———. 1959. Footsteps in the sand. Benn Publ., London, 254 pp.

———. 1960. A new species of pipistrelle bat (Chiroptera: *Pipistrellus*) from south Israel. Durban Mus. Novit., 5:261–267.

———. 1961. On Savi's pipistrelle (*Pipistrellus savii* Bonaparte 1837) in the Middle East, and a second record of *Nycticeius schlieffeni* Peters 1859 from Egypt. Senckenb. Biol., 42:41–44.

———. 1962. A new subspecies of the noctule bat (*Nyctalus noctula* Schreber 1774) from Lebanon. Proc. Zool. Soc. Lond., 139:337–339.

———. 1963*a*. A new bat for Israel, *Eptesicus innesi* Latasete, 1887 with notes on the affinities of this species. Z. Säugetierkd., 28:107–110.

———. 1963*b*. A note on the occurrence of the greater mouse-tailed bat *Rhinopoma microphyllum* Brunnich, 1782, in Lebanon. Mammalia, 27:305–307.

———. 1963*c*. On the occurrence in Israel of the pygmy gerbil, *Gerbillus* (*Hendecapleura*) *henleyi* de Winton. Mammalia, 27:352–355.

———. 1963*d*. Some observations on the white-toothed shrews (genus *Crocidura* Wagner, 1832) of Israel. Bull. Res. Counc. Isr. Sect. B, 11:177–182.

———. 1964*a*. The mammals of Arabia, Vol. I. Introduction, Chiroptera, Insectivora, Primates. Ernest Benn Ltd., London, xx + 192 pp.

———. 1964*b*. A new subspecies of Natterer's bat, *Myotis nattereri* Kuhl, 1818 (Mammalia: Chiroptera) from Israel. Z. Säugetierkd., 29:58–60.

———. 1968*a*. The mammals of Arabia, Vol. II. Carnivora, Hyracoidea, Artiodactyla. Ernest Benn Ltd., London, xiv + 193–381 pp.

———. 1968*b*. A new race of Baluchistan gerbil, *Gerbillus nanus* Blanford, 1875 (Rodentia, Cricetidae) from Oman, with remarks on the Arabian forms of the species. Mammalia, 32:60–63.

———. 1971. Observations on some notable Arabian mammals, with the description of a new gerbil (*Gerbillus*, Rodentia: Cricetidae). Mammalia, 35:111–125.

———. 1972. The mammals of Arabia, Vol. III. Lagomorpha and Rodentia. Ernest Benn Ltd., London, xvii + 384–670 pp.

———. 1973. Some comparative features of the skulls of wolves (*Canis lupus* Linn.) and pariah dogs (*Canis familiaris* Linn.) from the Arabian Peninsula and neighbouring lands. Bonn. Zool. Beitr., 24:185–191.

———. 1975. Desert coloration in rodents. Pp. 269–276, *in* Rodents in desert environments (I. Prakash and P. K. Ghosh, eds.). Dr. W. Junk, The Hague, 624 pp.

———. 1980. Occurrence of the noctule, *Nyctalus noctula* Schreber, 1774 (Chiroptera: Vespertilionidae) in Oman, Arabia. Mammalia, 44:409–410.

Harrison, D. L., and P. J. J. Bates. 1991. The mammals of Arabia. Harrison Zoological Museum, Kent, England, xvi + 354 pp.

Harrison, D. L., and R. E. Lewis. 1961. The large mouse-eared bats of the Middle East, with description of a new subspecies. J. Mammal., 42:372–380.

———. 1964. A note on the occurrence of the weasel (*Mustella nivalis* Linnaeus, 1766) (Carnivora: Mustellidae) in Lebanon. Z. Säugetierkd., 29:179–181.

Harrison, D. L., and D. Makin. 1988. Significant new records of vespertilionid bats (Chiroptera: Vespertilionidae) from Israel. Mammalia, 52:593–596.

Hart, H. 1885. A naturalist's journey to Sinai, Petra and south Palestine. Palest. Expl. Quart. Lond., 18:231–286.

———. 1891. Some accounts of the fauna and flora of Sinai, Petra and Wady Arabah. Palestine Exploration Fund, London, 255 pp.

Hasselquist, F. 1757. Iter Palaestinum. P. 619, *in* C. Linnaeus, ed., Stockholm (not seen, see Kumerloeve, 1975).

Hassinger, J. D. 1973. A survey of the mammals of Afghanistan resulting from the 1965 Street Expedition (excluding bats). Fieldiana Zool., 60:1–195.

Hatough, A., and D. El-Eisawi. 1988. The Arabian oryx in Jordan. J. Arid Environ., 14:291–300.

Hatough, A. M. A., D. M. H. El-Eisawi, and A. M. Disi. 1986. The effect of conservation on wildlife in Jordan. Environ. Conserv., 13:331–335.

Hatough-Bouran, A. 1990. The burrowing habits of desertic rodents *Jaculus jaculus* and *Gerbillus dasyurus* in the Shaumari Reserve in Jordan. Mammalia, 54:341–359.

Hatough-Bouran, A., and A. M. Disi. 1991. History, distribution, and conservation of large mammals and their habitats in Jordan. Environ. Conserv., 18:19–44.

Hatt, R. T. 1959. The mammals of Iraq. Misc. Publ. Mus. Zool. Univ. Mich., 106:113.

Heim de Balsac, H. 1936. Biogeographie des Mammiferes et des Oiseaux de l'Afrique du Nord. Bull. Biol. Fr. Belg., Ser. A, 1622:1–446.

———. 1965. Quelques enseignements d'odre Faunistique tires de l'etude du regime alimentaire du *Tyto alba* dans l'oust de l'Afrique. Alauda, 33:309–322.

Hellwing, S. 1971. Maintenance and reproduction of white toothed shrew, *Crocidura russula monacha* Thomas, in captivity. Z. Säugetierkd., 36:103–113.

———. 1973. The postnatal development of the white toothed shrew, *Crocidura russula monacha* in captivity. Z. Säugetierkd., 38:257–270.

Helversen, O. v. 1989. New records of bats (Chiroptera) from Turkey. Zool. Middle East, 3:5–18.

Hemmer, H. 1968. Ist die Hauskatze (*Felis catus* L.) polyphyletischen Ursprungs. Carniv. Genet. Newsl., 5:90–92.

———. 1978. Nachweis der Sandkatz (*Felis margarita harrisoni* Hemmer, Grubb und Groves, 1976) in Jordanien. Z. Säugetierkd., 43:62–64.

Hemmer, H., P. Grubb, and C. P. Groves. 1976. Notes on the sand cat, *Felis margarita* Loche, 1858. Z. Säugetierkd., 41:286–303.

Hemprich, F. W., and C. G. Ehrenberg. 1833. Symbolae physicae seu icones et descriptiones Mammalium Decade 1, 1828 (part of text published in 1833) (not seen, see Morrison-Scott, 1939).

Heptner, V. G., A. A. Nasimovich, and A. G. Bannikov. 1988. Mammals of the Soviet Union. I. Artiodactyla and Perissodactyla. Smithsonian Institution Libraries and the National Science Foundation, Washington, D.C., 1147 pp.

Herter, K. 1965. Hedgehogs: a comprehensive study. Phoenix House Publ. (J. M. Dent), London, 69 pp.

Heuglin, T. 1861. Einige Bemerkungen über die Wilbethiere des nördlichen Aegyptens und des Peträischen Arabiens. Petermanns Mitt. Gotha, (1861):310–312.

Hoogstraal, H. 1962. A brief review of the contemporary land mammals of Egypt (including Sinai). 1. Insectivora and Chiroptera. J. Egypt. Public Health Assoc., 37:143–162.

———. 1963. A brief review of the contemporary land mammals of Egypt (including Sinai). 2. Lagomorpha and Rodentia. J. Egypt. Public Health Assoc., 38:1–35.

Hoogstraal, H., M. N. Kaiser, R. A. Ormsbee, D. J. Osborn, I. Helmy, and S. Gaber. 1967. *Hyaloma rhipicephaloides*: its identity, hosts, ecology and *Rickettsia coroni*, *R. prowazekia*, and *Coxiella burnetti* infections in rodent hosts in Egypt. J. Med. Entomol., 4:391–400.

Hoogstraal, H., K. Wassif, and M. N. Kaiser. 1957. Results of the NAMRU-3 Southeastern Egypt Expedition, 1954. 6. Observations on non-domesticated mammals and their ectoparasites. Bull. Zool. Soc. Egypt, 13:52–75.

Hooijer, D. A. 1958. An early Pleistocene mammalian fauna from Bethlehem. Bull. Br. Mus. Nat. Hist. Zool., 3:267–292.

———. 1961. Middle Pleistocene mammals from Latamne, Orontes Valley, Syria. Ann. Archeol. Syrie, 11:117–132.

Horáček, I. 1991. Enigma of *Otonycteris*: ecology, relationship, classification. Myotis, 29:17–30.

Howells, V. 1956. A naturalist in Palestine. Andrew Melrose Publ., London, 180 pp.

Hsu, T. C., and F. E. Arrighi. 1966. The karyotype of 13 carnivores. Mamm. Chrom. Newsl., 21:155–160.

———. 1971. Distribution of constitutive heterochromatin in mammalian chromosomes. Chromosoma (Berl.), 34:243–253.

Hsu, T. C., and K. Benirschke. 1967–1971. An atlas of mammalian chromosomes. Springer Verlag, New York, loose leaf, numbered separately.

Hsu, T. C., H. H. Rearden, and G. F. Luquette. 1963. Karyological studies of nine species of Felidae. Am. Nat., 97:225–234.

Hungeford, D. L., and G. L. Snyder. 1966. The somatic chromosomes of a Syrian bear, *Ursus arctos syriacus*. Mamm. Chrom. Newsl., 21:150.

———. 1969. Chromosomes of the rock hyrax, *Procavia capensis* (Pallas). Experientia (Basel), 25:870.

Hutterer, R., and D. L. Harrison. 1988. A new look at the shrews (Soricidae) of Arabia. Bonn. Zool. Beitr., 39:59–72.

Ilani, G. 1975. Hyenas in Israel. Isr. Land Nat., 1:10–18.

———. 1987. Ibex on the Golan Heights. Isr. Land Nat., 12:126.

———. 1988a. New distribution areas of mongooses. Isr. Land Nat., 13(4):197.

———. 1988*b*. Otters still live on the Golan Heights. Isr. Land Nat., 14(1):245.

———. 1988*c*. Meanwhile, in the Judean Desert. Isr. Land Nat., 14:248–251.

Ilani, G., and A. Bouskila. 1982. Vole control by means of barn owls. Isr. Land Nat., 8:95.

Iljin, N. A. 1941. Wolf–dog genetics. J. Genet., 42:359–414.

IUCN. 1990. 1990 IUCN red list of threatened animals. IUCN, Gland, Switzerland, 228 pp.

Ives, R. 1950. The Palestinian environment. Am. Sci., 38:85–104.

Janecek, L. L., D. A. Schlitter, and I. L. Rautenbach. 1991. A genic comparison of spiny mice, genus *Acomys*. J. Mammal., 72:542–552.

Jordan, R. G., B. L. Davis, and H. Baccar. 1974. Karyotypic and morphometric studies of Tunisian *Gerbillus*. Mammalia, 38:667–680.

Jotterand-Bellomo, M. 1984. New developments in vertebrate cytotaxonomy VII Les Chromosomes des Rongeurs (order Rodentia Bowdich, 1821). Genetica (The Hague), 64:3–64.

Kadhim, A. H., A. M. Mustafa, and H. A. Jabir. 1979. Biological notes on jerboas *Allactaga euphratica* and *Jaculus jaculus* from Iraq. Acta Theriol., 24:93–98.

Kahmann, H. 1981. Zur Naturgeschichte des Löffelbilches, *Eliomys melanurus* Wagner, 1840 (Mammalia: Rodentia, Gliridae). Spixiana (Münch.), 4:1–37.

———. 1986. Jugendentwicklung und Erscheinungsbild des Löffelbilches, *Eliomys quercinus melanurus* (Wagner, 1839) ein Nachtrag. Säugetierkd. Mitt., 33:1–19.

Kaisila, J. 1966. The Egyptian fruit-bat, *Rousettus aegyptiacus* Geoffr. (Megachiroptera, Pteropodidae) visiting flowers of *Bombax malabaricum*. Ann. Zool. Fenn., 3:1–3.

Kamali, M. 1975. Sex-chromosome polymorphism in *Nesokia indica* from Iran. Mamm. Chrom. Newsl., 16:165–167.

Khalidi, W. 1992. All that remains: the Palestinian villages occupied and depopulated by Israel in 1948. The Institute for Palestinian Studies, Washington, D.C., 636 pp.

Kingdon, J. 1974. East African mammals: an atlas of evolution in Africa Vol. 2. Part A. Academic Press, London, 341 pp.

———. 1990. Arabian mammals: a natural history. Academic Press, London, 279 pp.

Kingswood, S. C., and A. T. Kumamoto. 1988. Research and management of Arabian sand gazelles in the USA. Pp. 212–226, *in* Conservation and biology of desert antelopes (A. Dixon and D. Jones, eds.). Christopher Helm, London, 238 pp.

Kirmiz, J. P. 1962. Adaptation to desert environment: a study on the jerboa, rat, and man. Butterworths, London, 168 pp.

———. 1965. The jerboa and the desert. Sci. J. (Lond.), 64–65:47–51.

Klein, J. M., A. R. Poulet, and E. Simonkovitch. 1975. Observations écologiques dans une zone en zootique de peste en Mauritanie. 2. Les rongeurs et en particulier *Gerbillus gerbillus* Olivier, 1810 (Rodentia, Gerbillinae). Off. Rech. Sci. Tech. Centr. Outre-Mer (Ent., Med., Par.), 13:13–28.

Kligler, I. J., and R. Comaroff. 1936. An epidemic outbreak of murine typhus in a labour group in an inland village in Palestine. Trans. R. Soc. Trop. Med. Hyg., 30:363–368.

Kock, D. 1969. Die Fledermaus-Fauna des Sudan. Abh. Senckenb. Naturforsch. Ges., 521:1–238.

———. 1983. Identifizierung der Palästina-Genetten von J. Aharoni als *Vormela peregusna* (Güldenstaedt, 1770). Z. Säugetierkd., 48:381–383.

Kock, D., and I. Nader. 1983. Pygmy shrew and rodents from the Near East. Senckenb. Biol., 64:13–23.

———. 1984. *Tadarida teniotis* (Rafinesque, 1814) in W-Palaearctic and a lectotype for *Dysopus rupelii* Temminck, 1826 (Chiroptera: Molossidae). Z. Säugetierkd., 49:129–135.

Kock, D., D. M. Shafei, and Z. S. Amr. 1993. The jungle cat, *Felis chaus* Güldenstaedt, 1776, in Jordan. Z. Säugetierkd., 58:313–315.

Kohane, M. J., and P. A. Parsons. 1988. Domestication—evolutionary change under stress. Evol. Biol., 23:31–48.

Kohle, I. 1981. Animal remains. Pp. 207, *in* Jawa, lost city of the Black Desert (S. W. Helms, ed.). Methuen and Co. Ltd., London, 237 pp.

Koopman, K. F. 1975. The bats of the Sudan. Bull. Am. Mus. Nat. Hist., 154:355–443.

Kosswig, G. 1955. Zoogeography of the Near East. Syst. Zool., 4:49–73.

Kratochvil, J. 1986. *Mus abbotti*—eine kleinasiatisch-balkanische Art (Muridae, Mammalia). Folia Zool., 35:3–20.

Kratochvil, J., and Z. Kratochvil. 1976. The origin of the domesticated forms of the genus *Felis* (Mammalia). Zool. Listy, 25:193–208.

Krishna, D., and I. Prakash. 1956. Hedgehogs of the desert of Rajasthan: part II; food and feeding habits. J. Bombay Nat. Hist. Mus., 53:362–366.

Kulzer, E. 1958. Untersuchungen über die biologie von Flughunden der Gattung *Rousettus* Gray. Z. Morphol. Oekol. Tiere, 47:374–402.

Kulzer, E., I. Helmy, and G. Necker. 1985. Untersuchungen über die Drüsen der Gesichtsregion der ägypischen Mausschwanz-Fledermaus *Rhinopoma hardwickei cystops* Thomas, 1903. Z. Säugetierkd., 50:57–68.

Kumerloeve, H. 1971. Zum stand des vorkommens von *Panthera pardus tulliana* Valenciennes 1856 in Kleinasien. Zool. Gart. N. F., Liepzig, 40:4–22.

———. 1973. In memorium Sana I. Atallah. Säugetierkd. Mitt., 21:82–84.

1975. Die säugetiere (Mammalia) Syriens und des Libanon. Veröff. Zool. Staatssamml. München, 18:159–225.

Kurtén, B. 1965. The Carnivora of the Palestine caves. Acta Zool. Fenn., 107:1–74.

Lamb, K. 1984. Oryx runs wild in Jordan after sixty years in exile. World Wildl. News, 1984:18–20.

Lataste, F. 1884. Description d'une espece nouvelle de Gerbilline d'Arabie (*Meriones longifrons*). Proc. Zool. Soc. Lond., 1884:88–109.

———. 1887. Notes prises au jour le jour sur différentes especes de l'ordre des rongeurs observées en captivité. Actes Soc. Linn. Bordeaux, 41(ser. 5, tome 1):201–536.

LaVal, R. K., and M. L. LaVal. 1980. Prey selection by the slit-faced bat *Nycteris thebaica* (Chiroptera: Nycteridae) in Natal, South Africa. Biotropica, 12:241–246.

Lavappa, K. S. 1977. Chromosomal banding patterns and idiograms of the Armenian hamster *Cricetulus migratorius.* Cytologia (Tokyo), 42:65–72.

Lawrence, T. E. 1926. Revolt in the Desert. George H. Doran Co., Garden City, New York, xii + 335 pp.

Lay, D. M. 1967. A study of the mammals of Iran, resulting from the Street Expedition of 1962–63. Fieldiana Zool., 54:1–282.

———. 1972. The anatomy, physiology, functional significance and evolution of specialized hearing organs of the gerbilline rodents. J. Morphol., 138:41–120.

———. 1983. Taxonomy of the genus *Gerbillus* (Rodentia: Gerbillinae) with comments on the applications of generic and subgeneric names and an annotated list of species. Z. Säugetierkd., 48:329–354.

Lay, D. M., K. Agerson, and C. F. Nadler. 1975. Chromosomes of some species of *Gerbillus* (Mammalia, Rodentia). Z. Säugetierkd., 40:141–150.

Lay, D. M., and C. F. Nadler. 1972. Cytogenetics and origin of north African *Spalax.* Cytogenetics (Basel), 11:279–286.

———. 1975. A study of *Gerbillus* (Rodentia: Muridae) east of the Euphrates River. Mammalia, 39:423–445.

Legge, A. J. 1972. Prehistoric exploitation of the gazelle in Palestine. Pp. 170-177 *in* Economic prehistory. Studies by members and associates of the British Academy Major Research Project in the early history of agriculture (E. S. Higgs, ed.). Cambridge Univ. Press, London, x + 219 pp.

Legge, A. J., and C. P. A. Rowley. 1987. Gazelle killing in Stone Age Syria. Sci. Am., 257:88–95.

Legouix, J. P., F. Petter, and A. Wisner. 1954. Etude de l'audition chez des mammiféres a bulles tympaniques hypertrophiées. Mammalia, 18:262–271.

Leshem, Y. 1979. Zoological treasure in Jerusalem. Isr. Land Nat., 5:56–61.

Lewis, R. E., and S. I. Atallah. 1966. A note on the occurrence of *Mellivora capensis* ssp. in northern Saudi Arabia (Mellivorinae: Mustellidae). Z. Säugetierkd., 31:390–392.

Lewis, R. E., and D. L. Harrison. 1962. Notes on the bats from the Republic of Lebanon. Proc. Zool. Soc. Lond., 138:473–486.

Lewis, R. E., J. H. Lewis, and S. I. Atallah. 1967. A review of Lebanese mammals. Lagomorpha and Rodentia. J. Zool. (Lond.), 153:45–70.

———. 1968. A review of Lebanese mammals, Carnivora, Pinnipedia, Hyracoidea and Artiodactyla. J. Zool. (Lond.), 154:517–531.

Lewis, R. E., J. H. Lewis, and D. L. Harrison. 1965. On a collection of mammals from northern Saudi Arabia. Proc. Zool. Soc. Lond., 144:61–74.

Lidicker, W. Z., Jr. 1962. The nature of subspecies boundaries in a desert rodent and its implications for subspecies taxonomy. Syst. Zool., 11:160–171.

Linnaeus, C. 1758. Systema Naturae par regna tria naturae, secundum classis, ordines, genera, species, cum chractereribus, differentiis, synomis, locis. Tenth ed. Vol. 1. Regnum Animale. pt. 1, 532 pp.

Linnaeus, C. D. 1762. Friedrich Hasselquists Reise nach Palästina in den Jahren von 1749 bis 1752. Auf Befehl Ihro Majestät der Königen von Schweden herausgegeben. J. C. Koppe, Rostock, not seen.

Lloyd, M. 1965. The Arabian oryx in Muscat and Oman. East Afr. Wildl. J., 3:124–127.

Madkour, G. 1961. The structure of the facial area in the mousetailed bat, *Rhinopoma hardwickei cystops*, Thomas. Bull. Zool. Soc. Egypt, 16:50–54.

———. 1977*a*. *Rousettus aegyptiacus* (Megachiroptera) as a fruit eating bat in the A. R. Egypt. Agr. Res. Rev. (Cairo), 55:167–172.

———. 1977*b*. Further observations on bats (Chiroptera) of Egypt. Agr. Res. Rev. (Cairo), 55:173–184.

Makin, D. 1977. Saqar atlaphi harqim b'eretz yisrael. Pp. 127–223, *in* Shamiret teva bisrael. Nature Reserve Authority, Jerusalem, 332 pp. (in Hebrew).

———. 1979. The quest for Israel's bats. Isr. Land Nat., 5:28–34.

Makin, D., and D. L. Harrison. 1988. Occurrence of *Pipistrellus ariel* Thomas, 1904 (Chiroptera: Vespertilionidae) in Israel. Mammalia, 52:419–422.

Makin, D., and H. Mendelssohn. 1986. Reconsidering fumigation of bat caves. Isr. Land Nat., 12:26–30.

———. 1989. A recent mass-kill of bats: who cares? Isr. Land Nat., 14:82–85.

Manfreddi-Romanini, M. G., C. Pellicciardi, F. Bolchi, and E. Capanna. 1975. Donnees nouvelles sur le contenu en ADN des noyaux postkinetique ches les chiropteres. Mammalia, 39:675–683.

Manson-Bahr, P. 1936. Wolves in Palestine and Egypt. Field, 167(4355):1448.

Markowitz, A. 1967. Desert game park. Wildlife (Lond.), 21:340.

Marshall, J. T., and R. D. Sage. 1981. Taxonomy of the house mouse. Symp. Zool. Soc. Lond., 47:15–25.

Maser, C. O. 1966. Commensal relationship between *Acomys* and *Rousettus*. J. Mammal., 47:153.

Matschi, V. P. 1912. Uber einige Rassen des Steppenluchses Felis (*Caracal caracal*). Sitzungsber. Ges. Naturf. Fr. Berl., 1912:55–67.

Matthey, R. 1953. Les chromosomes des Muridae: revision critique et materiaux nouveaux pour service a l' histoire de l'evolution chromosomique ches les rongeurs. Rev. Suisse Zool., 60:225–283.

———. 1954. Chromosomes et systematiques des Canides. Mammalia, 18:225–230.

Mayr, E. 1969. Principles of systematic zoology. McGraw-Hill Book Co., New York, 423 pp.

Mech, L. D. 1974. *Canis lupus*. Mamm. Species, 37:1–6.

Mendelssohn, H. 1965. Breeding the Syrian hyrax *Procavia capensis syriaca* Schreber, 1784. Int. Zoo Yearb., 5:116–125.

———. 1974. The development of the populations of gazelles in Israel and their behavioural adaptations. Pp. 722–743, *in* The behaviour of ungulates and its relation to management (V. Geist and F. Walker, eds.). International Union for the Conservation of Nature and Natural Resources, Morges, Switzerland. 940 pp.

————. 1982. Wolves in Israel. Pp. 173–195, *in* Wolves of the world: perspectives of behavior, ecology, and conservation (F. H. Harrington and P. C. Paquet, eds.). Noyes Publication, Park Ridge, New Jersey, 474 pp.

Mendelssohn, H., and Y. Yom-Tov. 1987. Mammals. Ministry of Defense Publishing House, Tel Aviv, 295 pp. (in Hebrew).

Mendelssohn, H., Y. Yom-Tov, G. Ilani, and D. Meninger. 1987. On the occurrence of Blanford's fox, *Vulpes cana* Blanford, 1877, in Israel and Sinai. Mammalia, 51:459–462.

Mermod, C. 1969. Les rongeurs d'une Daya au Sahara nordoccidental (Algerie). Terre Vie (Paris), 4:486–495.

Metaxas, C. C. 1891. Mémoire sur les animaux de la Mesopotamie. Rev. Sci. Nat. Apliquées Bull. Soc. Nat. Acclim. France, 9:321–328, 432–435.

Meylan, A. 1968. Formules chromosomiques de quelques petits mammiféres Nord-Américains. Rev. Suisse Zool., 75:691–696.

Miller, G. 1908. The recent voles of the *Microtus nivalis* group. Ann. Mag. Nat. Hist., 1(8):97–103.

Misonne, X. 1957. Mammiféres de la Turqui sud-orientale et du nord de la Syrie. Mammalia, 21:53–62.

Mohres, F. P., and E. Kulzer. 1956. Über die Orientierung der Flughunde (Chiroptera-Pteropodidae). Z. Vgl. Physiol., 38:1–29.

Morrison-Scott, T. C. S. 1939. Some Arabian mammals collected by Mr. Sgt. J. B. Philby, C.I.E. Novit. Zool., 41:181–211.

————. 1951. Exhibition of photograph of skin of Arabian cheetah. Proc. Zool. Soc. Lond., 121:201.

Morsy, T. A., A. M. Disi, and Z. S. Amr. 1981. Leishmania infection sought in rodents caught in Jordan. J. Egypt. Soc. Parasitol., 11:5–13.

Mountford, G. 1964. Disappearing wildlife and growing deserts in Jordan. Ibex, 7:229–232.

————. 1965. Portrait of a desert. Collins, London, 192 pp.

Müller, H. 1911. Beirage zur kenntnis der Stachelschweine Asiens, unsbesondere Palästinas I. Sitzungsber. Ges. Naturf. Fr. Berl., 1911:122.

————. 1920. Beirage zur kenntnis der Stachelschweine Asiens, unsbesondere Palästinas III. Zool. Anz., 51:195–200.

Murray, G. W. 1930. Egyptian mountains. Alpine J., 42:226–235.

Musil, A. 1927. The Middle Euphrates. American Geographic Society, New York, 426 pp.

Nadachowski, A., J. Smielowski, B. Rzebik-Kowalska, and A. Daoud. 1990. Mammals from the Near East in Polish collections. Acta Zool. Cracov., 33:91–120.

Nader, I. A. 1975. On the bats (Chiroptera) of the kingdom of Saudi Arabia. J. Zool. (Lond.), 176:331–340.

Nader, I. A., and W. Buttiker. 1980. Mammals of Saudi Arabia. Mammalia: fam. Canidae. Records of the Arabian wolf, *Canis lupus arabs* Pocock, 1934, from Saudi Arabia. Fauna Saudi Arabia, 2:405–411.

————. 1982. Mammals of Saudi Arabia. Recent records of the striped hyaena, *Hyaena hyaena* (Linnaeus 1758) from the kingdom of Saudi Arabia (Mammalia: Carnivora: Hyaenidae). Fauna Saudi Arabia, 4:503–508.

Nader, I. A., and D. Kock. 1983. Notes on some bats from the Near East (Mammalia: Chiroptera). Z. Säugetierkd., 48:1–9.

Nader, I. A., D. Kock, and A. D. Al-Khalili. 1983. *Eliomys melanurus* (Wagner 1839) and *Praomys fumatus* (Peters 1878) from the kingdom of Saudi Arabia (Mammalia: Rodentia). Senckenb. Biol., 63:313–324.

Nadler, C. F., and R. S. Hoffmann. 1970. Chromosomes of some Asian and South American squirrels (Rodentia, Sciuridae). Experientia (Basel), 26:1383–1386.

Nadler, C. F., and D. M. Lay. 1968. Chromosomes of some species of *Meriones* (Mammalia: Rodentia). Z. Säugetierkd., 32:285–291.

Naggan, L., A. Gunders, R. Dizian, Y. Dannon, S. Shibolet, A. Ronen, R. Schneeweiss, and D. Michaeli. 1970. Ecology and attempted control of cutaneous leishmaniasis around Jericho, in the Jordan Valley. J. Infect. Dis., 121:427–432.

Nehring, A. 1897. Über *Nesokia bacheri* sp. n. Zool. Anz., 20:503–505.

————. 1901. Über *Dipus schlüteri* n. sp. und einige andere Nager aus Palestina. Sitzungsber. Ges. Naturf. Fr. Berl., 8:163–176.

————. 1902*a*. Die geographische Verbreitung der Säugethiere in Palästina und Syrien. Mitt. Nachr. des Palästina-Ver., 1902:49–63.

————. 1902*b*. Über einen neuen Sumpffluchs (*Lyncus chrysomelanotis*) aus Palästina. Sitzungsber. Ges. Naturf. Fr. Berl., 1902:1–7 (127, 147).

Nelson, B. 1973. Azraq Desert oasis. Allen Lance, London, xix + 436 pp.

————. 1985. Return to Azraq. Oryx, 19:22–26.

Nelson, E. W. 1926. Bats in relation to the production of guano and the destruction of insects. U.S. Dept. Agriculture, Bull., 1395:1–12.

Neronov, V. M., M. E. Zvenigorodskaja, and I. F. Gachik. 1987. [A zoogeographical analysis of the rodent fauna of Arabia and Asia Minor.] Zool. Zh., 66:1055–1068 (In Russian, English summary).

Neuhäuser, G. 1936. Die Muriden von Kleinasien. Z. Säugetierkd., 11:161–236.

Neuville, R. 1951. La Paléolithique et la Mésolithique du Déserte du Judée. Arch. Inst. Paleontol. Hum. Mem., 24:1–233.

Nevo, E. 1961. Observations on Israeli populations of the mole rat, *Spalax e. ehrenbergi* Nehring, 1889. Mammalia, 25:127–144.

————. 1969. Mole rat *Spalax ehrenbergi*: mating behavior and its evolutionary significance. Science (Wash. D.C.), 163:448–456.

————. 1978. Genetic variation in natural populations: patterns and theory. Theor. Popul. Biol., 13:121–177.

————. 1982. Genetic structure and differentiation during speciation in fossorial gerbil rodents. Mammalia, 46:523–530.

————. 1985. Genetic differentiation and speciation in spiny mice, *Acomys*. Acta Zool. Fenn., 170:131–136.

————. 1988. Genetic diversity in nature—patterns and theory. Evol. Biol., 23:217–246.

————. 1989. Natural selection of body size differentiation in spiny mice, *Acomys*. Z. Säugetierkd., 54:81–99.

Nevo, E., and E. Amir. 1961. Biological observations on the forest dormouse, *Dryomys nitedula* Pallas, in Israel (Rodentia, Mucardinidae). Bull. Res. Counc. Isr. Sect. B, 9:200–201.

————. 1964. Geographic variation in reproduction and hibernation patterns of the forest dormouse. J. Mammal., 45:69–87.

Nevo, E., and H. Bar-El. 1976. Hybridization and speciation in fossorial mole rats. Evolution, 30:831–840.

Nevo, E., M. G. Filippucci, and A. Beiles. 1990. Genetic diversity and its ecological correlates in nature: comparisons between subterranean, fossorial, and above-ground small mammals. Pp. 347–366, *in* Evolution of subterranean mammals at the organismal and molecular levels (E. Nevo and O. A. Reig, eds.). A.R. Liss (Wiley-Liss), New York, 422 pp.

Nevo, E., S. Simson, A. Beiles, and S. Yahav. 1977. Adaptive variation in structure and function of kidneys of speciating subterranean mole rats. Toxicon, 15:541–548.

Newton, A. 1876. On the roebuck in Palestine. Proc. Zool. Soc. Lond., 1876:700–701.

Nissan, J. 1988. The early history of the Ancient Near East 9000–2000 BC. University of Chicago Press, Chicago, xiv + 215 pp.

Norberg, U. M., and M. B. Fenton. 1988. Carnivorous bats? Biol. J. Linn. Soc., 33:383–394.

Nowak, R. M., and J. L. Paradiso. 1983. Walker's mammals of the world. The Johns Hopkins University Press, Baltimore, Maryland, 1362 pp.

Noy, T., J. Schuldren, and E. Tchernov. 1980. Gilgal, a pre-pottery Neolithic A site in the Lower Jordan Valley. Isr. Explor. J., 30:63–82.

Ognev, S. I. 1928. The mammals of eastern Europe and northern Asia, Vol. I. Insectivora and Chiroptera. Glavnauka. Gosudarstrennoe Izdatel'stro, Moscow. (Published for the National Science Foundation by the Israel Program for Scientific Translation, 1962, 487 pp.)

————. 1931. The mammals of eastern Europe and northern Asia, Vol. II. Carnivora, Fissipidae. Glavnauka. Gosudarstrennoe Izdatel'stro, Moscow. (Published for the National Science Foundation by the Israel Program for Scientific Translation, 1962, 775 pp.)

————. 1935. The mammals of eastern Europe and northern Asia, Vol. III. Carnivora, Fissipidae, and Pinnipidae., Glavnauka. Gosudarstrennoe Izdatel'stro, Moscow. (Published for the National Science Foundation by the Israel Program for Scientific Translation, 1962, 752 pp.)

————. 1940. The mammals of eastern Europe and northern Asia, Vol. IV. Rodentia. Glavnauka. Gosudarstrennoe Izdatel'stro, Moscow. (Published for the National Science Foundation by the Israel Program for Scientific Translation, 1966, 429 pp.)

————. 1950. The mammals of eastern Europe and northern Asia, Vol. VII. Rodentia. Glavnauka. Gosudarstrennoe Izdatel'stro, Moscow. (Published for the National

Science Foundation by the Israel Program for Scientific Translation, 1964, 706 pp.)

Orsini, P., F. Bonhomme, J. Britton-Davidian, H. Croset, S. Gerasimov, and L. Thaler. 1983. Le complexe d'especes du genre *Mus* en Europe centrale et orientale. II. Criteres d'identification, répartition et caractéristiques écologiques. Z. Säugetierkd., 48:86–95.

Osborn, D. J., and I. Helmy. 1980. The contemporary land mammals of Egypt including Sinai. Fieldiana Zool. 5:1–579.

Oumish, Y., E. Saliba, and T. Alawi. 1982. Cutaneous leishmaniasis: an endemic disease in Jordan. Jordan Med. J., 16:55–61.

Owen, R. D., and M. B. Qumsiyeh. 1987. The subspecies problem in the trident leaf-nosed bat, *Asellia tridens:* homomorphism in widely separated populations. Z. Säugetierkd., 6:329–337.

Palmer, E. H. 1871. The desert of the Exodus. Deighton Bell and Co. Publ., London, 576 pp.

Panous, J. B. 1951. Les Chauves-souris du Maroc. Travaux de l'Institue Scientifique Chérifien, Tangier, Morocco, 122 pp.

Pathak, S., and D. Wurster-Hill. 1977. The distribution of constitutive heterochromatin in carnivores. Cytogenet. Cell Genet., 18:245–254.

Paz, U. 1982. The fauna of the Holy Land at the end of the Ottoman Period. Isr. Land Nat., 7:104–110.

Peshev, D. 1989. The mouse-like hamster (*Calomyscus bailwardi* Thomas, 1905), a new mammal for the Syrian fauna and the Arab Peninsula. Mammalia, 53:109–112.

Peshev, D. T., and K. Al Hossein. 1989. Karyology and biochemical characteristic of the pole cat (*Vormela peregusna syriaca* Pocock) (Carnivora, Mustelidae) from Syria. Acta Zool. Bulg., 38:54–57.

Peters, G., and R. Rödel. 1994. Blanford's fox in Africa. Bonn. Zool. Beitr., 45:99–111.

Peterson, R. L., and D. W. Nagorsen. 1975. Chromosomes of fifteen species of bats (Chiroptera) from Kenya and Rhodesia. R. Ont. Mus. Life Sci. Occas. Pap., 27:1–14.

Petter, F. 1954. Remarques biologiques sur des Rats Epineux de genere *Acomys*, repartition aus Sahara. Mammalia, 18:389–395.

———. 1956. Caractéres comparés de *Gerbillus allenbyi* et de deux autres espéces de sous-genre *Gerbillus*. Mammalia, 20:231–237.

———. 1957. Liste commentee des espéces de Gerbillidés de Palestine. Mammalia, 21:241–257.

———. 1961. Répartition géographique et écologie des rongeurs désertiques (du Sahara occidental a l'Iran oriental). Mammalia, 25(suppl.):1–219.

———. 1968. Retour au gite et nomadisme chez un rongeur a bulles tympaniques hypertrophées. Mammalia, 32:537–549.

———, 1971. Rodentia, Gerbillinae. *In* The mammals of Africa: an identification manual (J. Meester and H. W. Setzer, eds.). Smithsonian Inst. Press, Washington, D.C., Part 6.3, 4 pp., looseleaf.

————. 1973. Les animaux domestiques et leurs ancéstres. Collection Bordas Poche, Paris, 153 pp.

Pocock, R. I. 1934. On races of the striped and brown hyaenas. Proc. Zool. Soc. (Lond.), 1934:799–825.

————. 1936. The polecats of the genera *Putorius* and *Vormela* in the British Museum. Proc. Zool. Soc. Lond., 1936:691–723.

————. 1941. The fauna of British India, including Ceylon and Burma. Mammalia, Vol. I. Taylor and Francis Publ., London, 503 pp.

————. 1944. The wild cat (*Felis libyca*) of Palestine. Ann. Mag. Nat. Hist., ser. 11, 11:125–130.

————. 1951. Catalogue of the genus *Felis*. Trustees of the British Museum, London, 190 pp.

Popescu, N. C., and J. A. Di Paolo. 1980. Chromosome interrelationships of hamster species of the genus *Mesocricetus*. Cytogenet. Cell Genet., 28:10–24.

Poulet, A. R. 1970. Les Rhinopomatidae de Mauritanie. Mammalia, 34:237–243.

Prakash, I. 1960. Breeding of mammals in the Rajasthan Desert, India. J. Mammal., 4:386–389.

Prater, S. H. 1965. The book of Indian animals. Bombay Natural History Society, Bombay, 220 pp.

Qumsiyeh, M. B. 1980. New records of bats from Jordan. Säugetierkd. Mitt., 28:36–39.

————. 1985. The bats of Egypt. Spec. Publ. Mus., Texas Tech Univ. 23:1–102.

————. 1988. Pattern of heterochromatic variation and phylogeny in the rodent family Gerbillidae. Tex. J. Sci., 40:63–70.

————. 1989. Chromosomal fissions and phylogenetic hypotheses: cytogenetic and allozymic variation between species of *Meriones* (Rodentia, Gerbillidae). Occas. Pap. Mus., Texas Tech Univ., 132:1–16.

————. 1991. Karyotype of the east European hedgehog, *Erinaceus concolor* from Jordan. Z. Säugetierkd., 56:375–377.

————. 1992. Review of "Mammals of Arabia" by D. L. Harrison and P. J. Bates. J. Mammal., 73:228–229.

Qumsiyeh, M. B., Z. S. Amr, and D. M. Shafi. 1993. Status and conservation of the carnivores of Jordan. Mammalia, 57:55–62.

Qumsiyeh, M. B., and R. J. Baker. 1985. G- and C-band karyotypes of the Rhinopomatidae (Microchiroptera). J. Mammal., 66:541–544.

Qumsiyeh, M. B., and J. W. Bickham. 1993. Chromosomes and relationships of long-eared bats of the genera *Plecotus* and *Otonycteris*. J. Mamm., 74:376–382.

Qumsiyeh, M. B., and R. K. Chesser. 1988. Rates of protein, chromosome, and morphologic evolution in four genera of rhombomyine gerbils. Biochem. Syst. Ecol., 16:89–103.

Qumsiyeh, M. B., A. M. Disi, and Z. S. Amr. 1992. Systematics and distribution of bats (Mammalia, Chiroptera) of Jordan. Dirasat Nat. Sci (Amman), 19B(2):101–118.

Qumsiyeh, M. B., and J. K. Jones, Jr. 1986. *Rhinopoma hardwickei* and *R. muscatellum*. Mamm. Species, 263:1–5.

Qumsiyeh, M. B., R. D. Owen, and R. K. Chesser. 1988. Differential rates of genic and chromosomal evolution in bats of the family Rhinolophidae. Genome, 30:326–335.

Qumsiyeh, M. B., and D. A. Schlitter. 1981. Bat records from Mauritania, Africa (Mammalia: Chiroptera). Ann. Carnegie Mus. 50:345–351.

———. 1982. The bat fauna of Jabal al Akhdar, North East Libya. Ann. Carnegie Mus., 51:377–389.

Qumsiyeh, M. B., D. A. Schlitter, and A. M. Disi. 1986. New records and karyotypes of small mammals from Jordan. Z. Säugetierkd, 51:139–146.

Ragni, B., and E. Randi. 1986. Multivariate analysis of craniometric characters in European wild cat, domestic cat, and African wild cat (genus *Felis*). Z. Säugetierkd., 51:242–251.

Rahamat, O. 1982. The wild boar. Al-Reem, R. Soc. Cons. Nature (Jordan), 10:12–13 (in Arabic).

Ranck, G. L. 1968. The rodents of Libya: taxonomy, ecology, and zoogeographical relationships. U.S. Natl. Mus. Bull., 275:1–264.

Randi, E. and B. Ragni. 1991. Genetic variability and biochemical systematics of domestic and wild cat populations (*Felis silvestris:* Felidae). J. Mammal., 72:79–88.

Ranjini, P. V. 1966. The chromosomes of the Indian jackal (*Canis aureus*). Mamm. Chrom. Newsl., 19:5.

Rast, W. E. 1992. Through the ages in Palestinian archaeology: an introductory handbook. Trinity Press International, Philadelphia, 221 pp.

Raswan, C. R. 1935. The black tents of Arabia. Hutchinson Publ., London, 280 pp.

Rauwolff, L. 1582. Aigentliche bescreibung der Raiß, so er in die Morgenlander, furnemlich Syriam, Judeam, Arabiam, Mesopotamiam volbracht. Augspurg (not seen, see Kumerloeve, 1975).

Ray-Chaudhuri, S. P., S. Pathak, and T. Sharma. 1968. Chromosomes and affinities of Pteropidae (Megachiroptera) and Rhinopomatidae (Microchiroptera). Nucleus (Calcutta), (1968):96–101.

Reed, C. A. 1958. Observations on the burrowing rodent *Spalax* in Iraq. J. Mammal., 39:386–389.

———. 1959. Animal domestication in the prehistoric Near East. Science (Wash. D.C.), 130:1629–1639.

Rees, L. W. B. 1929. The Transjordan Desert. Antiquity, 3:389–407.

Renzoni, A. 1967. Chromosome studies in two species of rodents. Mamm. Chrom. Newsl., 8:111–112.

———. 1970. The karyotypes of two wild carnivores. Mamm. Chrom. Newsl., 11:26.

Renzoni, A., and P. Omodeo. 1972. Polymorphic chromosome system in the fox. Caryologia, 25:173–187.

Repenning, C. A., and O. Fejfar. 1982. Evidence for earlier date of 'Ubeidiya, Israel, hominid site. Nature (Lond.), 299:344–347.

Rieger, I. 1979. A review of the biology of the striped hyaena, *Hyaena hyaena* (Linné, 1758). Säugetierkd. Mitt., 27:81–95.

———. 1981. *Hyaena hyaena*. Mamm. Species, 150:1–5.

Rivon, N. 1957. "Wie der Panther erlegt wurde." H-Aretz, Jerusalem, 20–22 February 1957.

Roberts, T. J. 1977. Mammals of Pakistan. Ernst Benn Ltd., London, 361 pp.

Rotary, N. 1983. Shrew insights. Isr. Land Nat., 8:99–103.

Royen, C. E., J. E. Bowie, and K. S. Krikorian. 1944. Typhus research in Egypt, Palestine, Iraq, and Iran. Trans. R. Soc. Trop. Med. Hyg., 38:133–149.

Saint-Girons, M. 1962. Notes sur les dates de reproduction en captivité du Fennec, *Fennecus zerda* (Zimmermann, 1780). Z. Säugetierkd., 27:181–184.

Saleh, M. M. 1987. The decline of gazelles in Egypt. Biol. Conserv., 39:83–95.

Saliba, E. K., and Z. S. Amr. 1985. A contribution to the fleas of Jordan and their association with mammal hosts. Dirasat Nat. Sci. (Amman), 12:21–24.

Saliba, E. K., O. Y. Oumeish, J. Haddadin, Z. Amr, and R. K. Ashford. 1985. Cutaneous leishmaniasis in Mowaqqar area, Amman Governorate, Jordan. Ann. Trop. Med. Parasitol., 79:139–146.

Schauenberg, P. 1974. Données nouvelles sur le Chat des sables *Felis margarita* Loche, 1858. Rev. Suisse Zool., 81:949–969.

Schlawe, L. 1981. Material, Fundorte, Text- und Bildquellen als Grundlage für eine Artenliste zur Revision der Gattung *Genetta* G. Cuvier, 1816 (Mammalia, Carnivora, Viverridae). Zool. Abh. (Dres.), 37:85–182.

Schmidt-Nielson, K. 1964. Desert animals: physiological problems of heat and water. Oxford University Press and Clarendon Press, Oxford, London, xiv + 277 pp.

Schmiedeknecht, O. 1906. Die Wirbeltiere Europas mit Berucksichtigung der Faunen von Vorderasien und Nordafrika. Fischer Verlag, Jena (not seen, see Kumerloeve, 1975).

Schmitz, E. 1912. Eine Bärenjagel in Palästina (Hermon). Das Heilige Land (Berlin), 56:174–176.

Schoenfeld, M., and Y. Yom-Tov. 1985. The biology of two species of hedgehogs, *Erinaceus europaeus concolor* and *Hemiechinus auritus aegyptius*, in Israel. Mammalia, 49:339–355.

Searight, A. 1987a. The golden spiny mouse, *Acomys russatus lewisi*, in the wild and in captivity. Part 2. Ratel, 14:40–46.

———. 1987b. Some records of mammals from north-eastern Jordan. Pp. 311–317, *in* Proceedings of the Symposium on the Fauna and Zoogeography of the Middle East (F. Krupp, W. Schneider, and R. Kinzelbach, eds.). Dr. Ludwig Reichert Verlag, Wiesbaden, 365 pp.

Setzer, H. W. 1957. The hedgehogs and shrews of Egypt. J. Egypt. Public Health Assoc., 32:1–12.

———. 1958a. The mustelids of Egypt. J. Egypt. Public Health Assoc., 33:199–204.

———. 1958b. The gerbils of Egypt. J. Egypt. Public Health Assoc., 33:205–227.

———. 1959. The spiny mice (*Acomys*) of Egypt. J. Egypt. Public Health Assoc., 34:93–101.

———. 1960. Two new mammals from Egypt. J. Egypt. Public Health Assoc., 35:1–5.

————. 1961. The canids (Mammalia) of Egypt. J. Egypt. Public Health Assoc., 36:113–118.

Shalmon, B., S. M. MacDonald, and C. F. Mason. 1986. *Lutra lutra.* Oryx, 20:233–236.

Shkolnik, A., and A. Borut. 1969. Temperature and water relations in two species of spiny mice (*Acomys*). J. Mammal., 50:245–255.

Simons, E. L. 1990. Discovery of the oldest known anthropoidean skull from the Paleogene of Egypt. Science (Wash. D.C.), 247:1567–1569.

Simons, E. L., P. Andrews, and D. R. Pilbeam. 1978. Cenozoic apes. Pp. 120–146, *in* Evolution of African mammals (V. J. Maglio and H. B. S. Cooke, eds.). Harvard Univ. Press, Cambridge, Massachusetts, 641 pp.

Skinner, J. D. 1979. Feeding behaviour in caracal, *Felis caracal.* J. Zool. (Lond.), 189:523–557.

Skinner, J. D., and G. Ilani. 1981. The striped hyaena *Hyaena hyaena* of the Judean and Negev deserts of Israel and a comparison with the brown hyaena *Hyaena brunnea.* Paleobiology, 7:101–114.

Smith, R. H. 1985. Excavation at Pella of the Decapolis 1979–1985. Natl. Geogr. Res., 1:479–489.

Sokolov, V. J., V. N. Orlov, and V. M. Malygin. 1975. Chromosomnyje nabory dvuch vidov antilop iz Mongolii (Antilopinae, Bovidae). [Chromosome complements of two species of the Antilopinae from Mongolia.] Sist. Citogenet. Mlekop., Mat. Vsesoj. Simp., Moskra, 11:46–47.

Soldatovic, B., M. Tolksdorf, and H. Reichstein. 1970. Chromosomensatz bei verschieden Arten der Gattung *Canis.* Zool. Anz. (Jena), 184:155–167.

Southern, H. N. 1964. Handbook of British mammals, Mammal Society of British Isles. Blackwell Scientific Publications, Oxford, 320 pp.

Spitzenberger, F. 1970. Erstnachweise der Wimperspitzmaus (*Suncus etruscus*) für Kreta und Kleinasien und die Verbreitung der Art im südwestasiatischen Raum. Z. Säugetierkd., 35:107–113.

————. 1979. Die Säugetierfauna Zyperns Teil II: Chiroptera, Lagomorpha, Carnivora und Artiodactyla. Ann. Naturhist. Mus. Wien, 82:340–465.

Stencel, J. 1961. The distribution and bionomics of Kuhl's Bat, *Pipistrellus kuhli kuhli* (Natterer *in* Kuhl, 1819) in Lebanon (Chiroptera: Vespertilionidae). M.S. thesis, American University of Beirut, 30 pp.

Stewart, D. R. M. 1963. The Arabian oryx (*Oryx leucorys* Pallas). East Afr. Wild. J., 1:103–118.

Strelkov, P. P. 1969. Migratory and stationary bats (Chiroptera) of the European part of the Soviet Union. Acta Zoologica Cracov., 14:394–439.

Sykes, W. H. 1831. Catalogue of the Mammalia of Dukhum (Deccan); with observations on their habits, etc., and characters of new species. Proc. Zool. Soc. Lond., 1831:99–105.

Talbot, L. M. 1960. A look at threatened species: a report on some animals of the Middle East and southern Asia which are threatened with extinction. Oryx, 5:153–293.

Tchernov, E. 1968. Succession of rodent faunas during the upper Pleistocene of Israel. Paul Parey, Hamburg, 152 pp.

———. 1975. Rodent faunas and environmental changes in the Pleistocene of Israel. Pp. 331–362 *in* Rodents in desert environments (I. Prakash and P. K. Ghosh, eds.). Dr. W. Junk, The Hague, 624 pp.

———. 1979. Polymorphism, size trends and Pleistocene paleoclimatic response of the subgenus *Sylvamus* (Mammalia: Rodetia) in Israel. Isr. J. Zool., 28:131–159.

———. 1988. The biogeographical history of the southern Levant. Pp. 389–410 *in* The zoogeography of Israel (Y. Yom-Tov and E. Tchernov, eds.). Dr. W. Junk Publishers, Dordrecht, Netherlands, viii + 600 pp.

Tchernov, E., T. Dayan, and Y. Yom-Tov. 1987*a*. The paleogeography of *Gazella gazella* and *Gazella dorcas* during the Holocene of the southern Levant. Isr. J. Zool., 34:51–59.

Tchernov, E., L. Ginsburg, P. Tassy, and N. F. Goldsmith. 1987*b*. Miocene mammals of the Negev (Israel). J. Vertebr. Paleontol., 7:284–310.

Thalen, D. C. P. 1975. The caracal lynx (*Caracal caracal schmitzi*) in Iraq: Earlier and new records, habitat & distribution. Bull. Nat. Hist. Res. Cent., Univ. Baghdad, 6:1–24.

Theodor, O., and M. Costa. 1967. A survey of the parasites of wild mammals and birds in Israel. Part One. Ectoparasites. The Israel Academy of Sciences and Humanities, Jerusalem, 117 pp.

Theodor, O., and A. Moscona. 1954. On bat parasites in Palestine. Parasitology, 44:157–245.

Thesiger, W. 1964. The marsh Arabs. Longmans, Green and Co., London, 242 pp.

Thomas, O. 1892. Description of a new species of *Meriones* from Palestine. Ann. Mag. Nat. Hist., ser. 6, 9:147–149.

———. 1903. On the species of the genus *Rhinopoma*. Ann. Mag. Nat. Hist., (ser. 7), 11:496–499.

———. 1913. Some new feræ from Asia and Africa. Ann. Mag. Nat. Hist., ser. 8, 12:88–92.

———. 1918. The hedgehogs of Palestine and Asia Minor. Ann. Mag. Nat. Hist., ser 9, 2:211–213.

———. 1919*a*. Notes on gerbils referred to the genus *Meriones*, with description of new species and subspecies. Ann. Mag. Nat. Hist., ser. 9, 3:263–273.

———. 1919*b*. Some new mammals from Mesopotamia. J. Bombay Nat. Hist. Soc., 26:745–749.

———. 1919*c*. Two new gerbils from Sinai. Ann. Mag. Nat. Hist., ser. 9, 3:559–560.

———. 1919*d*. The white-toothed shrews of Palestine. Ann. Mag. Nat. Hist., ser. 9, 3:32–33.

———. 1920. A new shrew and two new foxes from Asia Minor and Palestine. Ann. Mag. Nat. Hist., ser. 9, 5:119–122.

———. 1922. A new jird (*Meriones*) from southern Palestine. Ann. Mag. Nat. Hist., ser. 9, 10:552.

Tohme, G., and H. Tohme. 1983. Quelques nouvelles données sur le statut actuel de l'hyene *Hyaena hyaena syriaca* Matschie, 1900 (Carnivora) au Liban. Mammalia, 47:345–351.

Tristram, H. B. 1866a. The land of Israel; a journal of travels in Palestine, undertaken with special reference to its physical character. Society for the Promotion of Christian Knowledge Publ., London, 649 pp.

———. 1866b. Report on the mammals of Palestine. Proc. Zool. Soc. Lond., 1866:84–93.

———. 1867. The natural history of the Bible. Society for the Promotion of Christian Knowledge Publ., London, 515 pp.

———. 1876. Notes on the discovery of the roebuck (*Cervus capreolus*) in Palestine. Proc. Zool. Soc. Lond., 1876:420–421.

———. 1877. Notes on *Eliomys melanurus* and on some other rodents in Palestine. Proc. Zool. Soc. Lond., 1877:40–42.

———. 1884. The survey of western Palestine. The fauna and flora of Palestine. Palestine Expl. Fund, London, Comm. 455 pp. (mammals on pp. 1–30.)

Twigg, G. I. 1978. The role of rodents in plague dissemination: a worldwide review. Mammal Rev., 8:77–110.

Vassart, M., L. Granjon, and A. Greth. 1991. Genetic variability in the Arabian oryx (*Oryx leucoryx*). Zoo Biol., 10:399–408.

Vassart, M., L. Granjon, A. Greth, J. F. Asmodé, and E. P. Cribiu. 1992. Genetic research in captive Arabian oryx in Saudi Arabia. Pp. 595–598 *in* Proceedings of the International Symposium "Ongulés/Ungulates 91" (F. Spitz, G. Janeau, G. Gonzalez, and S. Aulagnier, eds.). S.F.E.R.M.-I.R.G.M., Toulouse, Paris, 625 pp.

Vassart, M., L. Granjon, A. Greth, and F. M. Catzeflis. 1994. Genetic relationships of some *Gazella* species: an allozyme study. Z. Säugetierkd., 59:236–245.

Vassart, M., A. Greth, V. Durand, and E. P. Cribiu. 1993. Chromosomal polymorphism in sand gazelles (*Gazella subgutterosa marica*). J. Hered., 84:478–481.

Vesey-Fitzgerald, D. 1952. Wild life in Arabia. Oryx, 1(5):232–235.

———. 1953. Notes on some rodents from Saudi Arabia and Kuwait. J. Bombay Nat. Hist. Soc., 51:424–428.

Vogel, P. 1977. Neue Nachweise der *Rhinopoma hardwickei* (Chiroptera) aus Westafrika. Bonn. Zool. Beitr., 28:228–231.

Vogel, P., T. Maddalena, and F. Catzeflis. 1986. A contribution to the taxonomy and ecology of shrews from Crete and Turkey (*Crocidura zimmermani* and *C. suaveolens*). Acta Theriol., 31:537–545.

von Lehmann, E. 1965 [1966]. Über die Säugetiere im Waldgebiet NW-Syriens. Sitzungsber. Ges. Naturf. Fr. Berl. (N. F.), 5:22–38.

———. 1966. Taxonomische Bemerkungen zur Säugerausbeute der Kumerloeveschen Orientreisen 1953–1965. Zool. Beitr., 12:251–317.

Vorontsov, N. N., and N. A. Malygina. 1973. Karyological studies in jerboas and birch mice (Dipodoidea, Rodentia, Mammalia). Caryologia, 26:193–212.

Vujoševic, M. and J. Blagojevic. 1994. New localities with B chromosomes in *Apodemus flavicollis* (Rodentia, Mammalia). Arch. Biol. Sci. 46 (3–4):15.

Wagner, A. 1839. Mammals collected by von Schubert on his journey to Egypt and Palestine in 1836–37. Gelehrte Anzeigen (Munich), 8:297–300.

―――. 1840. Beschreibung einiger neuer Nager, welche auf der Herrn Hofrath v. Schubert gesammelt wurden, mit Bezugnahme auf einige andere werwandte Formen. Abh. Math.-Phys. Cl. Klg. Bayar. Akad. Wiss., München, 3:175–218.

Wahrman, J., and R. Goitein. 1972. Hybridization in nature between two chromosome forms of spiny mice (Rodentia: Murinae). Chromosomes Today, 3:228–237.

Wahrman, J., and P. Gourevitz. 1973. Extreme chromosome variability in a colonizing rodent. Chromosomes Today, 4:399–424.

Wahrman, J., C. Richler, R. Gamperl, and E. Nevo. 1985. Revisiting *Spalax*: mitotic and meiotic chromosome variability. Isr. J. Zool., 33:15–38.

Wahrman, J., C. Richler, and U. Ritze. 1988. Chromosomal considerations in the evolution of the Gerbillinae of Israel and Sinai. Chapter 16. Pp. 439–485 *in* The zoogeography of Israel (Y. Yom-Tov and E. Tchernov, eds.). Dr. W. Junk Publishers, Dordrecht, Netherlands, viii + 600 pp.

Wahrman, J., and A. Zahavi. 1955. Cytological contributions to the phylogeny and classification of the rodent genus *Gerbillus*. Nature, 175:600–602.

Wassif, K. 1949. Trident bat (*Asellia tridens*) in the Egyptian oasis of Kharga. Bull. Zool. Soc. Egypt, 8:9–12.

―――. 1954a. The bushy tailed gerbil, *Gerbillus calurus* Thomas, of south Sinai. J. Mammal., 35:243–248.

―――. 1954b. On a collection of mammals from northern Sinai. Bull. Inst. Deserte d'Egypte, 3 (for 1953):107–118.

―――. 1956. Studies on gerbils of the subgenus *Dipodillus* recorded from Egypt. Univ. Ain Shams Sci. Bull., 1:173–194.

Wassif, K., and H. Hoogstraal. 1959. On a collection of mammals from the Egyptian oases of Bahariya and Farafra. Univ. Ain Shams Sci. Bull, 6:137–146.

Wassif, K., R. G. Lutfy, and S. Wassif. 1969. Morphological, cytological and taxonomic studies of the rodent genera *Gerbillus* and *Dipodillus* from Egypt. Proc. Egypt. Acad. Sci., 22:77–96.

Wassif, K., and S. Soliman. 1979. The food of some wild rodents in the western desert of Egypt. Bull. Zool. Soc. Egypt, 29:43–51.

Weber, N. 1956. Notes on Iraq Insectivora and Chiroptera. J. Mammal., 36:123–126.

Weisbein, Y. 1988. The caracal: a study in Israel. Isr. Land Nat., 14:223–226.

Wetzel, R. M., and D. A. Schlitter. 1970. Sana I. Atallah. J. Mamm., 51:435.

Whybrow, P. J. 1984. Geological and faunal evidence from Arabia for mammal 'migrations' between Asia and Africa during the Miocene. Cour. Forschungsinst. Senckenb., 1984:189–198.

Wilson, D. E. and D. M. Reeder, eds. 1993. Mammal species of the world: a taxonomic and geographic reference. 2nd ed., Smithsonian Institution Press, Washington, D.C., xviii + 1206 pp.

Wisner, A., J. P. Legouix, and F. Petter. 1954. Etude histologique de l'orielle d'un Rongeur a bulles tympaniques hypertrophiées: *Meriones crassus*. Mammalia, 18:371–379.

Wooley, C. L. 1934. Ur excavations. II. The Royal Cemetery. British Mus. and Univ. Mus. Univ. Pennsylvania Publ., 604 pp.

Wurster, D. H. 1972. Sex-chromosome translocations and karyotypes in bovid tribes. Cytogenetics, 11:197–207.

Yaseen, A. E., H. A. Hassan, and L. S. Kawashti. 1994. Comparative study of the karyotypes of two Egyptian species of bats, *Taphozous perforatus* and *Taphozous nudiventris* (Chiroptera, Mammalia). Experientia 40:1111–1114.

Yerbury, J. W., and O. Thomas. 1895. On the mammals of Aden. Proc. Zool. Soc. Lond., 1895:542.

Yoffe, G. 1968. Large mammals in Israel. Oryx, 9:423.

Yom-Tov, Y. 1967. On the taxonomic status of the hares (genus *Lepus*) in Israel. Mammalia, 31:246–259.

———. 1988. 14. The zoogeography of the birds and mammals of Israel. Pp. 389–410 *in* The zoogeography of Israel (Y. Yom-Tov and E. Tchernov, eds.). Dr. W. Junk Publishers, Dordrecht, Netherlands, viii + 600 pp.

Yom-Tov, Y., D. Makin, and B. Shalmon. 1992a. The biology of *Pipistrellus bodenheimeri* (Microchiroptera) in the Dead Sea area of Israel. Z. Säugetierkd., 57:65–69.

Yom-Tov, Y., D. Makin, and B. Shalmon. 1992b. The insectivorous bats (Microchiroptera) in the Dead Sea area, Israel. Isr. J. Zool., 38:125–137.

Yom-Tov, Y., and B. Shalmon. 1989. First record of *Taphozous perforatus* in Israel. Mammalia, 53:661–662.

Yunker, C. E., and S. S. Guirgis. 1969. Studies on rodent burrows and their ectoparasites in the Egyptian desert 1: environment and microenvironment; some factors influencing acarine distributions. J. Egypt. Publ. Health Assoc., 44:498–542.

Zafriri, A., and S. Hellwing. 1973. The common shrew in Israel, *Crocidura russula monacha:* taxonomic aspects and data on reproduction under field conditions. Isr. J. Zool., 22:21.

Zahavi, A., and J. Wahrman. 1957. The cytotaxonomy, ecology and evolution of the gerbils and jirds of Israel (Rodentia: Gerbillinae). Mammalia, 21:341–380.

Zeuner, F. E. 1955. The goats of early Jericho. Palestine Expl. Quart., 86:70–86.

———. 1963. A history of domesticated animals. Hutchinson, London, 560 pp.

Zima, J. 1978. Chromosomal characteristics of Vespertilionidae from Czechoslovakia. Acta Sci. Nat. Acad. Sci. Bohemoslov. Brno, 12:1–38.

———. 1982a. [Karyotypes of three species of horsehoe bats (*Rhinolophus ferrumequinum, Rh. hipposideros, Rh. euryale*) from Czechoslovakia.] Lynx (Prague), 21:121–124 (in Czech, English summary).

———. 1982b. Chromosomal homology in the complements of bats of the family Vespertilionidae. II. G-band karyotypes of some *Myotis, Eptesicus* and *Pipistrellus* species. Folia Zool. (Zool. Listy), 31:31–36.

Zima, J., J. Cerveny, I. Horácccek, A. Cervena, and K. Prucha. 1991. Standard karyology of eighteen species of bats (Rhinolophidae, Vespertilionidae, Molossidae) from Eurasia. Myotis, 29:31–34.

Zima, J., and B. Král. 1984. Karyotypes of European mammals. Acta Sci. Nat. Acad. Sci. Bohemoslov. Brno, 18:(7):1–51, (8):1–62, (9):1–51.

Zima, J., M. Volleth, I. Horáček, J. Cerveny, A. Cervena, K. Prucha, and M. Macholàn. 1992a. Comparative karyololgy of rhinolophid bats (Chiroptera, Rhinolophidae). Pp. 229–236 in Prague studies in mammalogy (I. Horáček, and V. Vohral'k, eds.). Charles University Press, Praha, 245 pp.

Zima, J., M. Volleth, I. Horáček, J. Cerveny, and M. Macholàn. 1992b. Karyotypes of two species of bats, *Otonycteris hemprichi* and *Pipistrellus tramatus* (Chiroptera, Vespertilionidae). Pp. 237–242 in Prague studies in mammalogy (I. Horáček, and V. Vohral'k, eds.). Charles University Press, Praha, 245 pp.

Zohary, M. 1973. Geobotanical foundations of the Middle East. B. Fisher Verlag, Stuttgart, 739 pp.

Zulueta, J. D., S. Nasrallah, J. S. Karam, A. R. Anani, G. K. Sweatman, and D. A. Muir. 1971. Finding of tick-borne relapsing fever in Jordan by the Malaria Eradication Service. Ann. Trop. Med. Parasitol., 65:491–495.

Appendix

Localities and Coordinates

This locality index lists major localities from which mammals where obtained. It is to be used for the purposes of finding these localities in a map and is not intended for precise coordinates. Generic Arabic and Hebrew terms were used with locality names in many previous publications. This locality index is alphabetized by the main name followed by any generic modifiers or names. The following generic names are most commonly used. They are listed here with their English equivalents:

Ain/Ein/En (Ar, Heb): spring
Beit/Bet (Ar, Heb): house of
Beer/Bir (Ar, Heb): well
Deir (Ar): monastery
Faydhat (Ar): wadi, depression
Har/Hor (Heb): mountain
Horbat or Hurvat (Heb): ruin
Kfar/Kefar (Ar, Heb): village
Khirbet/Khorbat (Ar): ruin
Kiryat/Qiryat/Qaryat (Ar, Heb): village
Jebel or Jabal, pl. (Ar): mountain, hill
Jisr (Ar): bridge
Makhtesh/Hamakhtesh (Heb): depression
Mogharet (Ar): cave
Nahal (Heb): wadi
Nebi (Ar): prophet
Qasr (Ar): palace
Ramat/Ramot (Heb): hill, plateau
Sede (Ar, Heb): grazing area
Tel (Ar): hill
Wadi (Ar): wadi, a dry river valley

The Arabic modifier Al or El is frequently used by non-Arabic speakers in alphabetizing their localities. This modifier also changes pronunciation depending on the modified noun; thus Al Safi is pronounced As Safi Es Safi. Again, this becomes a nightmare for those not versed in the local language. In the index, I placed all modifiers and generic names *after* the main name. Alternative names or spellings are given in parentheses.

Locality	N Latitude	E Longitude
'Abde ('Avdat)	30° 48'	34° 46'
Abiad, Wadi El (Nahal Lavan)	30° 56'	34° 30'
Abu Anseer (not found)		
Abu Aweigila (SINAI)	30° 50'	34° 07'
Abu Durba (SINAI)	28° 29'	33° 20'
Abu Ghosh	31° 48'	35° 06'
Abu Hareireh, Tel	31° 23'	34° 36'
Abu Heran, Ain (N. Aqaba, not found)		
Abu Hureyra, Tel (see Abu Hareireh, Tel)		
Abu Kebir	32° 03'	34° 46'
Acco (Akko)	32° 55'	35° 05'
Acre Plain (see also Akko)	32° 55'	35° 08'
Adullam, Cave (Mogharet Khureitun)	31° 39'	35° 14'
Afaim, Nahal (Wadi Abu Ali)	33° 00'	35° 24'
'Afula	32° 38'	35° 20'
Agur, Sand Dunes	31° 00'	34° 24'
Ahuzam (see Shalwa)	31° 33'	34° 46'
'Ain Musa, Wadi	31° 48'	35° 40'
'Ajjul, Tel Al (not found)		
'Ajloun (Ajlun) 760 m	32° 20'	35° 45'
'Akaba (see 'Aqaba)	29° 31'	35° 00'
Akbara	32° 56'	35° 30'
Akko	32° 55'	35° 05'
Allenby Bridge (King Hussein Bridge)	31° 52'	35° 32'
Allonim, Kibbutz	32° 43'	35° 09'
Alma Cave	33° 03'	35° 30'
'Aluk, Al	32° 10'	35° 55'
Amatsya, Nahal	31° 32'	34° 55'
'Amman	31° 57'	35° 56'
'Amman Airport ca.	31° 50'	35° 57'
Ammur, Khirbet El	31° 47'	35° 06'
Amrah, Qasr	31° 48	36° 35
Amram, Nahal	29° 37'	34° 58'
Amud, Wadi (Nahal)	32° 51'	35° 32'
'Anan, Beit	31° 51'	35° 06'
Anavim, Kiryat (Qiryat Anavim)	31° 48'	35° 07'
'Aqaba (sea level)	29° 31'	35° 00'
'Aqraba	32° 44'	35° 48'
Aqua Bella	31° 47'	35° 07'
'Ara, Wadi	32° 30'	35° 05'
'Araba, Beit	31° 42'	34° 57'
'Araba, Wadi (Rift Valley section between the Dead Sea and 'Aqaba)		
Arad, Tel (Tel Shoket)	31° 17'	35° 08'
Arbel (Ruins)	32° 49'	35° 29'
'Arif, Har (Jebel Ureif)	30° 25'	34° 44'
'Arish, Al	31° 08'	33° 48'
Arnon, Wadi (see Mujib, Wadi)	31° 27'	35° 42'
Arod, Har and Nahal	30° 23'	34° 54'
Artuft (see Hartuv)	31° 46'	35° 00'
Arugot, Nahal	31° 28'	35° 24'
Arugot, Village	31° 44'	34° 46'

Locality	N Latitude	E Longitude
Aruma, Kfar	32° 09'	35° 19'
Arus, Wadi	33° 07'	35° 34'
Aseikhem, Jebel	31° 55'	36° 55'
Ashalim	31° 03'	35° 21'
Ashdod	31° 48'	34° 38'
Ashdot-Yaaqov (Nr.)	32° 40'	35° 35'
Ashona (not found)		
'Ashqelon	31° 40'	34° 35'
'Asluj, Bir (Beer Mashabim)	31° 01'	34° 45'
'Asluj, Wadi (see Wadi el Abiad)	30° 56'	34° 30'
'Assal, Wadi (Nahal Siyon)	32° 02'	35° 08'
Atmash, Ain Al	31° 48'	36° 48'
'Auja	30° 52'	34° 26'
'Auja, Ain	31° 56'	35° 17'
'Avdat ('Abde, 'Avedat)	30° 48'	34° 46'
Avraham, Sede (N.W. of Yotvata, Negev)	29° 53'	35° 03'
Awlad Ali (SINAI)	30° 52'	34° 04'
Ayelot Hashakhar (Hashahar)	33° 01'	35° 34'
Ayl	30° 13'	35° 32'
Ayyalon, Nahal (see Masrura)	31° 59'	34° 54'
Azraq ed Druz (Azraq Shamali)	31° 53'	36° 50'
Azraq Shishan (Azraq Janubi)	31° 50'	36° 49'
Bab el Wad	31° 49'	35° 02'
Baghak, Wadi Umm	31° 11'	35° 19'
Barha, Al	32° 34'	35° 50'
Barza, Khirbet	33° 01'	35° 19'
Bashan (not found; ? = Beisan)		
Bassa, El	31° 56'	35° 46'
Bayir	30° 46'	36° 41'
Beeri	31° 26'	34° 30'
Beerotayim, Nahal	30° 45'	34° 29'
Beersheba (Beer el Saba')	31° 15'	34° 47'
Beisan (Beit Shean)	32° 30'	35° 30'
Belka (approx)	32° 20'	37° 30'
Ben Shemen	31° 57'	34° 56'
Benei Beraq	32° 05'	34° 52'
Berurim	31° 46'	34° 47'
Betanan (see Beit Anan)	31° 51'	35° 06'
Bethlehem	31° 42'	35° 12'
Bezet, Nahal (see Wadi Karkara)	33° 05'	35° 06'
Bir Ein (not found)		
Bira, El	31° 54'	35° 13'
Bira, Wadi	32° 36'	35° 34'
Birak Sulayman (see Solomon's Quarries)	31° 41'	35° 10'
Birket Ramadan (approx.)	32° 15'	34° 50'
Bitan Aharon (= Beit Hayariyi)	32° 22'	34° 52'
Boqeq, Ain (Ain Um Baghak)	31° 08'	35° 22'
Boqeq, Nahal (Wadi Um Baghak)	31° 11'	35° 19'
Boqer, Sede	30° 52'	34° 46'
Bosmat, Nahal	29° 46'	34° 54'
Bteha, Jebel (see Tabgha)	32° 52'	35° 32'

Locality	N Latitude	E Longitude
Busayta, Al (SAUDI ARABIA)	29° 46'	40° 56'
Buweirdeh, Ayun	30° 35'	35° 19'
Buzieh (near Tel el Qadi)	33° 14'	35° 39'
Caesarea (see Qeisariya)	32° 30'	34° 53'
Capernaum (Kapharnaum, not found)		
Carmel Caves	32° 40'	34° 58'
Dab'ah	31° 36'	36° 04'
Dafna	33° 14'	35° 38'
Dagan, Beit	32° 00'	34° 53'
Dalam, Wadi	33° 04'	35° 10'
Damia, Jisr (Bridge)	32° 06'	35° 32'
Dan (Tel el Qadi)	33° 14'	35° 39'
Darajah, Wadi	31° 34'	35° 25'
Dareiya	32° 27'	35° 51'
Dawid, Nahal (Wadi Sudeir)	31° 28'	35° 24'
Dcheier, Ain (not found, see el Dschuheijir)		
Deir 'Alla, ppl.	32° 12'	35° 37'
Deir El Balah (GAZA, not found)		
Dekel	31° 10'	34° 21'
Devira	31° 25'	34° 49'
Dhana	30° 41'	35° 37'
Dhuhayba	31° 31'	35° 46'
Dhulayl, Wadi (not found)		
Dibbine Forest Park	32° 15'	35° 49'
Dimona	31° 04'	35° 01'
Dimra	31° 34'	34°
Dishon, Nahal (Wadi Ouba)	33° 05'	35° 31'
Disi, Al (60 mi E Dead Sea, ? = Disi)		
Disi, Wadi Rum Area	29° 37'	35° 33'
Djesi (see Al Disi)	29° 37'	35° 33'
Doleh, Beir El (Allen, 1915, not found)		
Doqeq	31° 00'	35° 22'
Dor (see Tantura)	32° 37'	34° 55'
Dorot	31° 30'	34° 39'
Dosi, Deir (Deir Ibn Ubaid)	31° 39'	35° 14'
Dschuheijir, Ain El, NW Dead Sea (Matschie, 1912, not found)		
Eilat (Elat)	29° 33'	34° 57'
Eilon (Elon)	33° 04'	35° 13'
Ekron	31° 51'	34° 50'
Elbiran (see El Bira)	31° 54'	35° 13'
Eliyahu, Sede	32° 27'	35° 31'
Elon	33° 04'	35° 13'
Elot	29° 35'	34° 58'
Emeq (by Theodor, not a specific locality)		
Emeq Yizrael (see Jezreel)	32° 36'	35° 13'
Emeq Zevulun	32° 52'	35° 05'
Emmaus Kubebe (see Qubeiba)	31° 50'	35° 08'
Eqron, Kfar	31° 52'	34° 50'
Esdraelon Plain (see Jezreel)	32° 36'	35° 13'
Eshel Hannassi	31° 19'	34° 42'
Even Sappir	31° 46'	35° 08'

Locality	N Latitude	E Longitude
Even Yehuda	32° 16'	34° 53'
Faidhat ed Dahikiya	31° 38'	37° 08'
Fajjar, Beit	31° 38'	35° 09'
Falik, Khirbet	32° 14'	34° 50'
Falik, Umm (upper Galilee, not found)		
Fallah, Wadi (not found)		
Fara, Ain	31° 50'	35° 18'
Faria, Wadi	32° 14'	35° 23'
Fashkhah, Ain (Ain Fashka)	31° 43'	35° 27'
Fauar, Wadi and Ain (Fawwar)	31° 18'	35° 40'
Feiran Oasis (SINAI)	28° 42'	33° 38'
Fidan, Wadi Al	30° 40'	35° 26'
Finan, Wadi Al (see Fidan)	30° 40'	35° 26'
Fo'ara (Faw'ara)	32° 37'	35° 46'
Fuhays, Al	32° 01'	35° 46'
Furtaga, Ain El	29° 03'	34° 33'
Ga'aton, Wadi or Nahal	33° 00'	35° 10'
Galim, Nahal and Cave (see Carmel)	32° 40'	34° 58'
Gaza	31° 30'	34° 28'
Gazzah, Wadi (Ghazzah)	31° 25'	34° 25'
Gebel Dhalfa (Dalfah, SINAI)	30° 45'	34° 12'
Gebel Egma (SINAI)	29° 12'	34° 02'
Gebel el Themed (Yithmid, SINAI)	29° 42'	34° 23'
Gebel Umm Afruth (SINAI)	29° 10'	34° 15'
Gedeirat, Ain El (Ain Gedeirat, SINAI)	30° 39'	34° 26'
Gedera	31° 49'	34° 46'
Gedi, Ain	31° 28'	35° 23'
Gerizim, Mt.	32° 12'	35° 16'
Gevim	31° 30'	34° 36'
Gevulot	31° 13'	34° 27'
Ghadyan, Ain (Ain Radian)	29° 53'	35° 03'
Gharandal	30° 05'	35° 12'
Ghazal, Ain	31° 42'	35° 37'
Ghazaleh, Jebel	30° 55'	35° 40'
Ghidyan, Ein (Ein Radian)	29° 53'	35° 03'
Ghor (the lower Jordan Valley)		
Gilat	31° 20'	34° 38'
Gilboa, Mt.	32° 31'	35° 24'
Gilead, Mt. and region (Jebel Osha)	32° 04'	35° 46'
Gileadi, Kefar	33° 15'	35° 34'
Gilgal (Jaljuliya)	32° 09'	34° 57'
Gimzo	31° 56'	34° 57'
Ginosar	32° 51'	35° 31'
Givat Ada	32° 31'	35° 00'
Givat Koah	32° 02'	34° 56'
Givat Ram	31° 47'	35° 12'
Givat Yeshayahu	31° 40'	34° 57'
Givatayim	32° 05'	34° 50'
Giv'ot Zed	32° 42'	35° 07'
Giyyora, Har	31° 45'	35° 05'
Gleidat, Ain (see Jaladat)	30° 42'	35° 37'

Locality	N Latitude	E Longitude
Glil Yam (Gelil Yam)	32° 09'	34° 50'
Greigroh, Al (not found)		
Grofit	29° 56'	35° 04'
Guierah (see Quweirah)	29° 48'	35° 18'
Guvrin, Beit (Beit Jibrin)	31° 36'	34° 54'
H4, station (Ruwayshid)	32° 30'	38° 12'
H5, station (Safawi)	32° 12'	37° 07'
Hadassim	32° 17'	34° 53'
Hadera	32° 26'	34° 55'
Hadid	31° 58'	34° 55'
Hadraj	30° 21'	37° 46'
Hahula Nature Reserve (see Hula)	33° 08'	35° 37'
Hai-Bar Reserve (see Yotvata)		
Haifa	32° 50'	35° 00'
Hakerem, Bet	31° 47'	35° 12'
Hallabat, Qasr Al	32° 06'	36° 20'
Halutsa (see Haluza)	31° 05'	34° 39'
Haluza	31° 05'	34° 39'
Hamadya	32° 31'	35° 31'
Hamapil	32° 22'	34° 59'
Hamda, Nahal (Wadi Abu Hamda)	30° 14'	35° 04'
Hammam, Wadi	32° 39'	35° 29'
Hammam Zarah	31° 35'	35° 34'
Hanan, Beit	31° 56'	34° 47'
Hananya, Beit (Hananyah)	32° 32'	34° 55'
Hanina, Beit	31° 50'	35° 12'
Hanita	33° 05'	35° 10'
Hanjurat al Gattar, Wadi	30° 47'	34° 58'
Hanna, Wadi (Nablus)	32° 13'	35° 16'
Haql	29° 17'	34° 57'
Harduf, Nahal and Har	31° 23'	35° 16'
Har'el	31° 49'	34° 57'
Harima	32° 38'	35° 53'
Harmul (see Sahil Harmul)		
Harod, En (Ain Herod)	32° 34'	35° 23'
Haroe, Kfar	32° 24'	34° 55'
Harta	32° 42'	35° 51'
Hartuv	31° 46'	35° 0'
Hasa, Al	30° 47'	35° 56'
Hasa, Wadi el	30° 57'	35° 45'
Hashofet, Ramat	32° 36'	35° 06'
Hassan, Tel (not located)		
Hatira, Wadi	30° 52'	35° 07'
Hatrura, Jebel	31° 14'	35° 20'
Hatseva (see Ein Husab)	30° 48'	35° 15'
Hawa	30° 43'	34° 55'
Hawfa, Ruin	32° 29'	35° 50'
Hawran	31° 33'	35° 37'
Hazera	31° 00'	35° 12'
Hazim, Al	31° 35'	37° 15'
Hazorea	32° 38'	35° 07'

Locality	N Latitude	E Longitude
Hebron	31° 32'	35° 06'
Helqum, Qasr Al (NE of H2, IRAQ)	33° 16'	40° 40'
Hemed, Ain	32° 01'	34° 50'
Hemma, El (ca.)	32° 38'	35° 38'
Hephzibah	32° 28'	34° 54'
Hermon, Mt. and Valley	33° 14'	35° 39'
Herod, Ain (not found)		
Herodes, Mt. (Hordos)	31° 40'	35° 14'
Herzelia	32° 10'	34° 50'
Hever, Nahal (see Wadi Khabra)	31° 24'	35° 23'
Hezme (see Hizma)	31° 50'	35° 16'
Hibar (not found)		
Hindis, Bir (Bir Ora)	29° 40'	34° 53'
Hiran, Umm Al (not found)		
Hittin	32° 48'	35° 28'
Hiyon, Nahal (Wadi Haiyani)	30° 11'	35° 03'
Hizma	31° 50'	35° 16'
Holon	32° 01'	34° 46'
Hor, Mt. (see Jebel Maderah)	30° 45'	34° 58'
Horbat Saadon	31° 02'	34° 36'
Hordos, Har (Mt. Herodes)	31° 40'	35° 14'
Horns of Hittin (see Hattin)	32° 48'	35° 28'
Horshat Ha 'arbaim (see Horshat, Tel)	33° 13'	35° 38'
Horshat, Tel	33° 13'	35° 38'
Horvot, Avedat	30° 48'	34° 46'
Hula (Lake and Nature Reserve)	33° 06'	35° 37'
Hulata	33° 03'	35° 37'
Huldah	31° 50'	34° 53'
Hummar, Al, ppl.	32° 01'	35° 49'
Hurfeish	33° 01'	35° 21'
Hurvat Bar Zeit (Khirbet Barza)	33° 01'	35° 18'
Husab, Ain	30° 48'	35° 15'
Husn, Al	32° 29'	35° 53'
Ibbin	32° 22'	35° 49'
Ibn Hammad, Wadi	31° 18'	35° 31'
Ibn Ubaid, Deir	31° 39'	35° 14'
Inab, Al	29° 58'	36° 54'
Iraq Al Amir	31° 55'	35° 45'
Irbid	32° 33'	35° 51'
Ishtafayna	32° 22'	35° 45'
Jabbok	32° 11'	35° 46'
Jaffa	32° 02'	34° 45'
Jafr, El	30° 16'	36° 11'
Ja'iz, Kafr	32° 37'	35° 49'
Jaladat, Ain	30° 42'	35° 37'
Jalama	32° 23'	35° 01'
Jarmaq, Jebel (=Har Meron)	33° 00'	35° 25'
Jarra, Wadi	31° 18'	35° 35'
Jawa, 1000 m	32° 20'	37° 02'
Jbeiha (also Jubeiha and Sweileh)	32° 02'	35° 50'
Jdeid, Bir	30° 28'	34° 44'

Locality	N Latitude	E Longitude
Jebbal Mountains	30° 50'	35° 35'
Jenin	32° 28'	35° 18'
Jerash (Jarash)	32° 17'	35° 53'
Jerash Refugee Camp	32° 16'	35° 52'
Jericho	31° 51'	35° 27'
Jerusalem	31° 47'	35° 13'
Jezreel Plain	32° 36'	35° 13'
Jimal, Umm Al (Ruin)	32° 20'	36° 22'
Jiza	31° 42'	35° 57'
Jubeiha (Jbeiha)	32° 02'	35° 50'
Kabri	33° 01'	35° 09'
Kadi, Tel El (Tel el Qadi, Dan)	33° 15'	35° 39'
Kafrein, Wadi	31° 50'	35° 35'
Kafrun (not found, ? = Kafrein)	31° 50'	35° 35'
Kallia	31° 46'	35° 30'
Kama, Bet (see Beit Qama)	31° 27'	34° 45'
Karak, El	31° 11'	35° 42'
Karantal (see Jebel Qarantal)	31° 52'	35° 26'
Karem, Tel (see Tulkarm)	32° 19'	35° 02'
Karkara, Wadi (Nahal Numer)	33° 05'	35° 06'
Karkom, Nahal	30° 22'	34° 50'
Karyatein (SYRIA)	33° 07'	36° 58'
Kasle, Khirbet (20 km W Jerusalem, not found)		
Kasr Hadschla	31° 49'	35° 24'
Kastal, Al	31° 45'	35° 56'
Katana (see Qatana)	31° 49'	35° 06'
Katrana, El (El Qatrana)	31° 14'	36° 03'
Kedron Gorge (see Qidron)	31° 43'	35° 19'
Kefren, Wadi (see Wadi Kufrein)	31° 52'	35° 42'
Kelekh, Wadi/Nahal	31° 31'	34° 48'
Kelt, Wadi (Qelt)	31° 51'	35° 25'
Kerak, Wadi	31° 15'	35° 38'
Kerem, Ain	31° 46'	35° 10'
Kerem Shalom	31° 12'	33° 17'
Kern, Wadi (see Wadi Kurn)	33° 03'	35° 06'
Keziv, Nahal (see Wadi Kurn)	33° 03'	35° 06'
Khabra Abu Guzour (Gazour, SINAI)	31° 00'	34° 20'
Khabra, Wadi	31° 24'	35° 23'
Khan Yunis	31° 21'	34° 18'
Khanzira	32° 28'	35° 42'
Kharana, Qasr Al	31° 44'	36° 28'
Khunayzir, Wadi and Village	30° 53'	35° 26'
Khureitun, Mogharet	31° 39'	35° 14'
Kineret (Kinereth)	32° 43'	35° 33'
King Hussein Bridge (see Allenby Bridge)		
Kishon River (see Qishon River)	32° 47'	35° 03'
Kufranjah	32° 18'	35° 42'
Kufrein, Wadi	31° 52'	35° 42'
Kuntila, Al (SINAI)	30° 00'	34° 41'
Kuraimah	32° 16'	35° 36'
Kurdani, Tel Al	32° 52'	35° 09'

Locality	N Latitude	E Longitude
Kurn, Wadi (= Khurn, Qarn)	33° 03'	35° 06'
Kurnub (Mamshit)	31° 01'	35° 04'
Kziv, Nahal (Wadi Qarn)	33° 03'	35° 06'
Lachish (Lachis, Nachla, Sukreir)	31° 34'	34° 51'
Lahav	31° 23'	34° 52'
Lahavot Habashan	33° 08'	35° 39'
Latrun	31° 50'	34° 59'
Lavan, Nahal (see Abiad, Wadi)	30° 56'	34° 30'
Lebweh, Wadi (not found)		
Ledjah (not found)		
Lejun	32° 34'	35° 11'
Lid, Beit	32° 19'	34° 55'
Lisan	31° 16'	35° 28'
Lod	31° 57'	34° 54'
Lubban	32° 02'	35° 02'
Ludd (see Lod)	31° 57'	34° 54'
Luzit Cave (near Beit Guvrin)	31° 41'	34° 53'
Ma'agan Mikhael (or Michael)	32° 33'	34° 55'
Ma'an	30° 11'	35° 43'
Madaba	31° 43'	35° 48'
Maderah, Jebel (Mt. Hor)	30° 45'	34° 58'
Madhkur, Bir	30° 24'	35° 21'
Mafraq	32° 20'	36° 12'
Magdabah (Magdaba, SINAI)	30° 53'	34° 02'
Magen	31° 18'	34° 25'
Maghara, Wadi El	30° 23'	34° 32'
Magor, Al	32° 55'	35° 36'
Mahammadia, Al (Amr, *M. libycus*, not found)		
Mahis	31° 59'	35° 46'
Majdal, Al	32° 14'	35° 51'
Makhtash haGadol	30° 56'	34° 59'
Makhtesh Haqatan	30° 58'	35° 11'
Makhtesh Hatsera (see Makhtesh Haqatan)	30° 58'	35° 11'
Makhtesh Ramon	30° 35'	34° 48'
Mallaha	33° 05'	35° 35'
Mamshit (Kurnub)	31° 01'	35° 04'
Manba, Nahal	33° 00'	35° 22'
Mansura, Al	32° 18'	35° 42'
Ma'oz Hayyim	32° 29'	35° 33'
Maqarim, Dam	32° 43'	35° 53'
Mar Jiryis, Deir (see Wadi Kelt)	31° 51'	35° 25'
Mar Saba, Deir (= monastery)	31° 42'	35° 20'
Marda, Tel	33° 01'	35° 14'
Maresha, Tel (nr. Beit Jibrin)	31° 35'	34° 54'
Marsa' (not found)		
Masada (north)	32° 41'	35° 36'
Masada (south) = (Sebbe)	31° 19'	35° 21'
Masaryk, Kfar	32° 53'	35° 06'
Mashabei Sade (Bir Asluj)	31° 00'	34° 47'
Mashabim, Beer (see Bir Asluj)	31° 01'	34° 45'
Mashari', N. Ghor	32° 15'	35° 35'

Locality	N Latitude	E Longitude
Mashash, Bir	31° 04'	34° 51'
Maslul	31° 19'	34° 35'
Masri, Wadi	29° 32'	34° 55'
Masrura, Wadi	31° 59'	34° 54'
Massua	31° 41'	34° 54'
Mata, Mogharet Al	30° 44'	35° 30'
Ma'yan Zevi	32° 34'	34° 57'
Mazar, Al (Al Shamali)	32° 28'	35° 48'
Mazra'ah	31° 23'	35° 38'
Me'arath Hateamim Cave (see Hartuv)	31° 46'	35° 00'
Megiddo	32° 35'	35° 11'
Meholla (not found,? = Mohila)		
Meir, Beit	31° 47'	35° 02'
Meiron, Mt. (Jarmak)	33° 00'	35° 25'
Melaha	33° 05'	35° 35'
Melikha, Bir (Malicha)	30° 18'	35° 07'
Menachem, Kfar (see Menahem)	31° 44'	34° 50'
Menahamya	32° 40'	35° 33'
Menahem, Kfar	31° 44'	34° 50'
Menayeh, Wadi	29° 47'	34° 58'
Meneya, Wadi (see Wadi Menayeh)	29° 47	34° 58'
Menuha, Beer (see Melikha)	30° 18'	35° 07'
Merhavya	32° 36'	35° 18'
Meron, Mt. (see Jarmak)	33° 0'	35° 25'
Messra, El (see El Mezra'ah)	31° 17'	35° 32'
Metsada (see Masada)	31° 19'	35° 21'
Metullah	33° 16'	35° 35'
Mezad Mukheila (see Mohila)	30° 38'	34° 55'
Mezra'ah, El	31° 17'	35° 32'
Midieh	31° 56'	35° 00'
Migdal Ha'Emeq	32° 41'	35° 14'
Milham, Bir	29° 49'	34° 57'
Miron, Mt. (see Jebel Jermak)	33° 00'	35° 25'
Mishash, Wadi (Nahal Sekher)	30° 09'	34° 44'
Mishmar Ha'Emeq	32° 34'	35° 09'
Mishmar Hanegev	31° 22'	34° 42'
Mitspe Ramon	30° 36'	34° 48'
Mivtahim	31° 15'	34° 24'
Mizpe Yodefat	32° 50'	35° 16'
Moab	31° 15'	35° 45'
Mohila, Mezad	30° 38'	34° 55'
Mogharet Al Roum	32° 16'	35° 52'
Moledot	32° 06'	34° 48'
Montfort	33° 03'	35° 13'
Mor, Nahal (Wadi Murra)	30° 50'	34° 48'
Motza (see Moza)	31° 47'	35° 09'
Mount Scopus	31° 48'	35° 14'
Mount of Temptation (see Jebel Qarantal)	31° 52'	35° 26'
Moza	31° 47'	35° 09'
Muhalla (see Mohila)	30° 38'	34° 55'
Mujib, Wadi	31° 27'	35° 42'

Locality	N Latitude	E Longitude
Mukallik, Wadi Al	31° 46'	35° 29'
Mukhmas	31° 52'	35° 17'
Munaysah, Al	32° 17'	36° 43'
Murra, Wadi	30° 50'	34° 48'
Musa Al Alami	31° 50'	35° 28'
Musa, Ayun (SINAI)	29° 53'	32° 39'
Musa, Nebi/Jebel	31° 47'	35° 26'
Musa, Wadi	30° 20'	35° 29'
Muwaqqar, Al	31° 49'	36° 06'
Nabi Musa	31° 47'	35° 26'
Nabi Rubin (Palmahim)	31° 56'	34° 42'
Nablus	32° 13'	35° 16'
Nadira	32° 19'	35° 39'
Nafkha, Wadi/Nahal (Nafkha)	30° 43'	34° 47'
Nahalal	32° 41'	35° 12'
Naharia	33° 01'	35° 05'
Nahr Rubin	31° 56'	34° 42'
Nahr, Wadi	31° 50'	35° 20'
Nahsholim	32° 37'	34° 55'
Nahum, Sede	32° 31'	35° 29'
Namir, Wadi (Wadi Dalam, Nahal Namer)	33° 04'	35° 10'
Naphekh, Wadi (see Nafkha)	30° 43'	34° 46'
Naqah, Al	31° 02'	35° 29'
Naqah, Al	31° 30'	35° 19'
Nasb, Wadi El, Sinai	28° 28'	34° 08'
Natanya	32° 20'	34° 51'
Na'ur	31° 53'	35° 50'
Navit, Berekhot (not found)		
Nathania (see Natanya)	32° 20'	34° 51'
Nazareth	32° 42'	35° 18'
Negba	31° 40'	34° 41'
Negev Desert	30° 30'	35° 00'
Neot Hakikar	30° 57'	35° 22'
Nes Ziyona	31° 55'	34° 48'
Netafim, Ain	29° 36'	34° 52'
Neve Ur	32° 35'	35° 33'
Nikrot, Nahal	30° 32'	34° 50'
Nimr, Wadi	31° 50'	35° 19'
Nimrin, Ghor	31° 54'	35° 34'
Nir, Bet	31° 39'	34° 52'
Nir David	32° 33'	35° 27'
Nir Gallim	31° 50'	34° 41'
Nir 'Oz	31° 19'	34° 24'
Nir Yitzhaq	31° 14'	34° 21'
Nirim	31° 20'	34° 24'
Nitsana (see Auja)	30° 52'	34° 26'
Nitsanim	31° 44'	34° 38'
Nizzana (see Auja)	30° 52'	34° 26'
Nizzana, Nahal (Wadi Auja)	30° 48'	34° 30'
Nizzanim	31° 43'	34° 30'
Nu'aymah	31° 54'	35° 26'

Locality	N Latitude	E Longitude
Nusariyat, Wadi (Kufrein, Wadi)	31° 52'	35° 42'
Oded, Beeret (see Jdeid)	30° 28'	34° 44'
Ora, Bir (see Bir Hindis)	29° 43'	34° 59'
Oranim	32° 43'	35° 06'
Oreima, Tel, near Tiberias (not found)		
Oren, Beit	32° 44'	35° 00'
Oren, Nahal, Etsba Cave (see Carmel)	32° 40'	34° 58'
Osha, Jebel	32° 04'	35° 46'
Ouba, Wadi	33° 05'	35° 31'
Palmahim (Nabi Rubin)	31° 56'	34° 42'
Pekin	32° 58'	35° 20'
Petah Tikvah	32° 05'	34° 53'
Petra	30° 19'	35° 26'
Qadi, Tel Al (Dan)	33° 15'	35° 39'
Qalqilya	32° 11'	34° 58'
Qama, Beit	31° 27'	34° 45'
Qarantal, Jebel	31° 52'	35° 26'
Qarn, Wadi (Nahal Kziv)	33° 03'	35° 06'
Qatana	31° 49'	35° 06'
Qatifa, Wadi	31° 49'	37° 23'
Qatrana, El (see El Katrane)	31° 14'	36° 03'
Qedesh, Tel and Wadi/Nahal	33° 07'	35° 32'
Qeisariya	32° 30'	34° 53'
Qelt, Wadi/Nahal/Gorge	31° 51'	35° 25'
Qeren, Nahal	31° 00'	34° 29'
Qidron, Gorge/Wadi	31° 43'	35° 19'
Qilt, Wadi (see Kelt, Wadi)	31° 51'	35° 25'
Qishon River	32° 47'	35° 03'
Qittayn, Umm al	32° 19'	36° 38'
Qu'alibah, Ain	32° 40'	35° 52'
Quayliba	32° 40'	35° 52'
Qubeiba	31° 50'	35° 08'
Quds, Al (Jerusalem, Yerushalaim)	31° 47'	35° 13'
Quetain, Umm el (see Qittayn)	32° 19'	36° 38'
Qumran	31° 44'	35° 27'
Quneitira (Golan, SYRIA)	33° 08'	35° 49'
Quneiya, Ain	32° 14'	36° 00'
Quraiqira	30° 40'	35° 26'
Quseimah, El (Kosseima, SINAI)	30° 40'	34° 22'
Quweira, Al	29° 48'	35° 18
Radian, Ein	29° 53'	35° 03'
Rafah	31° 17'	34° 14'
Rahaf, Nahal	31° 15'	35° 22'
Rahma (in Wadi Araba, not found)		
Rajil, Wadi	31° 49'	36° 53'
Ram, Wadi (Rum)	29° 34'	35° 24'
Ramallah	31° 54	35° 12'
Raman, Wadi (Nahal Ramon)	30° 37'	34° 54'
Ramatayim	32° 09'	34° 53'
Ramla	31° 56'	34° 52'
Ramleh	31° 56'	34° 52'

Locality	N Latitude	E Longitude
Ramon, Wadi/Makhtash	30° 37'	34° 54'
Ramtha	32° 34'	36° 00'
Ras El Ein	32° 6'	34° 56'
Ras El Feschiha (see Ras El Feshkhah)	31° 40'	35° 28'
Ras El Feshkhah	31° 40'	35° 28'
Ras En Nakura	33° 06'	35° 06'
Ras En Naqab	30° 00'	35° 29'
Ras En Naqura (see Ras En Nakura)	33° 06'	35° 06'
Ras En Negeb (Ras En Naqab)	30° 00'	35° 29'
Rasas, Um (Risas)	31° 30'	35° 55'
Rehavia Cave (see Jerusalem)	31° 47'	35° 13'
Rehobot (Biliramun)	31° 54'	34° 46'
Rehovot (see Rehobot)	31° 54'	34° 46'
Reishah (not found)		
Rekhme, Bir (Rachme, Yeroham)	31° 00'	34° 55'
Revivim	31° 02'	34° 43'
Risas, Um er	31° 30'	35° 55'
Rishah, Jabel Al	30° 15'	35° 14'
Rishon le Zion	31° 57'	34° 48'
Roded, Nahal	29° 35'	34° 57'
Rosh Haniqra (see Ras En Nakura)	33° 06'	35° 06'
Rosh Pinna	32° 58'	35° 32'
Rosh Zohar (Ras Zuweira)	31° 13'	35° 14'
Roshafim	32° 54'	35° 35'
Rotem, Mishor (Plat.)	31° 03'	35° 10'
Roum, Mogharet el (see Jerash Refugee Camp)		
Rtivon (not found)		
Rubin, Wadi/Nahr Al/Zahr Al	31° 56'	34° 42'
Ruhama	31° 30'	34° 41'
Rum, Wadi (Ram)	29° 34'	35° 24'
Ruman, Wadi	30° 37'	34° 54'
Ruppin, Kfar	32° 27'	35° 33'
Ruwayshid (see H4)	32° 30'	38° 12'
Sā	32° 34'	35° 54'
Sa'ad, Khirbet	31° 28'	34° 32'
Sa'ar	33° 02'	35° 06'
Saadim, Horbat	31° 45'	35° 08'
Sabalan	33° 01'	35° 20'
Sabalan, Jebel	33° 00'	35° 20'
Sabastiya	32° 17'	35° 12'
Safad	32° 58'	35° 30'
Safawi (H5 station)	32° 12'	37° 07'
Safi	31° 10'	35° 28'
Safi, Ghor as (Ghor as Safieh)	31° 03'	35° 30'
Safiah, Nahal	31° 24'	34° 43'
Safje (Bet Qama)	31° 27'	34° 46'
Safje (see Safi)	31° 10'	35° 28'
Sahab	31° 53'	36° 00'
Sahur, Beit	31° 42'	35° 13'
Saide, Ein and Qiryat	31° 45'	35° 09'
Sakib	32° 17'	35° 49'

Locality	N Latitude	E Longitude
Saleh, Nebi (see Ramleh)	31° 56'	34° 52'
Salem, Bir	31° 55'	34° 50'
Salfit	32° 05'	35° 10'
Salt	32° 03'	35° 44'
Samaria	32° 17'	35° 12'
Samuel, Nebi (see Nebi Samwili)	31° 50'	35° 11'
Samwili, Nebi	31° 50'	35° 11'
Sanasin, Khirbet (not found)		
Sanour	32° 31'	35° 15'
Sarafend	31° 57'	34° 51'
Sasa	33° 2'	35° 24'
Sataf	31° 46'	35° 08'
Sayad, Wadi (? = Siyal)	31° 21'	35° 24'
Schedschera (= Sejera)	32° 45'	35° 24'
Sdot Yam	32° 30'	34° 54'
Sebbe (Sebeh, Masada)	31° 19'	35° 21'
Sedom	31° 02'	35° 22'
Seisaban, Ghor	31° 51'	35° 35'
Sejera	32° 45'	35° 24'
Sekher, Nahal and Dunes (see Mishash)	30° 9'	34° 44'
Serbal (not found)		
Sha'ar Ha'amarqim	32° 43'	35° 07'
Shafaym	32° 13'	34° 49'
Shalwa	31° 33'	34° 46'
Sharon Plain	32° 16'	34° 52'
Shaumer, Umm (SINAI)	28° 22'	33° 56'
Shawbak (see Shobek)	30° 32'	35° 33'
Shawmari (area and Wildlife Reserve)	31° 47'	36° 49'
Shean, Beit (see Beisan)	32° 30'	35° 30'
She'arim, Bet	32° 42'	35° 11'
Sheikh Abreik	32° 40'	34° 56'
Sheikh Bureik (see Sheikh Abreik)	32° 40'	34° 56'
Sheikh Iskander	32° 31'	35° 09'
Shelomo, Nahal (Wadi Masri)	29° 32'	34° 55'
Shemaryahu, Kfar	32° 11'	34° 51'
Shemesh, Beit	31° 45'	35° 00'
Shen Ramon	30° 34'	34° 51'
Shetula	33° 05'	35° 17'
Shivta	30° 53'	34° 38'
Shobek	30° 32'	35° 33'
Shoqet, Tel (W. Arad)	31° 18'	34° 55'
Shorier, Tel El	31° 23'	34° 41'
Shtula	33° 05'	35° 17'
Shu'ayb, Wadi	32° 10'	35° 12'
Shuna Al Janubiya (South Shuna)	31° 53'	35° 34'
Shuna Ash Shamaliya (North Shuna)	32° 37'	35° 36'
Shunra, Holot	30° 58'	34° 32'
Silwan	31° 46'	35° 14'
Sir, As	31° 53'	35° 46'
Sirhan, Wadi (SAUDI ARABIA)	30° 45'	37° 57'
Siyal, Wadi	31° 21'	35° 24'

Locality	N Latitude	E Longitude
Solomon's Quarries	31° 41'	35° 10'
Som (Sawm), Kufr	32° 35'	35° 48'
Soreq, Nahal (Wadi Rubin)	31° 56'	34° 42'
St. Chariton	31° 39'	35° 14'
St. John's Monastery	31° 46'	35° 08'
Suad, Wadi (Nahal Ukam)	33° 04'	35° 18'
Sudeir, Wadi (Nahal Dawid)	31° 28'	35° 24'
Suelleh (see Suweilih)	32° 02'	35° 50'
Suf	32° 19'	35° 50'
Sukhn, Wadi	32° 08'	36° 04'
Sultan, Tel el (not found)		
Sur, Wadi Al	31° 54'	35° 43'
Surra Reserve Station	32° 24'	36° 09'
Suweilih	32° 02'	35° 50'
Suweima	31° 47'	34° 36'
Suweinit, Wadi (see Wadi Swenit)	31° 51'	35° 18'
Suweir, Bir El (Suwera)	29° 15'	34° 43'
Suweira	29° 15'	34° 43'
Swemi (see Suweima)	31° 47'	34° 36'
Swenit, Wadi	31° 51'	35° 18'
Taba, Wadi and° Ain el	29° 32'	34° 52'
Tabgha	32° 52'	35° 32'
Tabor, Mt.	32° 41'	35° 24'
Tabqat Fahl, ppl. and ruin	32° 27'	35° 37'
Tafilah	30° 52'	35° 36'
Tal'ah, Wadi	32° 10'	35° 28'
Tallah, Ain El	31° 02'	35° 38'
Tamar, Ein	30° 59'	35° 21'
Tantura	32° 37'	34° 55'
Tavor, Nahal (see Bira, Wadi)	32° 36'	35° 34'
Taymah (SAUDI ARABIA)	27° 38'	38° 29'
Tayyibah, At	32° 33'	35° 43'
Tel Aviv	32° 04'	34° 46'
Tel el Qadi	33° 14'	35° 39'
Tiberias, Lake	32° 48'	35° 35'
Timna, Nahal/Wadi and Mt. (Meneya)	29° 46'	34° 58'
Tin Fanis, Tel (not found)		
Tineh, Ain El (Ain El Tine)	32° 52'	35° 32'
Tira, Wadi	32° 46'	34° 58'
Tivon, Qiryat	32° 43'	35° 07'
Tsofit (Zofit)	32° 11'	34° 55'
Tsor'a	31° 45'	34° 58'
Tubeiq, Jebel El	29° 40'	37° 20'
Tulkarem (Tulkarm)	32° 19'	35° 02'
Turabah, Ain	31° 35'	35° 25'
TurabahTuraif (SAUDI ARABIA)	31° 43'	38° 35'
Turabi, Ain (Turaba, Terabeh, Mishor Yamin)	31° 35'	35° 25'
Tureibe Plain	31° 05'	35° 06'
Uchman, Mt.	33° 04'	35° 15'
Ukam, Nahal (Uchman, Wadi Suad)	33° 04'	35° 18'
Um Al Daraq, Jebel	32° 19'	35° 48'

Locality	N Latitude	E Longitude
UM Shomar (SINAI)	28° 22'	33° 55'
Unab, El	29° 57'	36° 57'
Urim	31° 18'	34° 31'
Usdum, Jebel (= Har Sedom)	31° 05'	35° 23'
Uweinid, Jebel and Wadi Al	31° 49'	36° 48'
Vitkin, Kfar	32° 23'	34° 52'
Yad Mordekhay	31° 35'	34° 33'
Yafo (See Jaffa)	32° 02'	34° 45'
Yagur (Yadjour)	32° 44'	35° 04'
Yahav, Ain	30° 36'	35° 10'
Yahav, ppl.	30° 38'	35° 15'
Yanani, Beit	32° 23'	34° 51'
Yanany, Bet	32° 23	34° 52'
Yarmouk Pumping Station	32° 38	35° 34'
Yarqon River	32° 07	34° 52'
Yassour	32° 54	35° 10'
Yavna, Holot (Yibna Dunes)	31° 52	34° 45'
Yeroham, Kfar (see Rekhme)	31° 0'	34° 55'
Yerushalaim (Jerusalem, Al Quds)	31° 47'	35° 13'
Yessod Hamaale	33° 04'	35° 37'
Yezrael Valley	32° 33'	35° 20'
Yiftah (Kibbutz)	33° 07'	35° 33'
Yohanan, Ramat	32° 47'	35° 08'
Yosef, Bet	32° 33'	35° 33'
Yotvata (Ein Ghidyan), area and Reserve	29° 53'	35° 03'
Zahala	32° 07'	34° 50'
Zakariya, Kefar (Zekharya)	31° 42'	34° 57'
Zalam, Wadi (see Dalam, Wadi)	33° 04'	35° 10'
Zamin, Wadi (not found)		
Zarqa (City)	32° 05'	36° 06'
Zarqa Ma'in, Wadi (Wadi Zerka Ma'in)	31° 36'	35° 38'
Zayy	32° 06'	35° 43'
Zedek, Migdal	32° 05'	34° 57'
Ze'elim	31° 12'	34° 32'
Zeelim, Nahal (see Siyal, Wadi)	31° 21'	35° 24'
Zeit, Bir	31° 58'	35° 12'
Zera, Bet	32° 41'	35° 34'
Zeved, Mt.	32° 59'	34° 25'
Zevul, Har (see Sabalan, Jebel)	33° 00'	35° 20'
Zikhron Ya'aqov, Wadi	32° 34'	34° 57'
Zin, Nahal	30° 57'	35° 19'
Ziqim	31° 36'	34° 31'
Ziza (see Jiza)	31° 42'	35° 57'
Zo'ar	31° 12'	35° 22'
Zora	31° 46'	34° 59'
Zubiya Forest	32° 26'	35° 46'

Index

A

Acinonyx, 157

Acinonyx jubatus, 156, 157 - 159; *A. j. venaticus*, 157

Acomys, 32, 33, 189, 233, 279, 280 - 281

Acomys cahirinus, 33, 39, 281 - 285, 286

Acomys dimidiatus, 33, 281

Acomys ignitus, 284

Acomys percivali, 286

Acomys russatus, 33, 39, 46, 281, 283, 285 - 286; *A. r. harrisoni*, 33, 285; *A. r. lewisi*, 33, 285, 286; *A. r. russatus*, 33, 285

Acomys wilsoni, 286

addax, 206 - 207. See also *Addax nasomaculatus*

Addax, 158, 206

Addax nasomaculatus, 206 - 207

Afghan fox, 149. See also *Vulpes cana*

afri, 213. See also *Gazella dorcas*

African wild ass, 54

Aharoni, Bathscheba, 10

Aharoni, Israel, 10

Alcelaphus, 206

Alcelaphus buselaphus, 46, 206

Allactaga, 32, 304, 306 - 307

Allactaga euphratica, 42, 46, 304 - 307

Ammotragus lervia, 206

Anderson's gerbil, 239 - 241. See also *Gerbillus andersoni*

Antrozous, 126

Apodemus, 25, 49, 233, 235, 280, 287

Apodemus flavicollis, 287 - 290, 291

Apodemus hermonensis, 287, 289

Apodemus mystacinus, 46, 280, 287, 288, 289, 290 - 291; *A. m. pohlei*, 290

Apodemus sylvaticus, 287, 290

Arabian camel, 321, 322. See also *Camelus dromedarius*

Arabian horseshoe bat, 98 - 99. See also *Rhinolophus clivosus*

Arabian sand gazelle, 193, 220 - 221. See also *Gazella subgutturosa*

Artiodactyla, 22, 198 - 223

Arvicola, 273, 278

Arvicola terrestris, 48, 273 - 275

Arvicolidae, 225, 229, 273

Asellia, 78, 104

Asellia tridens, 46, 105 - 107; *A. t. murriana*, 105, 106

Asiatic mouflon, 323

ass, 22, 191 - 193, 320 - 321. See also *Equus asinus*

Atallah, Sana Isa, 10

aurochs, ix, 207, 321 - 322. See also *Bos primigenius*

B

badgers, 53, 63, 177

Baluchistan gerbil, 42, 237, 240, 246, 247 - 249. See also *Gerbillus nanus*

bandicoot rat, 294

Barbastella, 109, 124

Barbastella barbastellus, 110, 111; *B. b. leucomelas*, 110

barbastelle, 109 - 110. See also *Barbastella barbastellus*

bats, ix, 8, 15, 17, 18, 19, 20, 24, 31, 35, 45, 53, 54, 56, 57, 76 - 140

bears, 140, 189 - 190

bicolored white-toothed shrew, 70 - 71, 74. See also *Croidura leucodon*

black rat, 296, 297 - 298, 317. See also *Rattus rattus*

H